Engineering Statistics Demystified

Demystified Series

Accounting Demystified
Advanced Statistics Demystified
Algebra Demystified
Alternative Energy Demystified
Anatomy Demystified
ASP.NET 2.0 Demystified
Astronomy Demystified
Audio Demystified
Biology Demystified
Biotechnology Demystified
Business Calculus Demystified
Business Math Demystified
Business Statistics Demystified
C++ Demystified
Calculus Demystified
Chemistry Demystified
College Algebra Demystified
Corporate Finance Demystified
Databases Demystified
Data Structures Demystified
Differential Equations Demystified
Digital Electronics Demystified
Earth Science Demystified
Electricity Demystified
Electronics Demystified
Environmental Science Demystified
Everyday Math Demystified
Forensics Demystified
Genetics Demystified
Geometry Demystified
Home Networking Demystified
Investing Demystified
Java Demystified
JavaScript Demystified
Linear Algebra Demystified
Macroeconomics Demystified
Management Accounting Demystified
Math Proofs Demystified
Math Word Problems Demystified
Medical Billing and Coding Demystified
Medical Terminology Demystified
Meteorology Demystified
Microbiology Demystified
Microeconomics Demystified
Nanotechnology Demystified
Nurse Management Demystified
OOP Demystified
Options Demystified
Organic Chemistry Demystified
Personal Computing Demystified
Pharmacology Demystified
Physics Demystified
Physiology Demystified
Pre-Algebra Demystified
Precalculus Demystified
Probability Demystified
Project Management Demystified
Psychology Demystified
Quality Management Demystified
Quantum Mechanics Demystified
Relativity Demystified
Robotics Demystified
Signals and Systems Demystified
Six Sigma Demystified
SQL Demystified
Statics and Dynamics Demystified
Statistics Demystified
Technical Math Demystified
Trigonometry Demystified
UML Demystified
Visual Basic 2005 Demystified
Visual C# 2005 Demystified
XML Demystified

Engineering Statistics Demystified

Larry J. Stephens

New York Chicago San Francisco
Madrid Mexico City Milan N
Seoul Singapore S

The McGraw-Hill Companies

Library of Congress Cataloging-in-Publication Data

Stephens, Larry J.
Engineering statistics demystified / Larry J. Stephens.—1st ed.
p. cm.
Includes index.
ISBN 0-07-146272-4 (alk. paper)
1. Engineering—Statistical methods. I. Title.
TA340.S74 2006
519.502'462—dc22

2006036984

1 2 3 4 5 6 7 8 9 0 DOC/DOC 0 1 0 9 8 7 6

ISBN-13: 978-0-07-146272-3
ISBN-10: 0-07-146272-4

The sponsoring editor for this book was Judy Bass, the editing supervisor was David E. Fogarty, and the production supervisor was Pamela A. Pelton. It was set in Times Roman by International Typesetting and Composition. The art director for the cover was Margaret Webster-Shapiro.

Printed and bound by RR Donnelley.

To my Mother and Father, Rosie and Johnie Stephens,
and to my Wife, Lana Christine Stephens

—Larry J. Stephens

ABOUT THE AUTHOR

Larry J. Stephens is a professor of mathematics at the University of Nebraska at Omaha. He has also taught at the University of Arizona, Gonzaga University, and Oklahoma State University. He has worked for NASA, Livermore Laboratory, and Los Alamos National Laboratory. He spent ten years as a consultant and conducted seminars for the engineering group at 3M in Valley, Nebraska. Dr. Stephens is the author of *Schaum's Outline of Beginning Statistics*, coauthor with Murray Spiegel of *Schaum's Outline of Statistics*, and author of *Advanced Statistics Demystified*. He has over 25 years of experience teaching statistical methodology, engineering statistics, and mathematical statistics.

CONTENTS

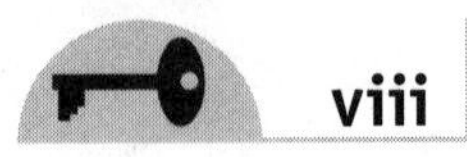

PREFACE

This book is intended for students currently in engineering statistics courses as well as engineers on the job who may not have had an engineering statistics course or who did have an engineering statistics course but need to brush up on the subject. The software available for statistics is truly outstanding. I have included output from several software packages throughout the book. In the Acknowledgments I have thanked the people who granted me permission to use this software output.

What background is required to understand engineering statistics?

The background required for engineering statistics is three semesters of Calculus.

What is different about this book?

One of the first differences that the reader will notice about the book is that it does not contain any statistical tables. "Why?" you may ask. If you have EXCEL (and everyone does) you will have the ability to find normal areas or percents. For example, nearly all statistics books contain the standard normal tables. By using the EXCEL functions =NORMDIST and =NORMINV, you can solve any of the statistical problems you encounter in the real world or university statistics course involving normally distributed variables. For example, adult males have heights that are normally distributed with a mean equal to 5 feet 10 inches, or 70 inches, and standard deviation equal to 3 inches. If you are trying to find how many adult males are 6 feet or shorter, the EXCEL answer is given by =NORMDIST(72,70,3,1), which gives 0.7475. Hence about 75% of adult males are 6 feet or shorter. Suppose you wish to know the 90th percentile of male heights. The answer, using EXCEL, is given by the expression =NORMINV(0.90,70,3) or 73.8447. The 90th percentile is 73.8 inches to the nearest tenth. That is, 90% of the males are 73.8 inches in height or shorter. In addition to normal probabilities, there are all the other information provided by the tables in statistics books. The functions =CHIDIST, =CHIINV, =FDIST, =FINV, =TDIST, =TINV, =BINOMDIST may be used in place of the chi-square tables, the F distribution tables, the student t tables, and the binomial tables. The EXCEL functions have the advantage that they give more accuracy and they are easier to use than tables in statistics texts. Many of the statistical packages are available on the internet for 30 day free trial downloadings. Go to the web addresses given above to download the packages.

Many of the problems that involve calculus may be solved either "by hand" or by using MAPLE. For example, suppose you were asked to find the value of c that would make the following function a probability density function: $f(x) = c(4 - x^2)$ for $-2 < x < 2$, and 0 for all other x values. The area under a probability density function is by definition equal to 1. If MAPLE is used to solve, the solution is `> evalf(int(c*(4-x^2),x=-2..2));` $10.66666667c$. The answer is $10.7c$ and if this is to equal 1, we have $c = 1/10.7 = 0.093$. We have the following by hand solution just like you did in your college calculus course:

$$c\int_{-2}^{2}(4-x^2)\,dx = 1 = c\left(4x-\frac{x^3}{3}\right)\Bigg|_{-2}^{2} = c\left[\left(8-\frac{8}{3}\right)-\left(-8+\frac{8}{3}\right)\right] = \frac{32c}{3} = 1 \text{ or } c = 3/32 = 0.093.$$

In the examples, exercises, and final examinations in this book, calculus problems are solved both ways.

It is recommended by the author that the user of this book keep EXCEL at his or her side when reading the book. If the user of the book is working at a job that has one of the other software packages available, then keep that software at your side.

Another difference that will be noticed about the book is that the reader is encouraged to use software in conjunction with the book to learn the subject of statistics.

Because there are so many different areas of engineering, it is difficult to know what examples to include in such a book. The author has attempted to use examples that are easy to understand, no matter which engineering area the user of the book is in. The author invites comments, feedback, and advice concerning the book. Please e-mail me at *LStephens@mail.unomaha.edu* with any comment you may have.

ACKNOWLEDGMENTS

I would like to thank the following people for granting me permission to use output from the various packages:

MAPLE: Dr. Tom Lee, Ph. D., Vice President Market Development and Executive Product Director, Maplesoft. I quote from the website: "Technical professionals, like engineers, research scientists and financial analysts, use Maplesoft products to save time and costs through increased productivity, rapid solution development and deployment, and easy technical knowledge capture." The web address for MAPLE is *www.maplesoft.com.*

MINITAB: Ms. Christine Bailey, Coordinator of the Author Assistance Program, Minitab Inc. 1829 Pine Hall Road, State College, PA 16801. I am a member of the author's assistance program that Minitab sponsors. This program has been extremely supportive of me as an author. "Portions of the input and output contained in this publication/book are printed with the permission of Minitab Inc. All material remains the exclusive property and copyright of Minitab Inc. All rights reserved." The web address for Minitab is *www.minitab.com.*

SAS: Ms. Sandy Varner, Marketing Operations Manager, SAS Publishing, Cary, NC. "Created with SAS software. Copyright 2005. SAS Institute Inc., Cary, NC, USA. All Rights Reserved. Reproduced with permission of SAS Institute Inc., Cary, NC." I quote from the website: "SAS is the leader in business intelligence software and services. Over the 30 years, SAS has grown—from seven employees to nearly 10,000 worldwide, from a few customer sites to more than 40,000—and has been profitable every year." The Web address for SAS is *www.sas.com.*

SPSS: Jill Rietema, Account Manager, Publications, SPSS. I quote from the website: "SPSS Inc. is a leading worldwide provider of predictive analytics software and solutions. Founded in 1968, today SPSS has more than 250,000 customers worldwide, served by more than 1200 employees in 60 countries." The web address for SPSS is *www.spss.com.*

STATISTIX: Dr. Gerard Nimis, President, Analytical Software, P O Box 12185, Tallahassee, FL 32317. I quote from the website: "If you have data to analyze—but you're a researcher, not a statistician—STATISTIX is designed for you. You'll be up and running in minutes without programming or using a manual. This easy-to-learn and simple-to-use software saves you valuable time and money. STATISTIX combines all the basic and advanced statistics and powerful data manipulation tools you need in a single inexpensive package." The web address for STATISTIX is *www.statistix.com.*

EXCEL: Microsoft Excel has been around since 1985. It is available to almost all college students and professional engineers. It is widely used in this book.

There are over 40 illustrations that the author has extracted from clip art in Microsoft Word. The author believes that the inclusion of the clip art adds to the understanding of the statistics. The illustrations are placed in the examples and the exercises throughout the book, and these add to the readability of the book.

I wish to thank Stanley Wileman for computer advice that he so unselfishly gave to me during the writing of the book. I wish to thank my wife, Lana, for her understanding while I was thinking about the best way to present some concept. Thanks also to Judy Bass, Senior Acquisitions Editor, and her staff at McGraw-Hill. Finally, I would like to thank Rasika Mathur, Project Manager, International Typesetting & Composition for her help with the final camera ready copy of the book.

—Larry J. Stephens

Engineering Statistics Demystified

CHAPTER 1

Treatment of Data Using EXCEL, MINITAB, SAS, SPSS, and STATISTIX

1.1 Graphical Displays of Data

This first chapter will deal with the treatment of data and introduce you to five of the six software packages that will be used in the book. EXCEL, MINITAB, SAS, SPSS, and STATISTIX are the primary statistical packages that will be referenced. MAPLE is a package that is more mathematically oriented. It will be used to assist with the calculus aspects of the course. The official Internet website for MAPLE is *www.Maplesoft.com*.

Output from different software will be contrasted to allow the student to compare the output and see the similarities and differences in the packages. The comparisons of the software will help the student learn the concepts. Downloads of

MINITAB, SPSS, and STATISTIX for approximately 30 days are available on the Internet at *www.minitab.com*, *www.spss.com*, and *www.statistix.com,* respectively. Another website where much of the software is available at reasonable rents is *www.e-academy.com*. EXCEL is readily available on university computers and on most home computers. SAS is available on most university computers. These six sources of statistical software will be used in the text to solve statistical problems. I am indebted to the software manufacturers for granting me the permission to use outputs in the book. When the engineer or the scientist goes to work, he or she will use the software to help solve his or her problems.

An article in the *USA Today* entitled "Average fuel efficiency rises slightly" (July 29, 2005) recently appeared. Among other facts, the article said that model 2005 cars averaged 24.7 miles per gallon (mpg)—unchanged for the third straight year. Vans averaged 20.4 mpg, SUVs 18.1 mpg, and pickups 17.1 mpg. Engineers, in an effort to improve mpg for pickups made many design changes. The mpg's were determined at the start of the project (2005) and then at the next year for the new trucks with the new design features (2006). The engineers wished to evaluate the effects of their changes on mpg. One hundred randomly selected 2005 pickup truck mpg's are shown in Table 1-1.

Engineers try to improve the mpg for pickup trucks.

Figure 1-1 is the MINITAB worksheet with the data from Table 1-1 in column 1. This data can either be entered directly from the keyboard or read from a file called mpg.MTW. A **histogram** is a graphical device that uses rectangles to show the distribution of data. The heights of the rectangles give the frequencies of the classes, and the bases represent the classes. It is one of the oldest constructs in use which shows distributions. The pull-down **Graph ⇒ Histogram** gives a dialog box in which a simple histogram is chosen. In the simple histogram dialog box, it is requested that a histogram for the variable mpg be prepared. The result is shown in Figure 1-2.

Table 1-1 mpg Values for 100 pickup trucks

19.7	17.9	15.7	16.2	15.9	15.7	17.9	20.2	15.2	18.4
20.8	18.8	17.4	17.1	16.3	16.9	12.6	16.1	17.5	16.6
22.9	20.7	20.1	18.8	18.0	13.5	15.3	17.9	16.3	12.6
17.8	16.3	16.2	16.0	15.3	16.4	16.2	18.9	19.8	14.7
14.4	18.1	19.1	14.2	21.1	18.8	11.4	18.0	17.5	19.0
14.0	14.4	18.2	15.5	15.4	19.5	15.2	17.6	16.6	16.3
16.1	16.3	19.9	15.6	15.3	18.6	18.9	14.6	19.7	20.1
19.6	20.8	15.5	19.3	17.1	14.8	17.0	21.2	14.1	15.3
17.8	15.5	17.7	10.4	18.8	18.1	20.5	16.0	18.3	17.0
19.2	12.9	19.5	12.5	14.7	14.9	13.1	19.0	16.8	21.5

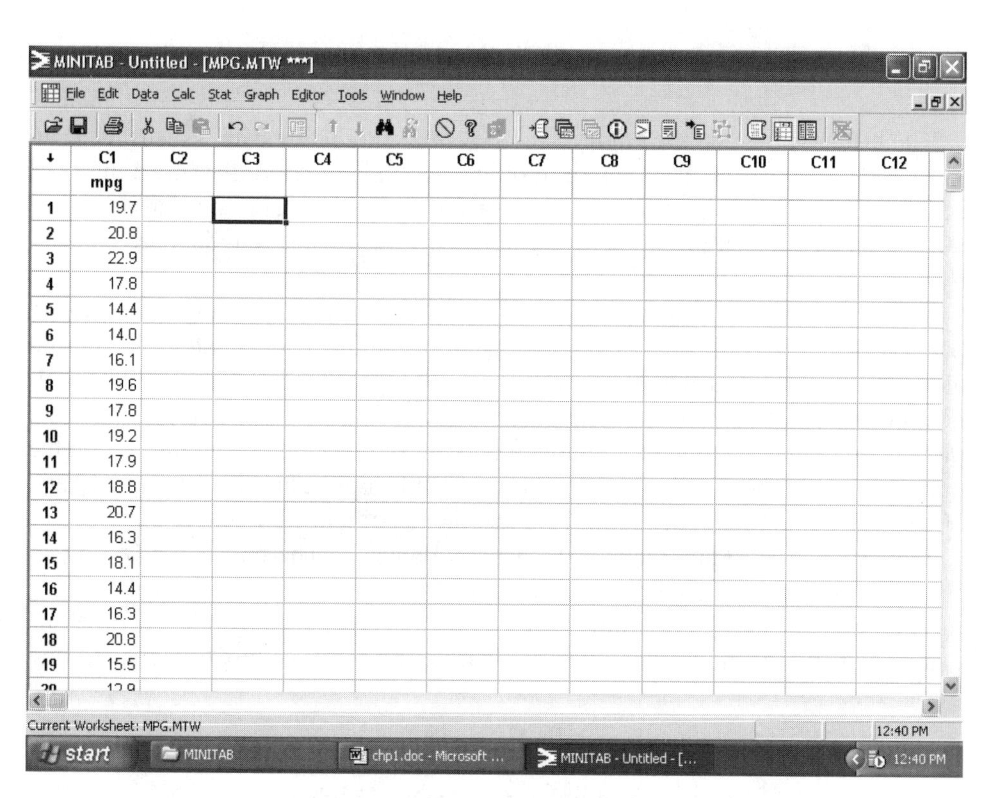

Figure 1-1 MINITAB worksheet containing the data from Table 1-1.

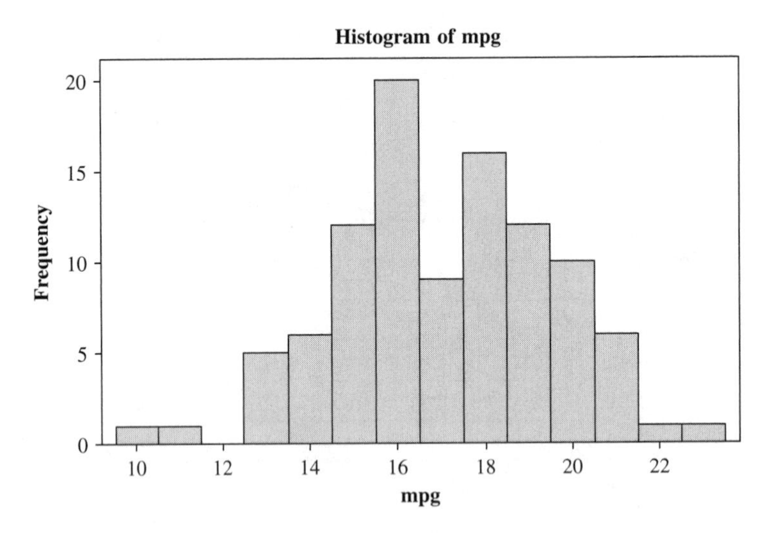

Figure 1-2 MINITAB histogram for the data in Table 1-1.

The histogram shows the engineer or the scientist how the data is distributed. In this case, the mpg data has a distribution, we call this **normal distribution**. Most of the data is near the middle. There are a few values that are down in the range 10–11, and a few that are up near the 22–23 range. Most of the data are concentrated in the middle between 13 and 21 mpg.

Another graphical picture called a **dot plot** is constructed using the software package MINITAB. Dot plots place dots atop numerical values that are placed along a horizontal axis. They are particularly useful with small data sets. The pull-down **Graph ⇒ Dot Plot** gives the graph shown in Figure 1-3. The mpg values are plotted along the horizontal axis and the dots are plotted above the numbers on the horizontal axis. The dot plot shows that most of the mpg values are between 12 and 20 with the minimum value being 10.4 and the maximum value being 22.9.

Figure 1-4 gives a box plot for the data in Table 1-1. The pull-down **Graph ⇒ Box Plot** gives the graph shown in Figure 1-4.

In its simplest form, a **box plot** shows a box that contains the middle 50% of the data values. It also shows two **whiskers** that extend from the box to the maximum

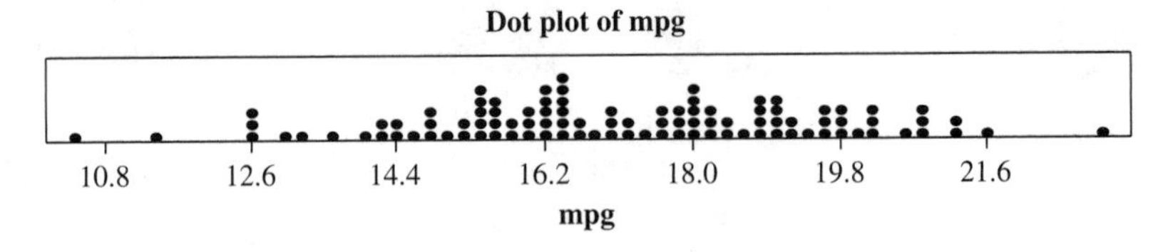

Figure 1-3 MINITAB dot plot for the data in Table 1-1.

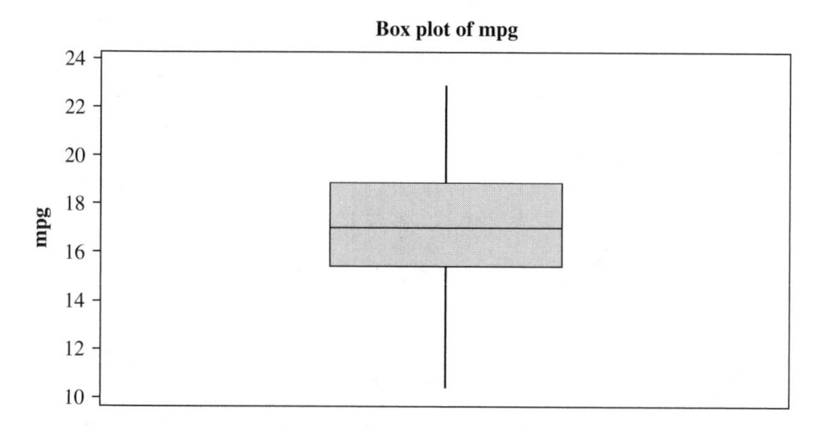

Figure 1-4 MINITAB box plot for the data in Table 1-1.

value and from the box to the minimum value. In Figure 1-4, the box extends from a number between 15 and 16 to a number between 18 and 19. The whiskers extend from a number a little larger than 10 to the box and from the box to a number a little less than 23. We shall have more to say about box plots as the text develops.

Figure 1-5 shows the worksheet for the software package SPSS containing the mpg data.

mpg.sav [DataSet1] - SPSS Data Editor

File Edit View Data Transform Analyze Graphs Utilities Add-ons Window Help

1 : mpg 19.7

	mpg	var	var	var	var	var	var	var	var	var
1	19.70									
2	20.80									
3	22.90									
4	17.80									
5	14.40									
6	14.00									
7	16.10									
8	19.60									
9	17.80									
10	19.20									
11	17.90									
12	18.80									
13	20.70									
14	16.30									
15	18.10									
16	14.40									
17	16.30									
18	20.80									
19	15.50									
20	12.90									
21	15.70									
22	17.40									

Data View / Variable View

SPSS Processor is ready

start Eng-Stat-Demystified chp1.doc - Microsoft ... mpg.sav [DataSet1] -... 2:56 PM

Figure 1-5 SPSS worksheet containing the data from Table 1-1.

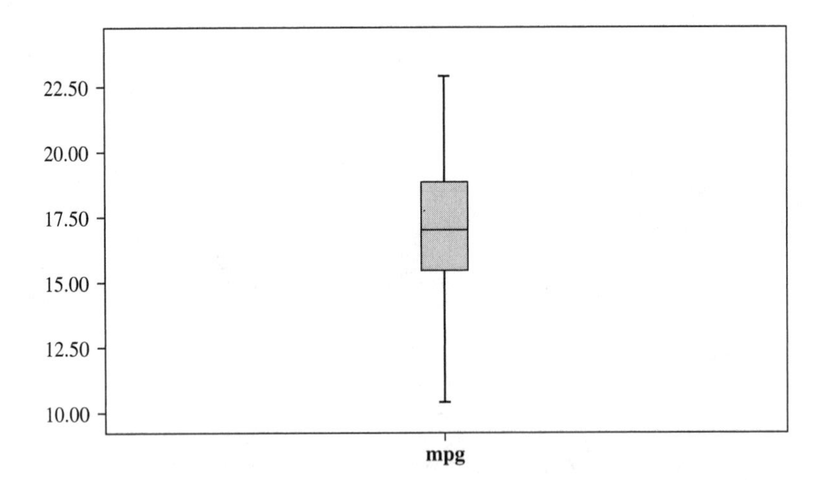

Figure 1-6 SPSS generated box plot.

Figure 1-6 gives an SPSS-created box plot. The pull-down **Graph ⇒ Box Plot** is used to create the box plot in SPSS. The box plot gives a box that covers the middle 50% of the data. The box in Figure 1-6 extends from 15.5 to 18.8. The whiskers extend to 10.4 on one end of the box and to 22.9 on the other end. If the data has a bell-shaped histogram, it will have a box plot like that shown in Figure 1-6. The middle of the data is shown by the heavy line in the middle of the box and will divide the box in half. Half of the data will be below the middle line and half will be above the middle line. The two whiskers will be roughly equal in length for a normal distribution.

The SAS pull-down **Graph ⇒ Histogram** gives the histogram of the mpg values and is shown in Figure 1-8. Figure 1-7 shows the data in the SAS worksheet.

A **stem-and-leaf** breaks the data down into stems and leaves. For example, 45 consists of stem 4 and leaf 5 or 4 tens and 5 ones. Figure 1-9 gives a STATISTIX created stem-and-leaf diagram of the data in Table 1-1. The smallest number is 104 and when multiplied by 0.1 becomes 10.4 mpg. Note that if the stem-and-leaf in Figure 1-9 is rotated 90 degrees, the shape of the histogram is shown. In addition to the shape being shown, the actual numbers that comprise the stem-and-leaf are also discernable. The pull-down **Statistics ⇒ Summary Statistics ⇒ Stem-and-leaf Plot** will give a stem-and-leaf diagram. For this stem-and-leaf, the first row represents 10.4; the third row represents 11.4; the sixth row represents 12.5, 12.6, 12.6, and 12.9; the seventh row represents 13.1; and the eighth row represents 13.5. The first column represents the cumulative count. The leaves (in column 3) for any row

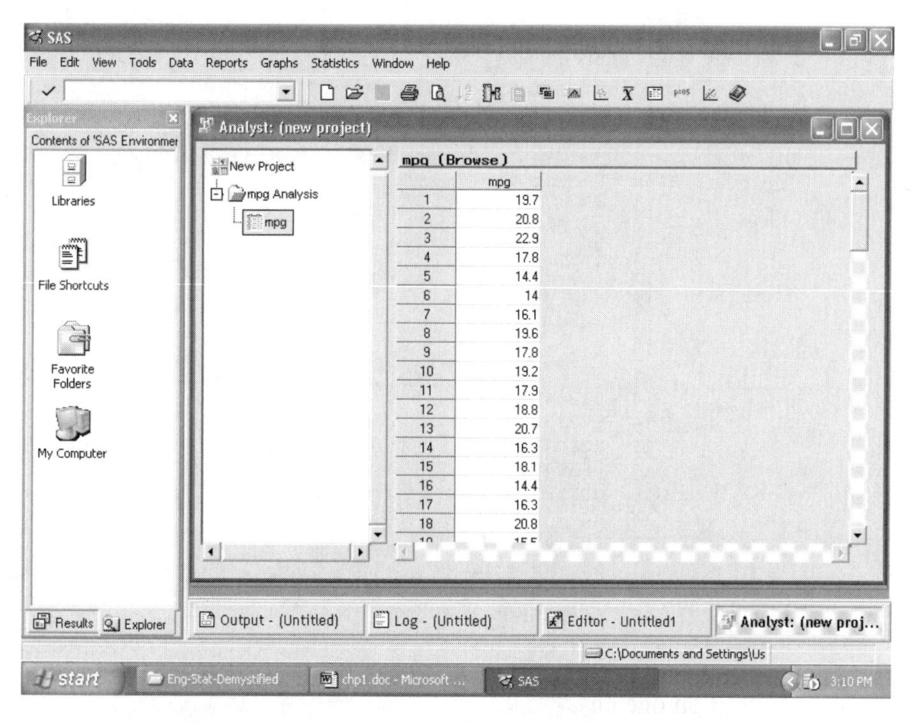

Figure 1-7 SAS worksheet for the data in Table 1-1.

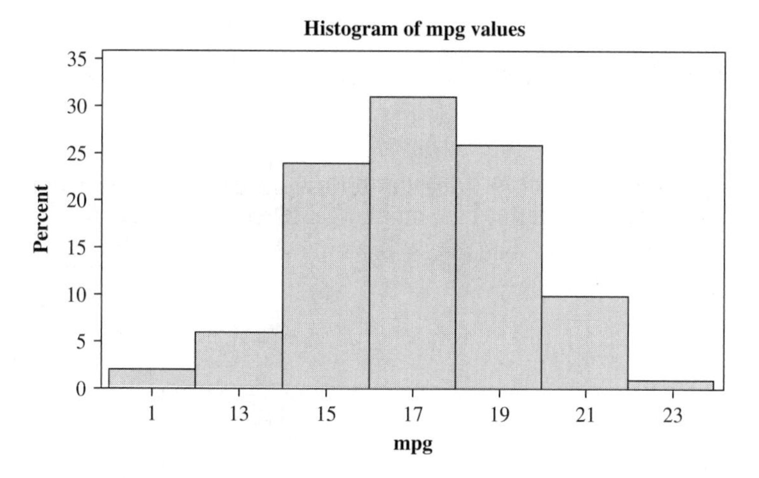

Figure 1-8 SAS histogram of the data in Table 1-1.

```
Statistix 8.0
Stem and Leaf Plot of mpg

  Leaf Digit Unit = 0.1                          Minimum   10.400
  10  4  represents 10.4                         Median    17.000
                                                 Maximum   22.900
       Stem  Leaves
    1   10   4
    1   10
    2   11   4
    2   11
    2   12
    6   12   5669
    7   13   1
    8   13   5
   13   14   01244
   18   14   67789
   25   15   2233334
   32   15   5556779
   45   16   0011222333334
   49   16   6689
  (5)   17   00114
   46   17   556788999
   37   18   0011234
   30   18   6888899
   23   19   00123
   18   19   5567789
   11   20   112
    8   20   5788
    4   21   12
    2   21   5
    1   22
    1   22   9

100 cases included   0 missing cases
```

Figure 1-9 STATISTIX stem-and-leaf plot of the data in Table 1-1.

are 0 through 4 or 5 through 9. (See the following from the upper portion of the stem-and-leaf in Figure 1-9.)

```
1    10   4
1    10
2    11   4
2    11
2    12
6    12   5669
7    13   1
8    13   5
```

The largest number is 22.9. The number in the third row from the bottom of the stem-and-leaf is 21.5, and the next two numbers in the fourth row from the bottom

are 21.1 and 21.2. The next four numbers are 20.5, 20.7, 20.8, and 20.8. The first column is the cumulative counts from the bottom. (See the following from the lower portion of the stem-and-leaf in Figure 1-9.)

```
8    20  5788
4    21  12
2    21  5
1    22
1    22  9
```

The class that contains the middle number contains the frequency for that class in parenthesis. In this case the middle number is 17.0. Note that the stem is the number of ones and is in the second column, and the leaf is the number of tenths and is in the third column.

Figure 1-10 shows the STATISTIX worksheet with the mpg data. The pull-down **Statistics ⇒ Summary Statistics ⇒ Histogram** is used to generate the histogram for the data in Table 1-1. This histogram is shown in Figure 1-11. Note that the classes used to form the histograms by the various packages differ. MINITAB uses 10, 11, ... , 23 at the center of the classes in Figure 1-2. SAS uses 11, 13, 15, 17, 19, 21, and 23 as the center of the classes in Figure 1-8. STATISTIX uses 10.3, 11.2,

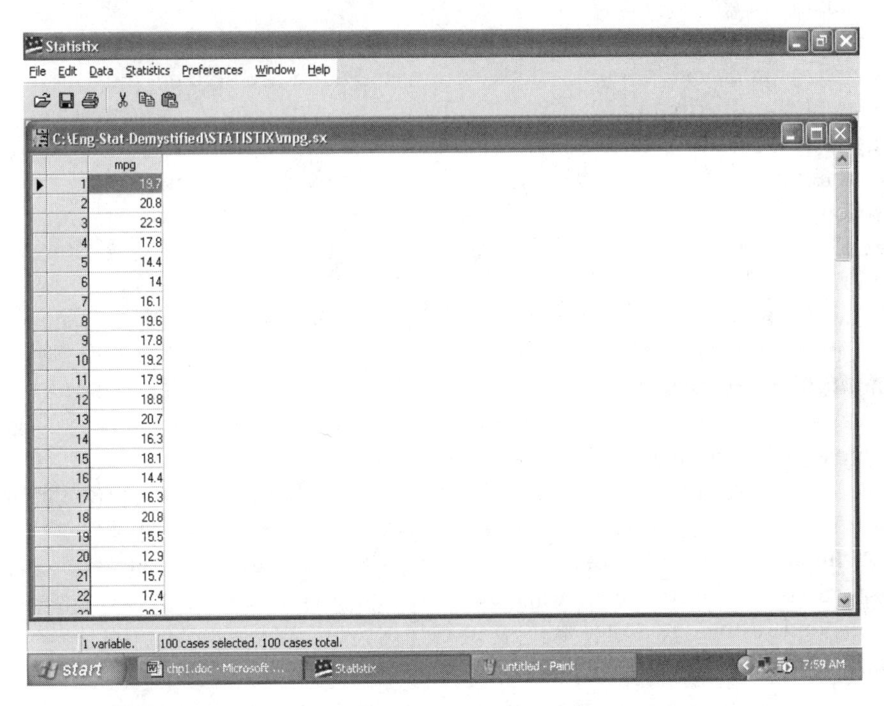

Figure 1-10 STATISTIX worksheet with data from Table 1-1.

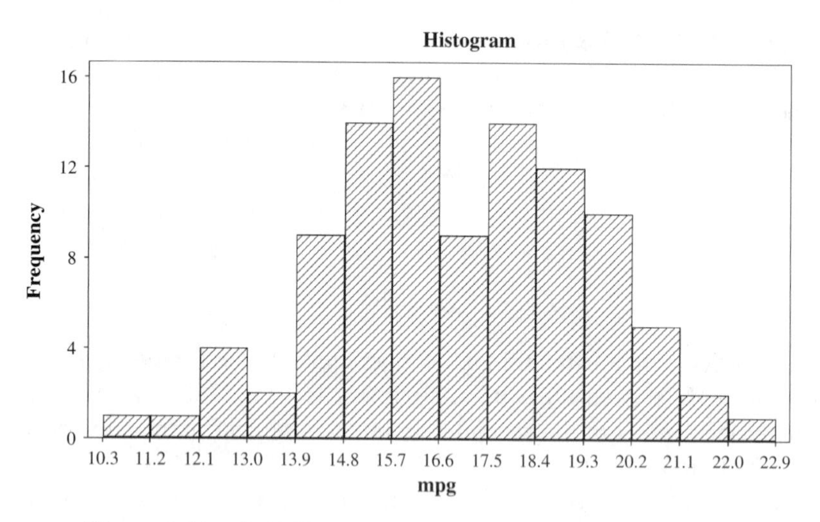

Figure 1-11 STATISTIX histogram of the data for Table 1-1.

12.1, … , 22.9 as the class limits in Figure 1-11. However, the distribution shape is the same with all the packages.

All of the statistical software given in this section reveal a set of data that is distributed symmetrically. A few values are low and a few are high but most are in the center of the distribution. Such a distribution is said to have a **normal** or **bell-shaped** distribution. Statistical graphics helps the engineer form some idea of the characteristic with which he or she is dealing. How variable is it? What type of distribution does it have? Are there some extreme values that occur? In all the spreadsheets shown, the data may be entered directly by the engineer, or it may be read from a file.

The histogram, the dot plot, the box plot, and the stem-and-leaf plot are used with quantitative data. **Quantitative engineering data** is obtained from measuring. **Qualitative engineering data** gives classes or categories for your data values. If your data is qualitative, other techniques are used to graphically display your data. Suppose daily inspection of some product results in rejection due to scratches, dents, smudges, failure to meet specifications, and other causes. Suppose the data shown in Table 1-2 was gathered during a given shift at a large company.

A **pie chart** is a plot of qualitative data in which the size of the piece of a pie is proportional to the frequency of a category. The angles of the pieces of a pie that correspond to the categories add up to 360 degrees. For example, the piece of a pie that represents dents has an angle equal to 0.40(360 degrees) = 144 degrees.

Table 1-2 Causes for rejection of items

Scratch	Spec	Dent	Spec	Dent
Dent	Scratch	Scratch	Spec	Dent
Smudge	Other	Scratch	Other	Scratch
Spec	Dent	Dent	Scratch	Scratch
Dent	Spec	Other	Spec	Dent
Scratch	Other	Spec	Dent	Scratch
Dent	Spec	Other	Spec	Other
Smudge	Dent	Dent	Spec	Spec
Spec	Smudge	Dent	Scratch	Dent
Other	Spec	Spec	Smudge	Dent

The piece of a pie that represents smudges has an angle equal to 0.08(360 degrees) = 11.5 degrees. The data in Table 1-2 may be summarized as follows:

Category	Frequency	Percentage
Dent	20	40
Scratch	10	20
Specifications	9	18
Other	7	14
Smudge	4	8

EXCEL may be used to form a pie chart of the data. The data, in summarized form, is entered into the worksheet as shown in Figure 1-12. The **chart wizard** is used to form the following exploded pie with 3-D visual effect.

A **bar chart** is a plot of qualitative data in which the frequency or the percent of the data is the height of a rectangle, and each category is the base of the rectangle. Figure 1-13 is an EXCEL horizontal bar chart illustrating the data in Table 1-2. The chart wizard is used to form the bar chart. The clustered bar with 3-D visual effect is chosen from the bar charts within the chart wizard.

A **Pareto chart** is a special form of a bar chart. It is a tool used by the engineers to prioritize problems for solution. It was named after the Italian economist Vilfredo Pareto (1848–1923). The bars of the Pareto chart are arranged in descending order. A Pareto chart for the data in Table 1-2 is shown in Figure 1-14. The curve above the histogram is the cumulative curve of frequencies. The curve starts at a count of 20 or

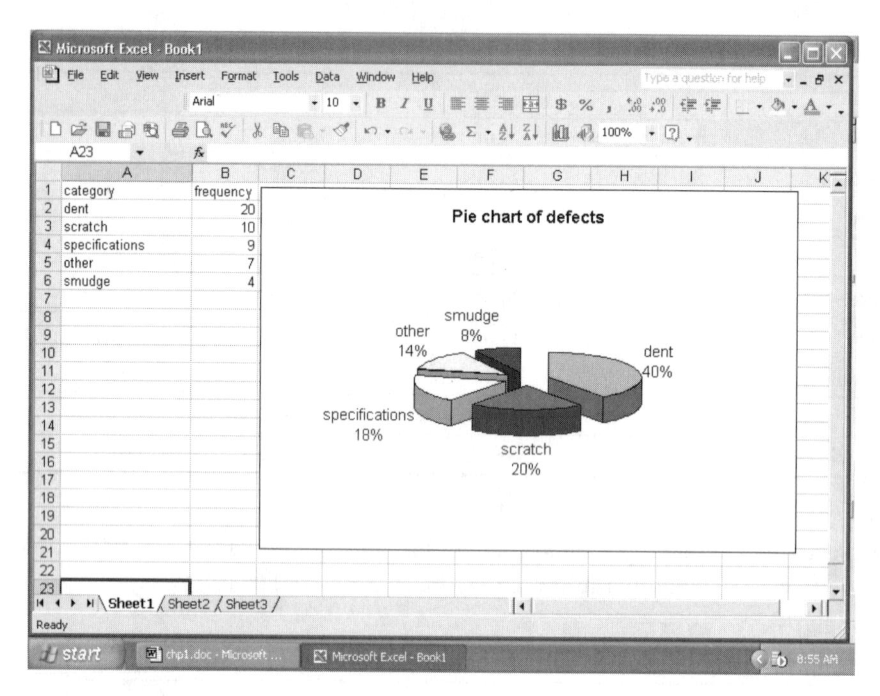

Figure 1-12 EXCEL pie chart of data in Table 1-2.

40% (20 out of 50 is 40%) above dents. The curve continues and above the tic mark for scratches. It passes through 30 or 60%. This means that 30 or 60% of the defects are either dents or scratches. Above spec (short for specifications) the curve passes through 39 or 78%. Seventy-eight percent of the defects are dents, scratches, or specifications not met. The curve continues in this manner. The pull-down **Graph ⇒ Pareto** is used to produce the Pareto chart shown in Figure 1-14.

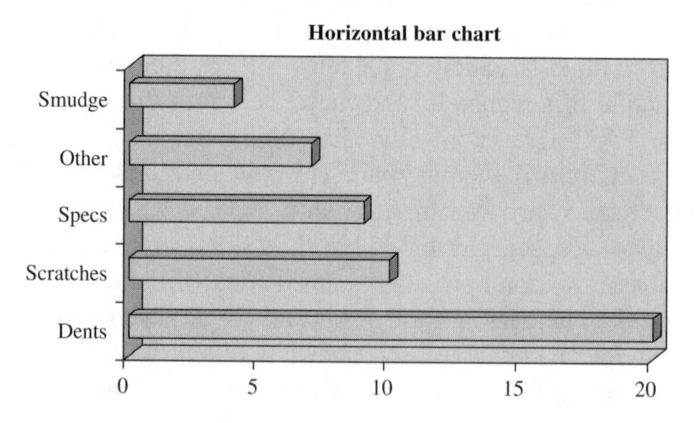

Figure 1-13 EXCEL horizontal bar chart of data in Table 1-2.

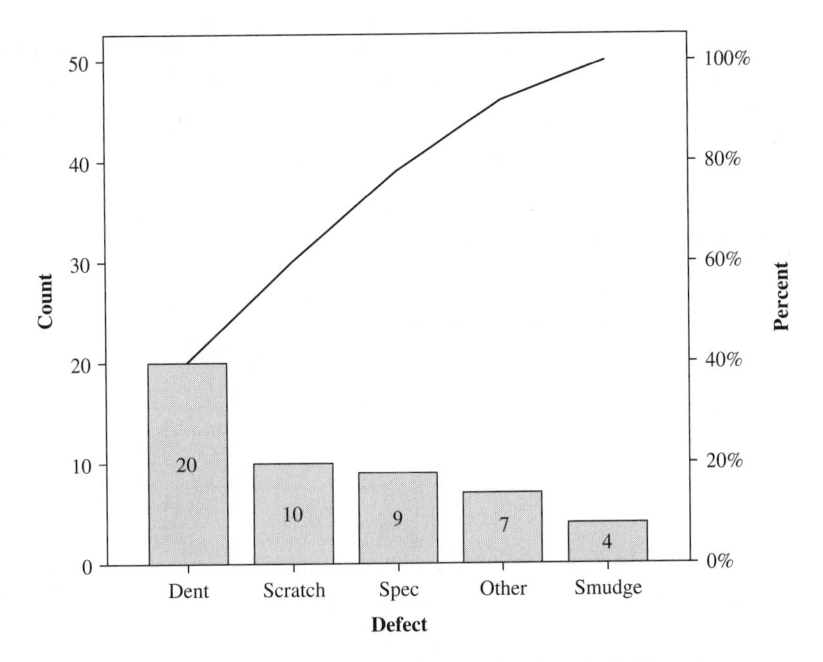

Figure 1-14 SPSS Pareto chart of the data in Table 1-2.

In this first section, we have considered various graphs from the different software packages for illustrating the distribution of the data.

1.2 Descriptive Measures of Data

Descriptive statistics is concerned with summarizing and presenting data concisely. The graphical techniques in Section 1.1 tell us about the distribution of the mpg values in Table 1-1. But, suppose we wanted one number that would in a sense represent all the data in the table. There are three descriptive statistical measures that we will discuss. They are mean, median, and mode. The **mean** is the sum of all the data divided by the number of observations. The **median** is the middle of the sorted data. The **mode** is the most frequently occurring value in the data set. The sum of the mpg values is 1704.9 and when this sum is divided by 100, the result is 17.049. If the data is sorted, the two values closest to the middle are 17 and 17. They are the 50th and the 51st values in the sorted array. They are averaged to obtain the median, 17. The most frequently occurring value is 16.3 mpg. It has a frequency of 5. We find that the mean, the median, and the mode are 17.049, 17, and

16.3, respectively. These values are found using EXCEL by entering the numbers in A2:A101. The label mpg is entered into A1. The pull-down **Tool ⇒ Data Analysis** is used to obtain the dialog box. Descriptive statistics is selected from the box. The results are as follows:

EXCEL output for descriptive statistics

mpg	
Mean	**17.049**
Standard error	0.239699
Median	**17**
Mode	**16.3**
Standard Deviation	2.396989
Sample Variance	5.745555
Kurtosis	–0.13851
Skewness	–0.17413
Range	12.5
Minimum	10.4
Maximum	22.9
Sum	1704.9
Count	100

The values of the mean, the median, and the mode are shown in bold.

When MINITAB is used, the pull-down is **Stat ⇒ Basic Statistics ⇒ Display Descriptive Statistics**. The results are as follows:

MINITAB output for descriptive statistics

Descriptive Statistics mpg								
Variable	N N*	**Mean**	SE Mean	St. Dev	Minimum	Q_1	**Median**	Q_3
mpg	100 0	**17.049**	0.240	2.397	10.400	15.425	**17.000**	18.875
Variable	Maximum							
mpg	22.900							

A measure of variation or dispersion that is simple to calculate and easy to understand is the **range**. The range is the largest value minus the smallest value in the data set. By consulting the EXCEL or the MINITAB output above, we see that the

maximum is 22.9, and the minimum value is 10.4. The range is, therefore, 22.9–10.4, or 12.5. If a data set is high in variability, the range is large. If a data set is low in variability, the range is small. The range, therefore, is representative of the variability of a set of data.

The SAS, SPSS, and STATISTIX output for the descriptive statistics are as follows. The SAS pull-down is **Statistics ⇒ Descriptive ⇒ Summary Statistics.**

SAS output for descriptive statistics

The MEANS Procedure					
Analysis Variable mpg					
Mean	Std. Dev	N	Minimum	Maximum	Median
17.0490000	2.3969886	100	10.4000000	22.9000000	17.0000000

The STATISTIX pull-down is **Statistics ⇒ Summary Statistics ⇒ Descriptive Statistics.**

STATISTIX output for descriptive statistics

Descriptive Statistics	
	mpg
N	100
Mean	17.049
SD	2.3970
Variance	5.7456
Minimum	10.400
Median	17.000
Maximum	22.900

The SPSS pull-down is **Analyze ⇒ Descriptive Statistics ⇒ Descriptive.**

SPSS output for descriptive statistics

Descriptive Statistics								
	N	Range	Minimum	Maximum	Sum	Mean	Std. Deviation	Variance
mpg	100	12.50	10.40	22.90	1704.90	17.0490	2.39699	5.746
Valid N (listwise)	100							

The **variance** and the **standard deviation** are easiest to understand if we select a smaller data set before finding these measures on a larger set, such as the mpg values. Suppose we wished to find the variance or the standard deviation of a set of golf scores for engineer, Jane Doe. The variance is defined to be

$$S^2 = \frac{\sum (X - \overline{X})^2}{n-1} \quad \text{or} \quad \frac{\sum X^2 - \frac{(\Sigma X)^2}{n}}{n-1}$$

The two forms can be shown to be algebraically equivalent.

Engineers play golf as well as solve problems.

Suppose Jane shot the following on five different occasions: 85, 90, 95, 88, and 92. First, note that the mean is 90. Using the first form of the equation for variance, we have the following. ($\overline{X}$ is the symbol that is used to represent the mean of a sample.)

X, score	$X - \overline{X}$, deviation	$[X - \overline{X}]^2$, square
85	–5	25
90	0	0
95	5	25
88	–2	4
92	2	4
	$\sum [X - \overline{X}] = 0$	$\sum [X - \overline{X}]^2 = 58$

The middle column in the table above illustrates that the sum of the deviations about the mean is always 0. The third column gives the numerator for the variance. When it is divided by $(n - 1)$, the variance is obtained. The variance is $S^2 = 58/4 = 14.5$. The standard deviation is the square root of $14.5 = 3.81$.

The variance and standard deviation may also be found using the equivalent form of the formula,

$$S^2 = \frac{\sum X^2 - \frac{(\Sigma X)^2}{n}}{n-1}$$

When using this formula, the sum and the sum of squares are found first. $\Sigma X = 85 + 90 + 95 + 88 + 92 = 450$ and $\Sigma X^2 = 7225 + 8100 + 9025 + 7744 + 8464 = 40{,}558$. Then,

$$S^2 = \frac{40558 - 40500}{4} = \frac{58}{4} = 14.5$$

Today, no one finds the variance or standard deviation by hand. A software package is used to find them.

The variance and standard deviation are given for the 100 mpg values by the five descriptive-statistics routines above. The variance is seen to be 5.75 in all five of the outputs, and the standard deviation is 2.40. When looking over the descriptive statistics, it is seen that the mean is 17.049, the median is 17.000, the minimum is 10.40, the maximum is 22.90, the range is 12.50, the variance is 5.75, and the standard deviation is 2.40. There are also other quantities in the descriptive statistics that we shall encounter later.

Table 1-3 Sorted mpg values for 100 pickup trucks

10.4	14.2	15.3	15.7	16.3	17.0	17.9	18.6	19.2	20.1
11.4	14.4	15.3	15.9	16.3	17.1	17.9	18.8	19.3	20.2
12.5	14.4	15.3	16.0	16.3	17.1	17.9	18.8	19.5	20.5
12.6	14.6	15.3	16.0	16.3	17.4	18.0	18.8	19.5	20.7
12.6	14.7	15.4	16.1	16.4	17.5	18.0	18.8	19.6	20.8
12.9	14.7	15.5	16.1	16.6	17.5	18.1	18.9	19.7	20.8
13.1	14.8	15.5	16.2	16.6	17.6	18.1	18.9	19.7	21.1
13.5	14.9	15.5	16.2	16.8	17.7	18.2	19.0	19.8	21.2
14.0	15.2	15.6	16.2	16.9	17.8	18.3	19.0	19.9	21.5
14.1	15.2	15.7	16.3	17.0	17.8	18.4	19.1	20.1	22.9

For data that is normally distributed, as is the data in Table 1-1, there is a special interpretation for the mean and the standard deviation. The **empirical rule** states that for any normal distribution the following relationships approximately hold true:

1. Approximately 68% of the distribution is between the (mean – standard deviation) and the (mean + standard deviation).
2. Approximately 95% of the distribution is between the (mean – 2 standard deviations) and the (mean + 2 standard deviations).
3. Approximately 99.7% of the distribution is between the (mean – 3 standard deviations) and the (mean + 3 standard deviations). The data in Table 1-1 is given in sorted form in Table 1-3.

One standard deviation below the mean is 17.049 – 2.397 = 14.652, and one standard deviation above the mean is 17.049 + 2.397 = 19.446. There are 68% of the values in Table 1-3 between 14.652 and 19.446. Two standard deviations below the mean is 17.049 – 4.794 = 12.255, and two standard deviations above the mean is 17.049 + 4.794 = 21.843. There is 97% of the data in Table 1-3 between these two values. All the data is within 3 standard deviations of the mean. The percents—68, 97, and 100—are approximately equal to 68, 95, and 99.7.

1.3 Measures of Relative Standing

A large group of engineering students took a standardized test in a course entitled "Calculus for Engineers." The scores attained by 225 students are shown in Table 1-4.

Isaac Newton (1643–1727) laid the foundation for differential and integral calculus.

Table 1-4 Scores made on a standardized test in calculus for engineers

526	600	492	577	480	528	304	525	522	549	386	358	565	565	574
488	397	327	448	420	624	606	417	487	539	417	467	444	620	498
608	546	502	573	449	438	688	468	613	506	517	518	532	427	491
493	502	488	453	546	517	721	775	310	532	437	340	446	413	502
535	656	426	319	527	397	457	311	298	719	594	709	521	441	487
553	468	435	666	305	365	543	486	472	525	522	378	483	379	539
370	506	536	496	606	539	661	444	520	676	662	404	449	413	417
716	546	447	612	479	359	584	533	460	503	478	498	581	542	537
724	570	438	587	342	407	446	622	605	553	529	637	384	495	504
619	500	327	460	586	314	569	562	588	375	645	381	455	580	434
431	525	642	537	615	712	519	529	630	536	409	629	566	526	576
593	404	471	482	364	582	416	452	480	619	495	267	401	792	388
622	798	488	425	512	702	406	507	460	559	426	658	565	274	664
679	534	645	579	530	381	667	477	433	472	632	552	641	440	574
477	590	489	402	326	651	619	355	445	577	743	521	607	574	599

Joe was told that his raw score was 650. This meant very little to Joe. What Joe really wanted to know was how did he "stack up" with the rest of the group. What percent of the scores were lower than his? Percentiles are the answer to Joe's question. There are 99 percentiles, 9 deciles, and 3 quartiles. For example, the 90th percentile is a score such that 90% of the scores are less than that score, and only 10% exceed that score. The ***p*th percentile** is the number that divides the lower $p\%$ of a set of data from the upper $(100 - p)\%$ of the data.

The first decile is the same as the 10th percentile, the second decile is the same as the 20th percentile, and so on. The first quartile is the same as the 25th percentile, the second quartile is the same as the 50th percentile, and the third quartile is the same as the 75th percentile. The notations $P_1, \ldots, P_{99}$ represent percentiles; the notations $D_1, D_2, \ldots, D_9$ represent deciles; and Q_1, Q_2 and Q_3 represent quartiles. We have the following equalities: $Q_1 = P_{25}$, $Q_2 = P_{50}$ = median, and $Q_3 = P_{75}$.

If the data in Table 1-4 is entered into the cells A1:O15 of the EXCEL worksheet, and the command =PERCENTILE(A1:O15,0.25) is entered into any empty cell, the number 441 is returned. Similarly, the command =PERCENTILE(A1:O15,0.50) gives 517, and the command =PERCENTILE(A1:O15,0.75) gives 580. We know that Joe, who scored 650 on the test, scored higher than 75% of those taking the test since the third quartile was 580. The ninth decile, given by the command =PERCENTILE(A1:O15,0.9), gives 645. This means that Joe's score is greater than Q_3 as well as D_9. The command =PERCENTILE(A1:O15,0.91) gives 655.2. Joe scored between D_9 and P_{91}.

If MINITAB is used to find some of these same measures, the data is put into column C1, and the descriptive statistics command gives basically the same values for the first, second, and third quartiles, as shown in the following:

Descriptive Statistics: C1								
Variable	N N*	Mean	SE Mean	St. Dev	Minimum	Q_1	Median	Q_3
C1	225 0	512.38	6.96	104.45	266.92	440.31	517.12	580.47
Variable	Maximum							
C1	797.80							

If Proc. Univariate of SAS is used, the following percentiles are given: 100% max is the largest value in the data set, $P_{99} = 775$, $P_{95} = 688$, $P_{90} = 645$, $Q_3 = 580$, $Q_2 = 517$, $Q_1 = 441$, $P_{10} = 379$, $P_5 = 327$, and $P_1 = 298$; and 0% min is the smallest number in the data set.

The UNIVARIATE procedure

Variable Score	
Quantile	**Estimate**
100% max	798
99%	775
95%	688
90%	645
75% Q_3	580
50% median	517
25% Q_1	441
10%	379
5%	327
1%	298
0% min	267

When SPSS is used and the pull-down **Analyze ⇒ Descriptive Statistics ⇒ Frequencies** is given, the following output is obtained:

Statistics

N	Valid	225
	Missing	0
Percentiles	1	280.24
	5	327.00
	10	378.60
	25	440.50
	50	517.00
	75	580.50
	90	647.40
	95	697.80
	99	787.58

When STATISTIX is used, the pull-down **Statistics ⇒ Summary Statistics ⇒ Percentiles** is used.

Percentiles

Variable	Cases	1.0	5.0	25.0	50.0	75.0
Score	225	280.24	327.00	440.50	517.00	580.50

It is seen that software packages give about the same values for the percentiles. However, differences do exist since the software programs do use slightly different techniques to find the percentiles.

One technique for finding percentiles "by hand" is the following:

1. Sort or order the n observations from the smallest to the largest. Suppose we are calculating the pth percentile.
2. Calculate $np/100$. If $np/100$ is not an integer, round it up to the next integer and find the corresponding ordered value. If $np/100$ is an integer, say k, calculate the mean of the kth and the $(k + 1)$th ordered observations.

For example, suppose we are looking for the 25th percentile of the data in Table 1-4. Then, $np/100$ is $225(25)/100 = 56.25$. Round this up to 57. The 57th observation when the 225 scores are sorted is 441. This is basically the same as given by all the software packages.

Another measure of dispersion is defined that depends on understanding percentiles. That measure is **Interquartile Range** (IQR). The IQR is defined to be

$$\text{IQR} = Q_3 - Q_1$$

The IQR measures the spread of the middle 50% of the data. The IQR for the spread in the calculus scores is 580.5 – 440.5 = 140 points. The IQR defines the width of the box in a box plot.

Another measure of relative standing for a measurement is the z-score for the measurement. The **z-score** for a measurement x is defined as

$$z\text{-score} = \frac{X - \overline{X}}{S}$$

For example, using the data in Table 1-4, the mean is 512.4, and the standard deviation is 104.5. Joe's score was 650. The z-score is $\frac{650 - 512.4}{104.5} = 1.32$. Joe's score is 1.32 standard deviations above the mean.

Using the rather simple techniques given in this first chapter, an engineer may analyze a set of data. They may find the distribution of the data by using software packages to construct a histogram, a box plot, a dot plot, and a stem-and-leaf plot. They may also find a representative measure such as mean, median, or mode. They may also find a measure that is reflective of the variability of the data. In addition, measures of location may be found.

Exercises

Seventy-five of these light bulbs were burned until they gave no more light. The time each burned was its lifetime.

An electrical engineer determined the lifetimes of 75 light bulbs in hours. The data is shown in Table 1-5.

Table 1-5 Lifetimes of ACE incandescent light bulbs

1491	1470	1564	1487	1575
1475	1600	1483	1460	1473
1499	1660	1464	1502	1433
1433	1508	1526	1463	1485
1470	1522	1507	1517	1547
1484	1560	1529	1494	1545
1496	1512	1485	1526	1547
1546	1502	1560	1512	1476
1449	1494	1504	1494	1457
1522	1448	1459	1453	1515
1470	1457	1533	1515	1553
1705	1462	1670	1499	1625
1529	1548	1547	1462	1485
1475	1497	1497	1489	1534
1505	1509	1541	1403	1525

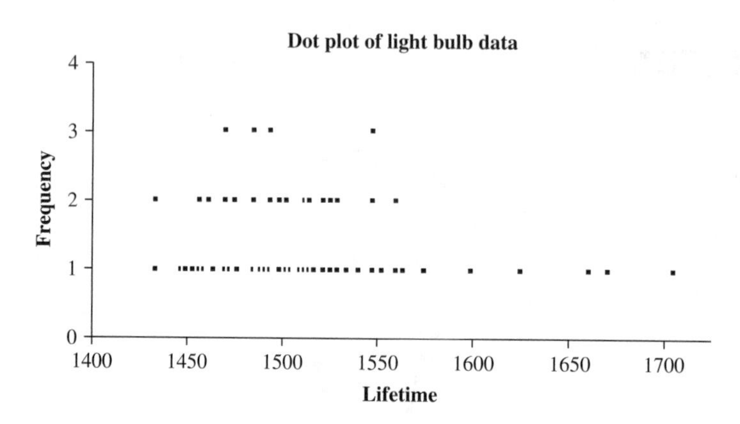

Figure 1-15 EXCEL dot plot of lifetime of bulbs.

1. Even though EXCEL does not have a dot plot as one of its built in plots, the dot plot shown in Figure 1-15 was created using scatter plot of the chart wizard in EXCEL and the data in Table 1-5. Can you explain how the dot plot was created using scatter plot of chart wizard? What does the dot plot tell you about the lifetimes of the light bulbs?
2. Looking at the SPSS histogram of the data from Table 1-5, in Figure 1-16, how wide is each class of this histogram? What is the frequency of the class, 1700–1725 hours? The distribution of the data is not normal or bell-shaped. How would you describe the data?

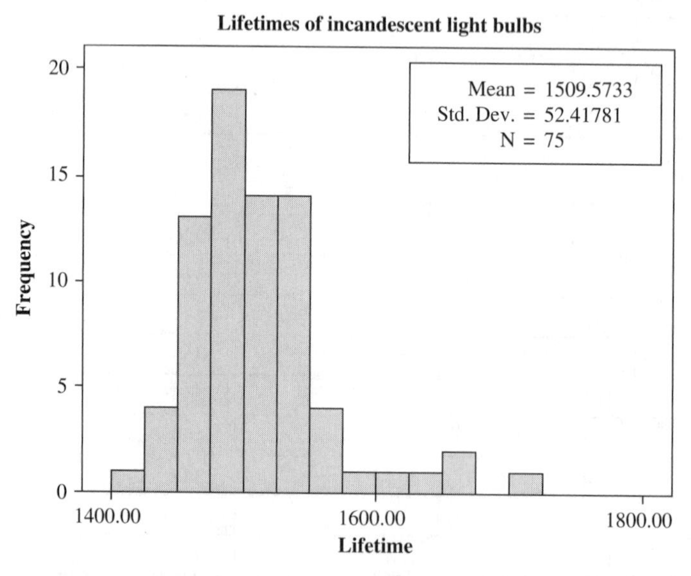

Figure 1-16 SPSS histogram of lifetime of light bulbs.

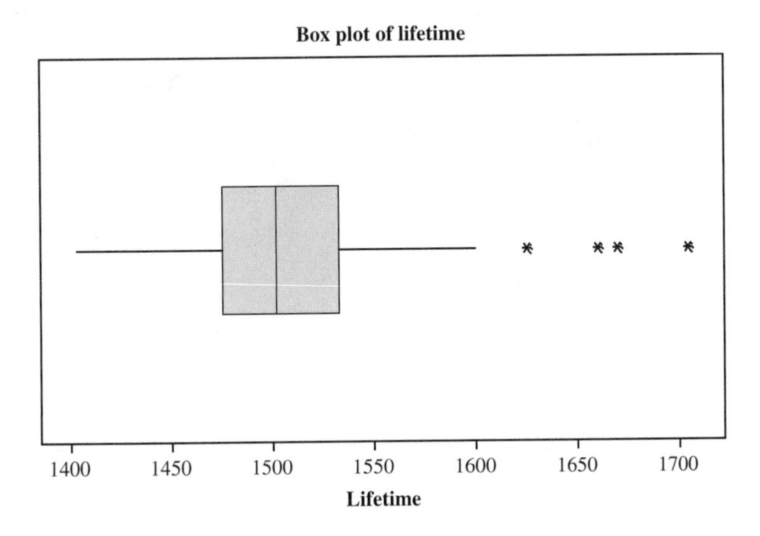

Figure 1-17 MINITAB box plot of the data in Table 1-5.

3. A MINITAB box plot for the data in Table 1-5 is shown in Figure 1-17.
   (a) Identify the four outliers as shown by the asterisks.
   (b) Identify the median as shown by the vertical line inside the box.
   (c) Give the value of the left-hand side of the left whisker.
   (d) Give the value of the right-hand side of the right whisker.
4. The stem-and-leaf shown in Figure 1-18 is for the data in Table 1-5. The stems are hundreds, and the leaves are tens. The first row represents 14 hundreds and 0 tens, or 1400. The second row represents two numbers: 14 hundreds and 3 tens, or two 1430s. The number in the first column is the cumulative frequency. The STATISTIX output gives the cumulative frequency, the stem, and the leaf. It accumulates from the top and from the bottom until the class containing the medium is encountered. When this class is reached, the number in that class is printed in parenthesis. This class is called the **median class**, since it contains the median.
   (a) How many 1470s are indicated?
   (b) How many numbers are 1520 or larger?
   (c) How many numbers are 1450 or smaller?
5. The SPSS descriptive statistics are given for the data in Table 1-5 in Figure 1-19.

```
Stem-and-leaf Plot of lifetime

  Leaf Digit Unit = 10                        Minimum  1403.0
  14  0  represents 1400.                     Median   1502.0
                                              Maximum  1705.0
              Stem   Leaves
        1      14    0
        3      14    33
        9      14    445555
       21      14    666667777777
       37      14    8888888999999999
     (12)      15    000000011111
       26      15    222222233
       17      15    44444445
        9      15    6667
        5      15
        5      16    0
        4      16    2
        3      16
        3      16    67
        1      16
        1      17    0

75 cases included    0 missing cases
```

Figure 1-18 STATISTIX stem-and-leaf for the data in Table 1-5.

($\overline{X}$ is the symbol that represents the mean and S represents the standard deviation.)

(a) What percent of the values in Table 1-5 are between $\overline{X}-S$ and $\overline{X}+S$?

(b) What percent of the values in Table 1-5 are between $\overline{X}-2S$ and $\overline{X}+2S$?

(c) What percent of the values in Table 1-5 are between $\overline{X}-3S$ and $\overline{X}+3S$?

(d) Why do these percents differ considerably from 68%, 95%, and 99.7%?

6. The STATISTIX descriptive statistics are given for the data in Table 1-5 in Figure 1-20.

   (a) Refer to Figure 1-20 and give the z-score for the minimum value of the data in Table 1-5.

   (b) Refer to Figure 1-20 and give the z-score for the maximum value of the data in Table 1-5.

	N	Minimum	Maximum	Mean	SD
lifetime	75	1403.00	1705.00	1509.5733	52.41781
Valid N (listwise)	75				

Figure 1-19 SPSS descriptive statistics for the data in Table 1-5.

Statistix 8.0					
Descriptive Statistics					
Variable	N	Mean	SD	Minimum	Maximum
Lifetime	75	1509.6	52.418	1403.0	1705.0

Figure 1-20 STATISTIX output for descriptive statistics and the data in Table 1-5.

7. The following SAS output is part of the output obtained from the pull-down **Statistics ⇒ Descriptive ⇒ Distributions** applied to the lifetime data in Table 1-5.

The Univariate Procedure	
Quantiles (Definitions)	
Quantile	**Estimate**
100% max	1705
99%	1705
95%	1625
90%	1560
75% Q_3	1533
50% median	1502
25% Q_1	1475
10%	1457
5%	1448
1%	1403
0% min	1403

The **lower inner fence** is $Q_1 - 1.5$ (IQR), the **lower outer fence** is $Q_1 - 3$ (IQR), the **upper inner fence** is $Q_3 + 1.5$ (IQR), and the **upper outer fence** is $Q_3 + 3$ (IQR). Data between $Q_1 - 1.5$ (IQR) and $Q_1 - 3$ (IQR) are called **mild outliers**. Data between $Q_3 + 1.5$ (IQR) and $Q_3 + 3$ (IQR) are also called mild outliers. Data less than $Q_1 - 3$ (IQR) or greater than $Q_3 + 3$ (IQR) are called **extreme outliers**. Refer to the above SAS output. Find the four fences. List all mild and extreme outliers.

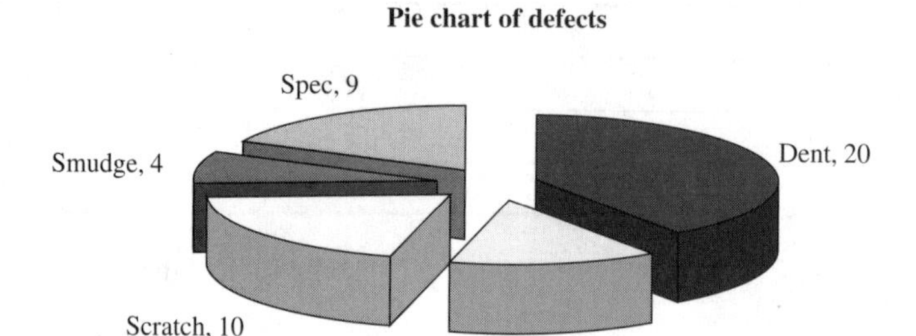

Figure 1-21 EXCEL pie chart.

8. The file defect.xls is read into EXCEL, and the 50 causes of rejections of items are read into the EXCEL worksheet. The 50 causes are sorted, and the following summary table is formed.

Dent	20
Other	7
Scratch	10
Smudge	4
Spec	9

   Using this summary table and the chart wizard, the pie chart in Figure 1-21 is formed. Find the angle that goes with each of the five slices of the pie chart.

9. Refer to Figure 1-22. What is the most frequently occurring defect? What are the two most frequently occurring defects? What are the three most frequently occurring defects?

10. Use the technique given in section 1.3 and EXCEL to find P_5, Q_1, Q_2, Q_3, and P_{95} for the data in Table 1-5. Input the data from Table 1-5 into the EXCEL worksheet. Then use the sort routine to sort the 75 values from the smallest to the largest. Then, apply the technique given in this chapter to find the percentiles. Next, apply the built in function =PERCENTILE (array, *k*) to find the same percentiles and compare your values.

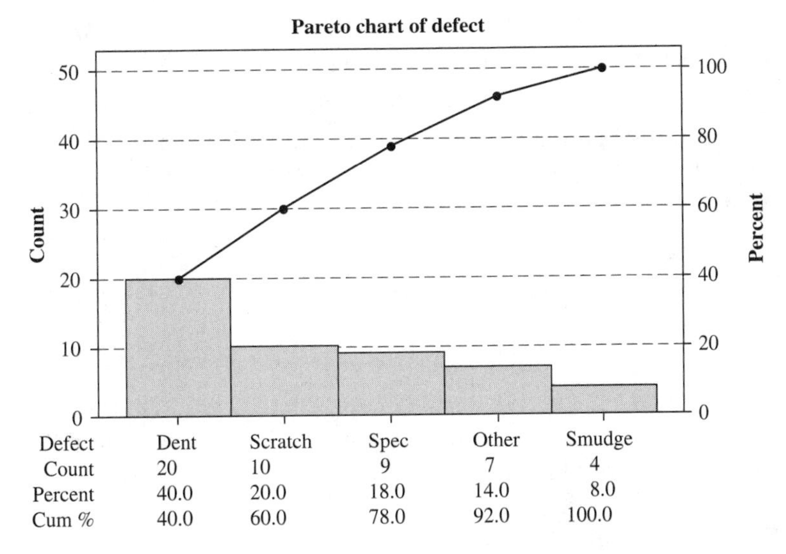

Figure 1-22 MINITAB Pareto chart of defects.

Summary

1. A **histogram** is a graphical device that uses rectangles to show the distribution of data. The heights of the rectangles give the frequency of the classes, and the bases represent the classes.
   MINITAB **Graph ⇒ Histogram**
   SAS **Graph ⇒ Histogram**
   STATISTIX **Statistics ⇒ Summary Statistics ⇒ Histogram**
2. **Dot plots** place dots atop numerical values that are placed along a horizontal axis. They are particularly useful with small data sets.
   MINITAB **Graph ⇒ Dot Plot**
3. In its simplest form, a **box plot** shows a **box** which contains the middle 50% of the data values. It also shows two **whiskers** that extend from the box to the maximum value and from the box to the minimum value.
   MINITAB **Graph ⇒ Box Plot**
   SPSS **Graph ⇒ Box Plot**
4. A **stem-and-leaf** breaks the data down into stems and leaves. For example, 45 might consist of stem 4 and leaf 5, or 4 tens and 5 ones.
   STATISTIX **Statistics ⇒ Summary Statistics ⇒ Stem-and-Leaf Plot**

5. **Quantitative engineering data** is obtained from measuring. **Qualitative engineering data** gives classes or categories for your data values.
6. A **pie chart** is a plot of qualitative data in which the size of the piece of a pie is proportional to the frequency of a category. The angles of the pieces of a pie that correspond to the categories add up to 360 degrees.
   EXCEL **chart wizard**
7. A **bar chart** is a plot of qualitative data in which the frequency or the percent of the data is the height of a rectangle and each category is the base of the rectangle.
   EXCEL **chart wizard**
8. A **Pareto chart** is a special form of a bar chart. It is a tool used by the engineers to prioritize problems for solution. It was named after the Italian economist Vilfredo Pareto (1848–1923). The bars of the Pareto chart are arranged in descending order.
   SPSS **Graph ⇒ Pareto**
9. **Descriptive Statistics** is concerned with summarizing and presenting data concisely.
   EXCEL **Tools ⇒ Data Analysis** is used to obtain the dialog box. **Descriptive Statistics** is selected from the box.
   MINITAB **Stat ⇒ Basic Statistics ⇒ Display Descriptive Statistics**
   SAS **Statistics ⇒ Descriptive ⇒ Summary Statistics**
   STATISTIX **Statistics ⇒ Summary Statistics ⇒ Descriptive Statistics**
   SPSS **Analyze ⇒ Descriptive Statistics ⇒ Descriptive**
10. The **mean** is the sum of all the data divided by the number of observations. The **median** is the middle of the sorted data. The **mode** is the most frequently occurring value in the data set.
11. A measure of variation or dispersion that is simple to calculate and easy to understand is the **range**. The range is the largest value minus the smallest value in the data set.
12. The **variance** is defined to be

$$S^2 = \frac{\sum(X-\overline{X})^2}{n-1} \quad \text{or} \quad \frac{\sum X^2 - \frac{(\Sigma X)^2}{n}}{n-1}$$

The two forms can be shown to be algebraically equivalent. The **standard deviation** is the square root of the variance.

13. The **empirical rule** states that, for any normal distribution, the following relationships approximately hold true:
    (a) Approximately 68% of the distribution is between the (mean – standard deviation) and the (mean + standard deviation).
    (b) Approximately 95% of the distribution is between the (mean – 2 standard deviations) and the (mean + 2 standard deviations).
    (c) Approximately 99.7% of the distribution is between the (mean – 3 standard deviations) and the (mean + 3 standard deviations).
14. The *p*th percentile is the number that divides the lower *p*% of a set of data from the upper (100 – *p*) % of the data.
    EXCEL **=PERCENTILE(array,k)**
    SPSS **Analyze ⇒ Descriptive Statistics ⇒ Frequencies**
    STATISTIX **Statistics ⇒ Summary Statistics ⇒ Percentiles**
15. The first decile is the same as the 10th percentile, the second decile is the same as the 20th percentile, and so forth. The first quartile is the same as the 25th percentile; the second quartile is the same as the 50th percentile; and the third quartile is the same as the 75th percentile. The notations $P_1, \ldots, P_{99}$ represent percentiles, the notations $D_1, D_2, \ldots, D_9$ represent deciles; and Q_1, Q_2, and Q_3 represent quartiles. We have the following equalities: $Q_1 = P_{25}$, $Q_2 = P_{50}$, and $Q_3 = P_{75}$.
16. One technique for finding percentiles "by hand" is the following:
    (a) Sort or order the *n* observations from the smallest to the largest. Suppose we are calculating the *p*th percentile.
    (b) Calculate *np*/100. If it is not an integer, round it up to the next integer and find the corresponding ordered value. If it is an integer, say *k*, calculate the mean of the *k*th and the (*k* + 1)th ordered observations.
17. The **Interquartile Range** (IQR) is defined to be IQR = $Q_3 - Q_1$.
18. The *z*-score for a measurement *x* is defined as *z*-score = $\frac{X - \overline{X}}{S}$.

CHAPTER 2

Probability

2.1 Sample Space, Events, and Operations on Events

Before discussing probability, some definitions are needed. An **experiment** is any operation or procedure whose outcome cannot be predicted with certainty. The **sample space** associated with an experiment consists of all possible outcomes associated with the experiment. An **event** is some subset of the sample space.

EXAMPLE

The simplest example of an experiment is flipping a coin. The sample space is a head or a tail, or in the shorthand of mathematics, $S = \{H, T\}$. Another example is the toss of a die.

Rolling a die is an experiment. You cannot predict its outcome with certainty.

The sample space for this experiment may be represented as $S = \{1, 2, 3, 4, 5, 6\}$.

EXAMPLE

Consider the following experiment. A student is to be selected from a course entitled differential equations for engineers, to be interviewed concerning the engineering program at Midwestern University. The students in the course may be classified according to their sex and their area of engineering. Table 2-1 gives the makeup of the class according to sex and major.

Define the following events on the experiment of selecting one student from this class. Event A is that a female is selected and event B is that the student is majoring in chemical or electrical engineering. These events may be represented in the form of a **Venn diagram**. In a Venn diagram, the sample space is represented as a rectangle and events are rectangles inside the sample space. Figure 2-1 shows the sample space S and events A and B. The events A and B are composed of the following outcomes:

A = {Eve, Heather, Sue, Lana, Beverly, Jody, Brittany, Sandy, Lisa, Kristi, Courtney, Marie}

B = {Scott, Hassan, Todd, Jim, Sam, Eve, Heather, Sue, Lana, Beverly}

Table 2-1 Composition of the class differential equations for engineers

	Chemical Engineering	**Electrical Engineering**	**Civil Engineering**	**Industrial Engineering**
Male	Scott, Hassan	Todd, Jim, Sam	Tom, Joe	Ted
Female	Eve, Heather, Sue	Lana, Beverly	Jody, Brittany, Sandy, Lisa	Kristi, Courtney, Marie

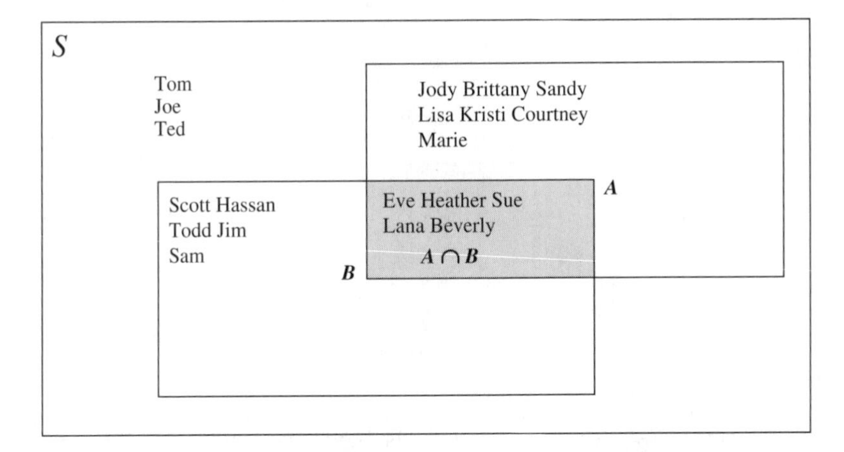

Figure 2-1 There are five outcomes in the intersection.

Operations on the sets may be then defined. The main operations are intersection, union, and complement.

The **intersection** of A and B are the outcomes they have in common. The intersection is written as $A \cap B$. The intersection is $A \cap B = \{$Eve, Heather, Sue, Lana, Beverly$\}$. It is shown in Figure 2-1 in gray.

The **union** of two events, A and B, is the set of outcomes in A or B, or both. It is represented as $A \cup B$. In Figure 2-1, $A \cup B = \{$Jody, Brittany, Sandy, Lisa, Kristi, Courtney, Marie, Eve, Heather, Sue, Lana, Beverly, Scott, Hassan, Todd, Jim, Sam$\}$

The **complement** of A is all the outcomes in S that are not in A. It is represented as A^c. In Figure 2-1, $A^c = \{$Scott, Todd, Sam, Hassan, Jim, Tom, Joe, Ted$\}$. The complement of B is $B^c = \{$Jody, Sandy, Kristi, Marie, Brittany, Lisa, Courtney, Tom, Joe, Ted$\}$.

EXAMPLE

The following may be seen by considering Figure 2-1. (The reader should verify these.)

$A \cap B^c = \{$Jody, Sandy, Kristi, Marie, Brittany, Lisa, Courtney$\}$

$B \cap A^c = \{$Scott, Todd, Sam, Hassan, Jim$\}$

$(A \cup B)^c = \{$Tom, Joe, Ted$\}$

The general Venn diagram representation of the intersection of events A and B is shown in Figure 2-2.

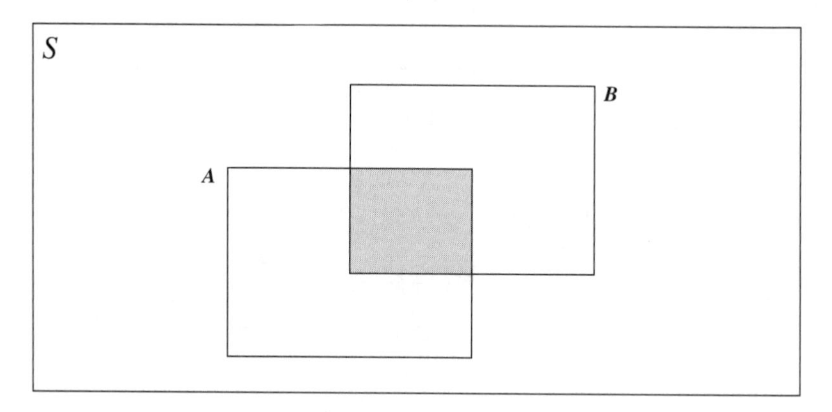

Figure 2-2 Intersection of events A and B is shown in gray.

The union is shown in Figure 2-3. Event A is shaded vertically, event B is shaded horizontally, and $A \cap B$ is crosshatched.

The complement of A, represented as A^c, is the gray portion of Figure 2-4.

Union means the same as the word **or**, intersection means the same as the word **and**, and the complement is the same as the word **not**. That is, the intersection of two events is the outcomes that are in A and B, the union is the outcomes that are in A or B. The complement of A is the outcomes that are in S that are not in A.

Venn diagrams are often used to verify relationships among sets.

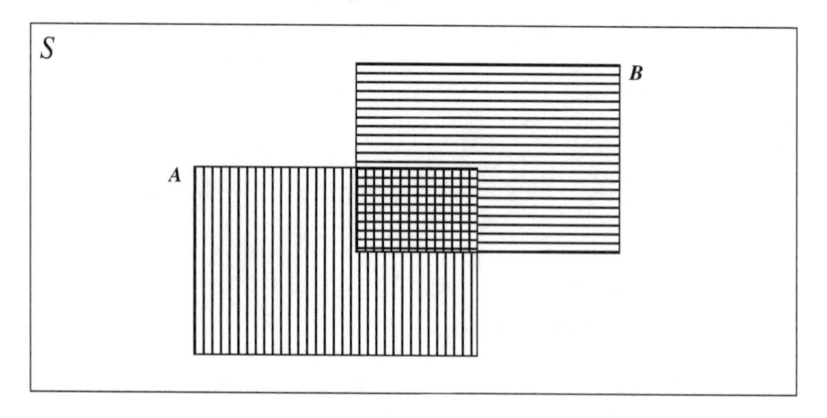

Figure 2-3 Union of events A and B are lined horizontal (B), vertical (A), or crosshatched (A and B).

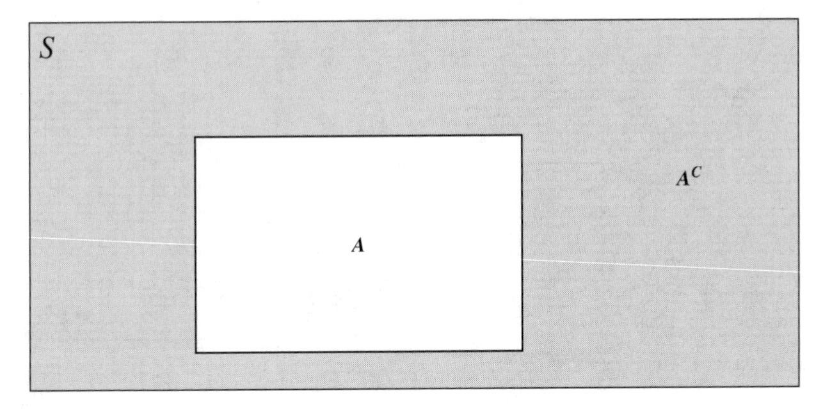

Figure 2-4 Complement of event A is shown in gray.

EXAMPLE

For example, suppose we wished to show that $(A \cap B)^c = A^c \cup B^c$. First, draw a Venn diagram representing $(A \cap B)^c$. This is everything outside of the intersection of A and B. This is shown in Figure 2-5. $(A \cap B)^c$ is shown in gray.

In Figure 2-6, B^c is shaded horizontally.

In Figure 2-7, A^c is shaded vertically.

When the union of the set in Figure 2-6 and the set in Figure 2-7 is created, the result is Figure 2-5. Hence, it is seen that $(A \cap B)^c = A^c \cup B^c$.

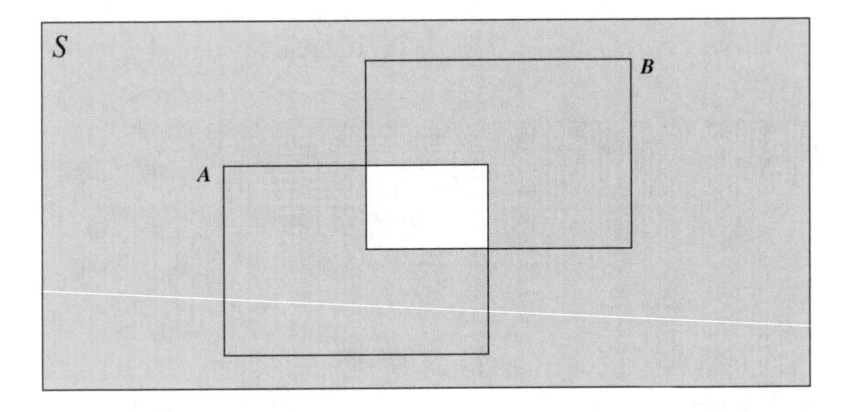

Figure 2-5 $(A \cap B)^c$ is shaded gray.

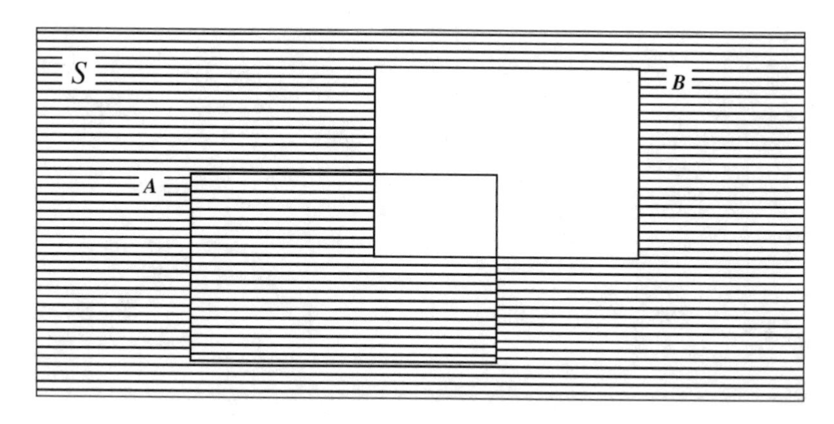

Figure 2-6 B^c is shaded horizontally.

2.2 The Multiplicative Rule, Permutations, and Combinations

Suppose an engineer is asked to rate the new Toyota Yaris according to steering, shifting, handling, and comfort. The Yaris is due to be available in 2007. Each characteristic may be rated as satisfactory or unsatisfactory. A **tree diagram** may be used to list the number of different ratings that the engineer can give the Yaris. This tree diagram can be built using an EXCEL worksheet. The results are shown in Figure 2-8.

There are two choices for each category. Each set of four is called a **branch** of the tree. This tree diagram has $2 \times 2 \times 2 \times 2 = 16$ branches.

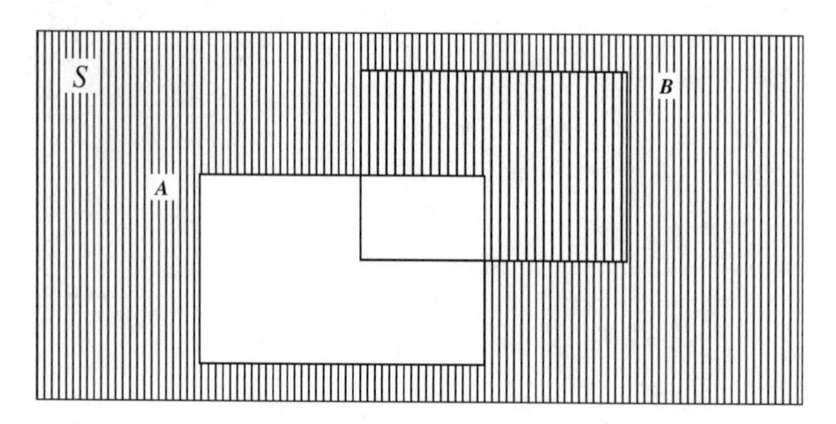

Figure 2-7 A^c is shaded vertically.

Steering	Shifting	Handling	Comfort
Satisfactory	Satisfactory	Satisfactory	Satisfactory
Satisfactory	Satisfactory	Satisfactory	Unsatisfactory
Satisfactory	Satisfactory	Unsatisfactory	Satisfactory
Satisfactory	Satisfactory	Unsatisfactory	Unsatisfactory
Satisfactory	Unsatisfactory	Satisfactory	Satisfactory
Satisfactory	Unsatisfactory	Satisfactory	Unsatisfactory
Satisfactory	Unsatisfactory	Unsatisfactory	Satisfactory
Satisfactory	Unsatisfactory	Unsatisfactory	Unsatisfactory
Unsatisfactory	Satisfactory	Satisfactory	Satisfactory
Unsatisfactory	Satisfactory	Satisfactory	Unsatisfactory
Unsatisfactory	Satisfactory	Unsatisfactory	Satisfactory
Unsatisfactory	Satisfactory	Unsatisfactory	Unsatisfactory
Unsatisfactory	Unsatisfactory	Satisfactory	Satisfactory
Unsatisfactory	Unsatisfactory	Satisfactory	Unsatisfactory
Unsatisfactory	Unsatisfactory	Unsatisfactory	Satisfactory
Unsatisfactory	Unsatisfactory	Unsatisfactory	Unsatisfactory

Figure 2-8 Possible satisfactory/unsatisfactory ratings for the Yaris.

Suppose there are k sets of elements—n_1 in the first set, n_2 in the second set, … , and n_k in the kth set. A sample of k elements is formed by taking one element from each of the k sets. The number of different samples that can be formed is the product $n_1n_2 \dots n_k$. This is called the **multiplicative rule**.

There are several ways to go from one place to another.

EXAMPLE

There are three flights from Houston to Chicago, four flights from Chicago to Memphis, and five flights from Memphis to Atlanta. How many choices of flights include the Houston-Chicago-Memphis-Atlanta connection? The multiplicative rule gives the answer as $3 \times 4 \times 5$, or 60.

Suppose a sample of two is selected from four, and the order of selection is not important. For simplicity, assume the items are the letters a, b, c, and d. There are six possible choices, $\{(a,b), (a,c), (a,d), (b,c), (b,d), (c,d)\}$. If the order of selection makes a difference, then there are 12 choices, $\{(a,b), (b,a), (a,c), (c,a), (a,d), (d,a), (b,c), (c,b), (b,d), (d,b), (c,d), (d,c)\}$. When order of selection is of no concern, we say there are 6 **combinations** possible when selecting two from four. When the order of selection is important, we say there are 12 **permutations** possible when selecting two from four.

Generally, when r objects are selected from a set of n distinct objects, there are $P_r^n = \frac{n!}{(n-r)!}$ permutations possible. There are $\binom{n}{r} = \frac{n!}{r!(n-r)!}$ combinations possible. Applying this formula to the above example, $n = 4$ and $r = 2$. The number of permutations possible is $P_2^4 = \frac{4!}{(4-2)!} = \frac{24}{2} = 12$ and the number of combinations possible is

$$\binom{n}{r} = \frac{n!}{r!(n-r)!} = \frac{4!}{2!(4-2)!} = \frac{24}{2 \times 2} = 6$$

Twelve permutations and six combinations were the answers obtained above when all the combinations and permutations were worked out by listing all of them.

When n and r become larger, the number of combinations and permutations can be quite large.

One of 2,598,960 poker hands that are possible.

EXAMPLE

Suppose a poker hand is dealt to an engineer at a casino, and the engineer would like to know how many poker hands are possible when 5 cards are dealt from 52. The answer can be worked out using EXCEL. Enter the expression =COMBIN(52,5) in any empty cell. The number 2,598,960 is returned. This means that there are 2,598,960 different poker hands that the engineer could be dealt.

EXAMPLE

The engineering club at Midwestern University is to select a president, a vice president, and a treasurer from the 38 members of the club. Note that order is important in this case. The answer is $P_3^{38} = \frac{38!}{35!} = 38 \times 37 \times 36 = 50{,}616$. The EXCEL solution is obtained by entering the expression =PERMUT(38,3) in any cell. The number 50,616 is returned.

EXAMPLE

A calibration study needs to be conducted to see if 20 scales are giving the same weights. How many ways may 5 be selected to perform a preliminary study. The answer is =COMBIN(20,5). The value 15,504 is returned.

EXAMPLE

Industrial engineers are timing workers as to there speed at assembling computers. There are 10 workers. How many different permutations are possible? When the rankings are finished, there will be a number 1 (the fastest), a number 2, … , and a number 10 (the slowest). The answer is $P_{10}^{10} = \frac{10!}{0!} = 10! = 3{,}628{,}800$. This is given in EXCEL by =PERMUT(10,10).

2.3 Probability

The probability that an event will occur is the likelihood of the occurrence of the event. When walking on campus, the likelihood of spotting a student with a cell phone in his hand is very high. The likelihood that a student has been on the Internet within the last 24 hours is high. The likelihood that a student has been to a BMW-dealership within the last 24 hours is low. The likelihood that the student is majoring in computer engineering is low. Probability is a way of assigning a numerical value to these events. There are many different ways of defining probability. We shall discuss three of these. These three definitions are the **classical** definition, the **relative frequency** definition, and the **personal or subjective** definition. There are some definitions such as the measure theoretic approach that are very mathematical and that require more mathematics background than most of the readers of this book have.

The classical definition might also be called the equally likely outcome definition. Suppose an experiment has n equally likely outcomes, and an event E occurs if anyone of k of these outcomes occurs as an outcome of the experiment. The classical definition of probability defines the probability of event E as $P(E) = k/n$.

EXAMPLE

Suppose, a lot consisting of 100 items contains 5 defectives. If one item is randomly selected, the probability that it is defective is P(item is defective) = 5/100 or 0.05. There are $k = 5$ outcomes out of $n = 100$ that satisfy the event that the item is defective. It is common terminology to say that there is a 5% chance that a defective item will be chosen. Most people would regard this as a low probability.

EXAMPLE

Suppose 3 items are inspected and if at least 1 defective is found, the lot will be 100% inspected. Otherwise, the lot will be passed on. How likely is it that a lot containing 5 defectives will be passed on? The lot will be passed on if no defectives are found in the sample, or if all 3 come from the 95 nondefectives. There are $k = \binom{95}{3}$ such samples possible. There are $n = \binom{100}{3}$ samples that could be drawn. The probability that the lot will be passed on, using EXCEL to do the calculations, is =COMBIN(95,3)/COMBIN(100,3) = 0.855999.

The **frequency probability definition** states that if an experiment is repeated many times, the probability of an event is the proportion of times the event occurs in n repetitions.

EXAMPLE

If a jet from Chicago to Omaha is on time 80% of the time, the probability, 0.80, is assigned to the event that the jet will arrive on time when flying from Chicago to Omaha. If an engineer communicates 50% of the time, by e-mail, with a company with whom he consults, then the probability is 0.50 that he will communicate by e-mail with the company.

A third definition is that of **personal or subjective probability evaluations**. An aeronautical engineer is assigning subjective probability when they assign 0.80 as the probability that a manned spaceflight to Mars in 2010 will be successful. A civil engineer states that the probability that the dikes will hold in New Orleans is 0.75. This is a subjective probability assignment. An army engineer estimates that the probability a terrorist organization will get their hands on weapons of mass destruction to be 0.10 is another example of personal probability. Sometimes personal or subjective probability is the only kind available. The events mentioned in this paragraph do not lend themselves to repeatability.

Regardless of the source of the probability assignment, there are probability laws that the probability of events must satisfy. These will be considered in the following sections.

2.4 The Axioms of Probability

A probability function is actually a set function. The domain is composed of sets, and the range is real numbers.

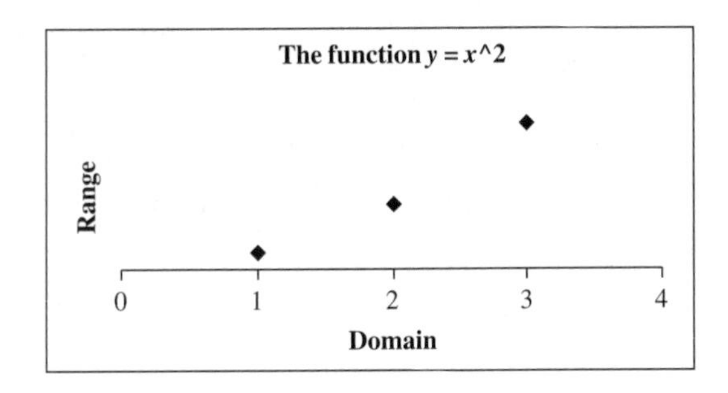

Figure 2-9 A function is a mapping from the real numbers to the real numbers.

The usual mathematical function maps a real number into a real number. For example, $y = x^2$ is a real valued function. If the domain is 1, 2, and 3, then the range is 1, 4, and 9. The domain and the range are real numbers. See Figure 2-9.

Contrast the following example with the above example. Toss a coin twice. The sample space is $S = \{HH, HT, TH, TT\}$. Let E_1 be the event that the outcome has no heads, E_2 be the event that the outcome has 1 head, and E_3 be the event that the outcome has 2 heads. $P(E_1) = 1/4$, $P(E_2) = 2/4$, and $P(E_3) = 1/4$. Note that the probability function has domain— E_1, E_2, E_3; and range 1/4, 2/4, 1/4. In other words, the domain is composed of sets, and the range is real numbers.

The axioms of probability for a finite sample space are as follows. Suppose you have a finite sample space S and an event A in S. Define $P(A)$, the probability of A, to be a value of a set function that satisfies the following three conditions:

Axiom 1: $0 \leq P(A) \leq 1$ for each event A in S.

Axiom 2: $P(S) = 1$.

Axiom 3: If A and B are mutually exclusive events in S, then $P(A \cup B) = P(A) + P(B)$.

Mutually exclusive events have empty intersections. The three axioms are consistent with the classical and the frequency definition of probability.

EXAMPLE

Suppose an experiment has a sample space that is made up of four mutually exclusive events, that is $S = \{A, B, C, D\}$. Check whether the following assignment of probabilities are permissible. (See if the axioms allow the assignment.)

a. $P(A) = 0.25$, $P(B) = 0.25$, $P(C) = 0.25$, and $P(D) = 0.25$
b. $P(A) = 0.25$, $P(B) = 0.35$, $P(C) = 0.45$, and $P(D) = -0.05$
c. $P(A) = 0.25$, $P(B) = 0.25$, $P(C) = 0.35$, and $P(D) = 0.25$
d. $P(A) = 0.20$, $P(B) = 0.25$, $P(C) = 0.25$, and $P(D) = 0.25$

a. All the probabilities are between 0 and 1, and their sum is 1. This assignment is allowed.
b. The assignment is not allowed because $P(D) < 0$.
c. The sum of the probabilities is not valid because it exceeds 1.
d. The sum of the probabilities is not valid because it is less than 1.

2.5 Some Elementary Theorems (Rules) of Probability

The third axiom of the three axioms in the last section can be extended from two mutually exclusive events to n mutually exclusive events by the use of mathematical induction. It says that if $A_1, A_2, \ldots, A_n$ are mutually exclusive events within a sample space, then $P(A_1 \cup A_2 \cup \cdots \cup A_n)$ is the sum of the probabilities of the n events.

EXAMPLE

An industrial engineer has found that, when she takes a sample of size 5 of a product, 90% of the time there is no defective in the sample, 3% of the time there is 1 defective, 2% of the time there are 2 defectives, 2% of the time there are 3 defectives, 2% of the time there are 4 defectives, and 1% of the time there are 5 defectives. What is the probability of taking a sample that has at least 3 defectives in the 5? The event with at least 3 defectives in the 5 is composed of three mutually exclusive events. The mutually exclusive events are 3 defectives, 4 defectives, and 5 defectives. The probability of at least 3 defectives is $(0.02 + 0.02 + 0.01) = 0.05$.

Suppose there are two events that are not mutually exclusive. What is $P(A \cup B)$? $(A \cup B)$ can be expressed as the union of three mutually exclusive events as follows: $(A \cup B) = (A \cap B^c) \cup (A \cap B) \cup (B \cap A^c)$. This relationship is shown in Figure 2-10.

From this mutually exclusive decomposition, we have the following from Axiom 3.

$$P(A \cup B) = P(A \cap B^c) + P(A \cap B) + P(B \cap A^c).$$

Now add and subtract $P(A \cap B)$ to the right side to obtain

$$P(A \cup B) = P(A \cap B^c) + P(A \cap B) + P(B \cap A^c) + P(A \cap B) - P(A \cap B)$$

$$P(A \cup B) = P(A) + P(B) - P(A \cap B)$$

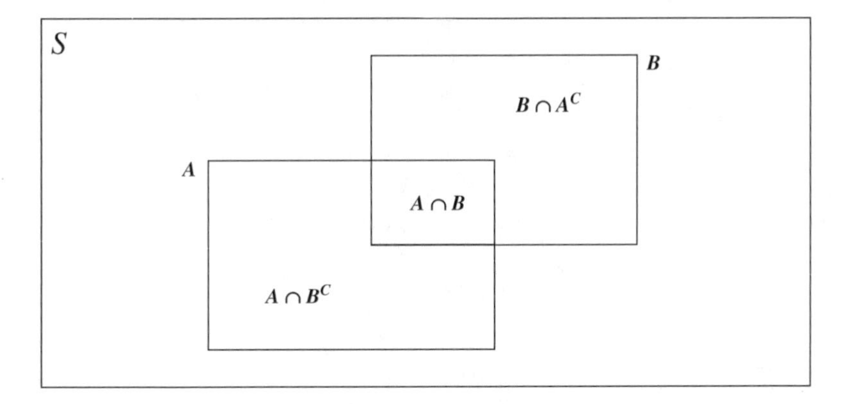

Figure 2-10 $(A \cup B) = (A \cap B^c) \cup (A \cap B) \cup (B \cap A^c)$.

This relationship is called the **general probability addition rule** for two events, and if A and B are mutually exclusive $P(A \cup B) = P(A) + P(B)$ is called the **special addition rule**.

EXAMPLE

It is known that 60% of a class is female, that 50% are majoring in chemical or electrical engineering, and 25% are female and are majoring in chemical or electrical engineering. What percent are female, or majoring in chemical or electrical engineering? A is the event that a member of the class is female, and B is the event that a member of the class is majoring in chemical or electrical engineering. The general probability addition rule gives $P(A \cup B) = P(A) + P(B) - P(A \cap B) = 0.60 + 0.50 - 0.25 = 0.85$. Thus, 85% are female, or majoring in chemical or electrical engineering. See the example at the beginning of section 2.1 to check this out.

When three or more nonmutually exclusive events are involved, Venn diagrams become more complicated. Figure 2-11 shows three such events.

Event A is composed of regions 1, 2, 3, and 6. Event B is composed of regions 1, 3, 4, and 7. Event C is composed of regions 1, 2, 4, and 5. Region 8 is $(A \cup B \cup C)^c$, or everything that is in S but outside the union of A, B, and C.

EXAMPLE

Suppose event A is that an engineering student owns a cell phone, event B is that an engineering student has a personal digital assistant (PDA), and event C is that an engineering student lives at home. Region 1 consists of all those students who own a cell phone, own a PDA, and live at home. Region 2 consists of those students who own a cell phone, live at home, but does not own a PDA. Region 3 consists of those students who own a cell phone, have a PDA, but do not live at home. Region 4 consists of those students who own a PDA, live at home, but do not have a cell phone.

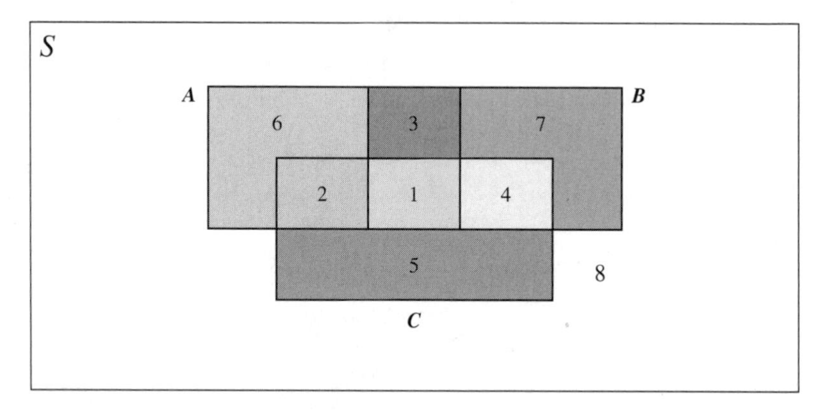

Figure 2-11 Three nonmutually exclusive events A, B, and C.

Region 5 consists of those students who live at home, do not have a cell phone and do not own a PDA. Region 6 consists of those students who own a cell phone, do not own a PDA, and do not live at home. Region 7 consists of those students who have a PDA, do not own a cell phone, and do not live at home. Region 8 consists of all those students who do not own a cell phone, do not own a PDA and do not live at home. It can be shown that the following relationship holds: (*This is called the general addition rule for three events.*)

$$P(A \cup B \cup C) = P(A) + P(B) + P(C) - P(A \cap B) - P(A \cap C) \\ - P(B \cap C) + P(A \cap B \cap C)$$

Suppose the following probabilities are known.

Region	1	2	3	4	5	6	7	8
Probability	0.02	0.04	0.03	0.05	0.16	0.45	0.20	0.05

From Figure 2-11, $P(A \cup B \cup C)$ must equal 0.95. If the additions and the subtractions on the right hand side of the equation for $P(A \cup B \cup C)$ are performed, the answer is $(0.54 + 0.30 + 0.27 - 0.05 - 0.06 - 0.07 + 0.02) = 0.95$. This confirms the equation for this example.

Suppose A and A^c are complementary events. Then $A \cap A^c$ is empty, and $A \cup A^c = S$. $P(A \cap A^c) = P(A) + P(A^c)$ and since $A \cup A^c = S$, $P(A) + P(A^c) = 1$ or $P(A^c) = 1 - P(A)$. The rule of complements states that

$$P(A^c) = 1 - P(A).$$

EXAMPLE

An engineer selects a sample of 5 iPods from a shipment of 100 that contains 5 defectives. Find the probability that the sample contains at least 1 defective. At least 1 defective means 1 or 2 or 3 or 4 or 5 defectives. Remember **or** and union mean the same. The probability $P(1 \text{ or } 2 \text{ or } 3 \text{ or } 4 \text{ or } 5 \text{ defectives}) = P(1) + P(2) + P(3) + P(4) + P(5)$. The events—1 defective, 2 defectives, 3 defectives, 4 defectives, and 5 defectives—are mutually exclusive, and we add the probabilities. However, the whole sample space consists of 0 defectives, 1 defective, 2 defectives, 3 defectives, 4 defectives, or 5 defectives. And since the probability of a certain event is 1, we know that

$$P(0) + P(1) + P(2) + P(3) + P(4) + P(5) = 1$$

The complement of at least 1 defective is no defective, and therefore the slick and easy solution is

$$P\{\text{at least 1 defective}\} = 1 - P\{0 \text{ defectives}\}$$

$$P\{\text{at least 1 defective}\} = 1 - \frac{\binom{95}{5}}{\binom{100}{5}} = 1 - \frac{57940519}{75287520} = 1 - 0.77 = 0.23$$

EXCEL is used to evaluate the expression that is subtracted from 1. There are 57,940,519 ways that 5 nondefectives can be chosen from the 95 nondefectives and 75,287,520 ways that 5 nondefectives can be chosen from the 100 nondefectives. It is interesting to consider the work that the rule of complements saves in this problem. If the rule of complements is not used, the following must be evaluated.

$$\frac{\binom{5}{1}\binom{95}{4}}{\binom{100}{5}} + \frac{\binom{5}{2}\binom{95}{3}}{\binom{100}{5}} + \frac{\binom{5}{3}\binom{95}{2}}{\binom{100}{5}} + \frac{\binom{5}{4}\binom{95}{1}}{\binom{100}{5}} + \frac{\binom{5}{5}\binom{95}{0}}{\binom{100}{5}}$$

If EXCEL is used to evaluate this expression, the following is in cell A1: =COMBIN(5,1)*COMBIN(95,4)/COMBIN(100,5). Similar expressions are in cells A2:A5. The numerical values obtained are shown in the following table. The last number is the expression =SUM(A1:A5).

A	B
0.211425811	=COMBIN(5,1)*COMBIN(95,4)/COMBIN(100,5)
0.018384853	=COMBIN(5,2)*COMBIN(95,3)/COMBIN(100,5)
0.00059306	=COMBIN(5,3)*COMBIN(95,2)/COMBIN(100,5)
6.30915E-06	=COMBIN(5,4)*COMBIN(95,1)/COMBIN(100,5)
1.32824E-08	=COMBIN(5,5)*COMBIN(95,0)/COMBIN(100,5)
0.230410047	=SUM(A1:A5)

Note that the sum agrees with the answer found above. The amount of computation that can be saved using the rule of complements in problems is considerable.

2.6 Conditional Probability

The sample space that a probability is computed with respect to has a great bearing upon the probability. Consider the following simple example.

EXAMPLE

If a die is tossed, the probability that the face 2 turns up is 1/6. If a die is tossed, and it is known that the face is an even number, then the probability that the face 2 turns up is 1/3. This is twice what the probability is if nothing is known. If it is known that the face that turns up is an odd number, then the probability that the face 2 turns up is 0. We see that the probability of an event is conditioned by what other events, we know, have occurred.

EXAMPLE

Reconsider the example in section 2.1. A class consists of twenty engineering students. One student is to be selected to discuss the engineering program at Midwestern University. Figure 2-12 shows two events. Event A is that a female is selected, and event B is that the student is majoring in chemical or electrical engineering. The names of those students composing the two events, and those students not falling into either event are given. The probability of event A is $P(A) = 12/20 = 0.6$. The probability of event B is $P(B) = 10/20 = 0.5$. Suppose, it is known that event B has occurred, and we wish to know the probability of A. In other words, it is known that the student is majoring in chemical or electrical engineering. We want to know the probability that it is a female. The notation for such a conditional probability is $P(A \mid B)$. Looking at Figure 2-12, we know that we are in rectangle B, and we want to find the probability that the chosen student is a female. The condition that the student is an electrical or chemical engineer limits us to rectangle B. Within rectangle B, there

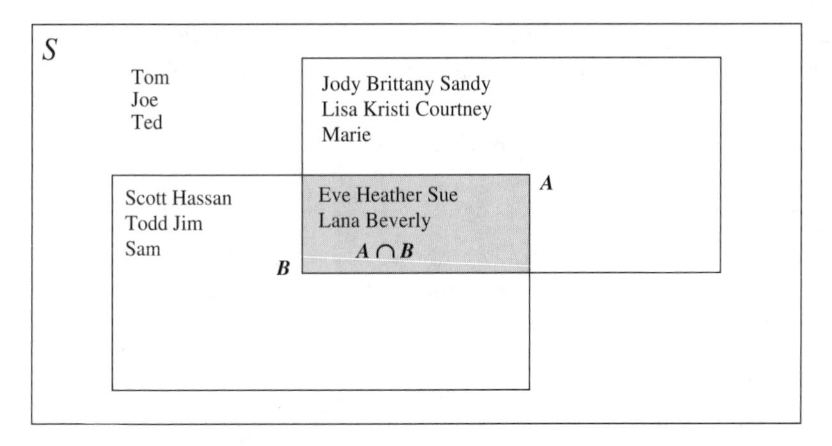

Figure 2-12 Conditional probability illustration.

are five females. The probability is 5/10 = 0.5. The conditional probability was found by dividing the number in event $A \cap B$ by the number in B. If the number in $A \cap B$ is divided by the number in S, and the number in B is also divided by the number in S, the following formula results:

$$P(A|B) = \frac{P(A \cap B)}{P(B)}$$

The formula

$$P(A|B) = \frac{P(A \cap B)}{P(B)}$$

is called the formula for conditional probability.

EXAMPLE

One percent of parts have type 1 defect, 2% have type 2 defect, and 0.5% have both types of defects. A part is known to have type 1 defect. What is the probability that it has type 2 defect? Let A represent the event that a part has a type 1 defect, and B represent the event that a part has also a type 2 defect. We are asked to find $P(B|A)$. $P(B|A) = \frac{P(A \cap B)}{P(A)} = \frac{0.005}{0.01} = 0.5$. Now suppose, a part is known to have a type 2 defect. What is the probability it also has a type 1 defect? We are now asked to find $P(A|B) = \frac{P(A \cap B)}{P(B)} = \frac{0.005}{0.02} = 0.25$.

Take the conditional probability formula and solve it for $P(A \cap B)$ by multiplying both sides by $P(B)$. The following form of the formula is obtained.

$$P(A \cap B) = P(A \mid B)\, P(B)$$

EXAMPLE

A box contains 5 defective and 195 nondefective cell phones. A quality control engineer selects 2 cell phones at random **without replacement**. What is the probability that

a. Neither is defective?

b. Exactly 1 is defective?

c. Both are defective?

a. P(first is nondefective $\cap$ second is nondefective) = P(first is nondefective) P(second is nondefective | first is nondefective) $= \frac{195}{200}\frac{194}{199} = 0.9505$

b. P{(first is nondefective $\cap$ second is defective) $\cup$ (first is defective $\cap$ second is nondefective)}. The union of two mutually exclusive events tells us to add.

= P(first is nondefective $\cap$ second is defective) + P(first is defective $\cap$ second is nondefective)

= P(first is nondefective) P(second is defective | first is nondefective) + P(first is defective) P(second is nondefective | first is defective)

$= \frac{195}{200}\frac{5}{199} + \frac{5}{200}\frac{195}{199} = 0.0490.$

c. P(first is defective $\cap$ second is defective) = P(first is defective) P(second is defective | first is defective) $= \frac{5}{200}\frac{4}{199} = 0.0005.$

Notice that the three probabilities add to 1. This tells us that we are certain to obtain 0, 1, or 2 defectives when we select two from the 200.

Another concept associated with conditional probabilities is the concept of independence of events. If $P(A \mid B) = P(A)$ then we say that A and B are **independent events**. Note that when A and B are independent events $P(A \cap B) = P(A \mid B)\, P(B) = P(A)\, P(B)$.

EXAMPLE

A coin is tossed twice. A is the event that the first toss is a head, and B is the event that the second toss is a head. $P(A) = 0.5$ and $P(B \mid A) = 0.5 = P(B)$. The coin has no idea what the outcome on the first toss was. The two outcomes are said to be independent of one another.

EXAMPLE

A box contains 5 defective and 195 nondefective cell phones. A quality control engineer selects 2 cell phones at random **with replacement**. What is the probability that

a. Neither is defective?

b. Exactly 1 is defective?

c. Both are defective?

Note that this example is just like the one preceding the last example, except the sampling is done with replacement. This makes the second draw independent of the first draw.

a. $P(\text{first is nondefective} \cap \text{second is nondefective}) = \frac{195}{200}\frac{195}{200} = 0.950625$

b. $P(\text{exactly one defective}) = \frac{195}{200}\frac{5}{200} + \frac{5}{200}\frac{195}{200} = 0.04875$

c. $P(\text{both are defective}) = \frac{5}{200}\frac{5}{200} = 0.000625$

Note that, in the case of sampling with replacement, the three probabilities also add to 1. $(0.950625 + 0.04875 + 0.000625 = 1)$

If n events are independent, then $P(A_1 \cap A_2 \cap \cdots \cap A_n)$ equals the product of the probabilities of the events A_i, $i = 1, 2, \ldots, n$.

EXAMPLE

If a die is rolled five times, what is the probability of obtaining five 6s in a row? The rolls are independent of one another. The probability of a 6 on any one roll is 1/6. The probability of five 6s in a row is $\left(\frac{1}{6}\right)^5$.

EXAMPLE

A machine has a probability of producing a defective equal to 0.05, every time it produces a product. If three of its products are selected randomly and independently during a shift, what is the probability that a quality engineer will find:

a. No defectives in the three

b. One defective in the three

c. Two defectives in the three

d. Three defectives in the three

a. P(nondefective **and** nondefective **and** nondefective) = (.95) (.95) (.95) = 0.857375

b. P(defective **and** nondefective **and** nondefective **or** nondefective **and** defective **and** nondefective **or** nondefective **and** nondefective **and** defective) = .05(.95)(.95) + .95(.05)(.95) + .95(.95)(.05) = 0.135375

c. P(defective **and** defective **and** nondefective **or** defective **and** nondefective **and** defective **or** nondefective **and** defective **and** defective) = .05(.05)(.95) + .05(.95)(.05) + .95(.05)(.05) = 0.007125

d. P(defective **and** defective **and** defective) = .05(.05)(.05) = 0.000125

Note *When you see and ($\cap$), multiply and when you see or ($\cup$), add. Also note that the probabilities add up to 1.*

2.7 Bayes' Theorem

We now combine many of the ideas in the first six sections to come up with two very important results. One is known as the rule of total probability and the other is known as Bayes' theorem. They are developed with the help of an example.

EXAMPLE

A plant has three suppliers. S_1 supplies 30% of the parts to the plant, S_2 supplies 50% of the parts, and S_3 supplies the remaining 20%. One percent of the parts that S_1 supplies are defective, 2% of the parts supplied by S_2 are defective, and 3% of S_3 parts are defective. The event of interest is that a part received by the plant is defective. The plant has defective parts in inventory if they came from supplier S_1 and are defective, or if they came from supplier S_2 and are defective, or if they came from supplier S_3 and are defective. Let D be the event that a part received by the plant is defective. D may be expressed as

$$D = (D \cap S_1) \cup (D \cap S_2) \cup (D \cap S_3)$$

This represents a disjoint union, and the probability of the disjoint union is the sum of the probabilities.

$$P(D) = P(D \cap S_1) + P(D \cap S_2) + P(D \cap S_3)$$

$$P(D) = P(S_1)P(D \mid S_1) + P(S_2)P(D \mid S_2) + P(S_3)P(D \mid S_3)$$

$$P(D) = (0.3)(0.01) + (0.5)(0.02) + (0.2)(0.03) = 0.019$$

The percent of parts at the plant that are defective is 1.9%. Expanding this idea, we have the following **rule of total probability**.

If $S_1, S_2, \ldots, S_n$ are mutually exclusive events, and one of the n events must occur, then

$$P(D) = P(S_1)P(D \mid S_1) + P(S_2)P(D \mid S_2) + \cdots + P(S_n)P(D \mid S_n)$$

EXAMPLE

To carry this example one step further, suppose we choose a defective item at the plant and ask the question, "What is the probability that it came from supplier S_3?" This may be formulated as follows:

Find $P(S_3 \mid D)$. Consider the following derivation:

$$P(S_3 \mid D) = \frac{P(D \cap S_3)}{P(D)} = \frac{P(S_3)P(D \mid S_3)}{P(S_1)P(D \mid S_1) + P(S_2)P(D \mid S_2) + P(S_3)P(D \mid S_3)}$$

$$P(S_3 \mid D) = \frac{(0.20)(0.03)}{(0.30)(0.01) + (0.50)(0.02) + (0.20)(0.03)} = \frac{0.006}{0.019} = 0.3158$$

Given that the part was defective, the probability it came from supplier S_1 is $P(S_1 \mid D) = (0.30)(0.01)/0.019 = 0.1579$

Given that the part was defective, the probability it came from supplier S_2 is $P(S_2 \mid D)\ (0.50)(0.02)/0.019 = 0.5263$

Note that $(0.3158 + 0.1579 + 0.5263) = 1$.

If this last result is extended from 3 suppliers to n suppliers, it is called Bayes' theorem.

2.8 Mathematical Expectation

Now that we have established the concepts of probability and its properties, we turn our attention to one of the many uses of probability. What do we expect to happen if we know the probabilities concerning an event, and what action do we take as a result of that expectation?

EXAMPLE

The following game is played. A coin is flipped. If it lands heads, you are paid \$5. If it lands tails, you pay \$5. If you perform this experiment many times, what do you expect your winnings to be? The answer is simple. Your expectation is 0. We arrive at this in the following way. Consider the following table:

Winning	\$5, if heads	–\$5, if tails
Probability	1/2	1/2

The expected winning is 5(1/2) – 5(1/2) = 0. Such a game is called **fair**. If your expected winnings are zero, the game is said to be fair.

The definition of **mathematical expectation** is as follows. If you win amount a_1 with probability p_1, amount a_2 with probability p_2 , … , and amount a_n with probability p_n your mathematical expectation is $\text{Exp} = \Sigma a_i p_i$. The amounts are real numbers; and the probabilities are between 0 and 1, and sum to 1.

EXAMPLE

The game of **chuck-a-luck** is played as follows. Three dice are thrown. You pay \$5 to play. You bet on the number of times the number 6 appears in the toss of the three dice. Let x represent the number of times 6 occurs when the three dice are tossed. The following table gives the probabilities associated with different values that x may assume.

x	0	1	2	3
$P(x)$	125/216	75/216	15/216	1/216

The rules for this game are summarized in the following table:

Winning	–\$5	\$1	\$15	\$25
Probability	125/216	75/216	15/216	1/216

What the table says is: If no 6s occur, you loose the \$5 you paid to play. If one 6 occurs, you get your \$5 back plus \$1. If two 6s occur you get your \$5 back plus \$15. If your toss results in three 6s occurring, you get your \$5 back plus \$25. Your mathematical expectation of the game is Exp = (–5)(125/216) + (1)(75/216) + (15)(15/216) + (25)(1/216) = –300/216 = –1.38. If you play the game 100 times, you expect to loose about \$138. You will loose \$1.38 per game on the average. This is not a fair game. It favors the "house".

NOTES CONCERNING THE CHUCK-A-LUCK EXAMPLE

1. How can the distribution of x be derived? By using EXCEL as well as copy and paste techniques, the 216 possible outcomes associated with rolling the

three dice may easily be obtained. The outcomes are placed in the range A1:C216. Then the function =COUNTIF(A1:C1,"6") is entered in D1 and a click-and-drag is performed across the 216 observations. Then, a count shows that 6 occurred zero 125 times, 6 occurred once 75 times, 6 occurred twice 15 times, and 6 occurred 3 times once. The table giving x and $P(x)$ are easily obtained by knowing the number of times 0, 1, 2, and 3 occurred. The reader is encouraged to use EXCEL to verify this.

2. The game could be made a fair game in many different ways. One way is to leave it alone, except for one change. Give the $5 back as well as a grand prize of $325 if 6 occurs 3 times. This gives the following table:

Winning	–$5	$1	$15	$325
Probability	125/216	75/216	15/216	1/216

The mathematical expectation is Exp = $(-5)(125/216) + (1)(75/216) + (15) \times (15/216) + (325)(1/216) = 0$. The reader is encouraged to give another version of the game that will result in a fair game.

Note that the mathematical expectation does not tell you what will happen on one performance of the experiment, but rather what to expect from the experiment in the long run.

EXAMPLE

A game is played as follows. You pay $1 to play. A coin is flipped four times. If four tails or four heads are obtained, you get your $1 back plus $5 more. Otherwise you forfeit your $1. What is the mathematical expectation? You win $5 with probability 1/16 + 1/16. You win –$1 with probability 14/16. The mathematical expectation is $5(2/16) - 1(14/16) = -4/16$ or -0.25. There is an average of a quarter loss per player per game. The probabilities are arrived at as follows. You win $5 if you toss all heads or all tails. Because of independence, $P(HHHH) = (1/2)^4 = 1/16$. Similarly $P(TTTT) = (1/2)^4 = 1/16$. You add the two because they are mutually exclusive. You loose $1 if you get the complement of *HHHH* or *TTTT*. The probability of the complement of all heads or all tails is $1 - 2/16 = 14/16$.

EXAMPLE

An engineering company prepares an estimate for a job. The cost of preparing the estimate is $10,000. The amount of profit over and above the $10,000, to be made, is $25,000 if their estimate is accepted. The probability that their estimate will be accepted is 0.7, and the probability that their estimate will not be accepted is 0.3. What is the expected profit? The expected profit is Exp = $25,000(0.7) – $10,000(0.3) = $14,500.

EXAMPLE

An engineering consulting company must decide between two jobs. The decision is based on the following information. Which job has the greater expected profit?

Job 1

Probability	0.2	0.8
Outcome	Loss of \$30,000	Profit of \$100,000

Job 2

Probability	0.4	0.6
Outcome	Loss of \$20,000	Profit of \$125,000

For Job 1, Exp = −30,000(0.2) + 100,000(0.8) = \$74,000, and for Job 2 Exp = −20,000(0.4) + 125,000(0.6) = \$67,000. The engineering consulting company should take Job 1 because it has the greater expected profit. (This, of course, assumes all other factors are the same.)

What is the difference between the first three examples and the last two examples of this section? The difference is in the computation of the probabilities. In the games of chance, the probabilities are computed using the classical definition of equal outcomes. In the engineering examples, the probabilities are computed from limited observations, or from personal or subjective sources. The probabilities in the engineering examples are not as accurate as those in the examples that use the classical method of computing probabilities. As long as the dice are balanced and the coins are fair, the mathematical expectation will accurately reflect what occurs on the average. There are also some questions about the monetary values that are assigned in the real world examples.

Exercises

1. An experiment consists of selecting and classifying three items as either conforming (C) or nonconforming (N). The items are randomly selected from a large batch. They are all similar items. The sample space may be represented as $\{CCC, CNC, CCN, CNN, NCC, NNC, NCN, NNN\}$. The probability an item is conforming is 0.97, and the probability that an item is nonconforming is 0.03.

   (a) Assign a probability to each point in the sample space. (Give six decimal places.)

Define the following events: $A = \{\text{at least two items conform}\}$
$B = \{\text{at most two items conform}\}$. Find the following probabilities.

(b) $P(A)$ (c) $P(B)$ (d) $P(A \cap B)$ (e) $P(A \cup B)$ (f) $P(A^c)$ (g) $P(B^c)$

2. An experiment consists of selecting and classifying four items, each as either conforming or nonconforming. The items are randomly selected from a large batch. They are all similar items. Give the sample space for this experiment, and the probabilities associated with each outcome. The probability an item is conforming is 0.9, and the probability that an item is nonconforming is 0.1.
   Define the following events:
   $A = \{\text{two or three items conform}\}$
   $B = \{\text{one or two items conform}\}$

   Draw a Venn diagram and show the outcomes from the sample space that make up A and B, and those outcomes outside of A or B.

   Find the elements in the following sets.

   (a) $A \cap B^c$ (b) $B \cap A^c$ (c) $A \cap B$ (d) $(A \cup B)^c$

   Find the following probabilities:

   (e) $P(A)$ (f) $P(B)$ (g) $P(A \cap B)$ (h) $P(A \mid B)$ (i) $P(B \mid A)$

3. Use Venn diagrams to show that

   (a) $(A^c \cup B^c)^c = A \cap B$ (b) $(A \cap B^c)^c = A^c \cup B$

4.
   (a) A group of engineers are asked to rank the top four from a list of 10 design characteristics of automobiles. How many rankings of the top four from the 10 are possible?
   (b) An engineer randomly selects, for testing, a sample of size 10 from a batch of 50 items. How many different samples are possible?
   (c) An engineering quiz consists of two true/false questions, and eight multiple choice questions with three choices for each question. In how many different ways may the test be answered?
   (d) If a tree diagram is constructed using EXCEL having three choices for flights from Chicago to London, four choices from London to Paris, and three from Paris to Rome, how many branches would the tree have if we were constructing flights from Chicago to London to Paris to Rome?

5. The following probabilities are known: $P(A) = 0.35$, $P(B) = 0.45$, and $P(A \text{ and } B) = 0.25$. Find the following probabilities:

   (a) $P(A \text{ or } B)$ (b) $P(\text{not } A)$ (c) $P(A \cup B)^c$ (d) $P(B \mid A)$ (e) $P(A \cap B)^c$

6. Three plants manufacture iPods. Plant A manufactures 60% of the iPods, plant B manufactures 15%, and plant C manufactures 25%. Plant A has a defective rate of 0.5 per hundred, plant B has a defective rate of 1 per hundred, and plant C has a defective rate of 0.75 per hundred.
   (a) What is the defective rate of the three plants when taken together?
   (b) If a defective is found in a store, find the probability that it was manufactured by each of the plants.
7. A quality control engineer selects a sample of size 4 from a batch having 30 items of which 5 have a minor discoloration. Find the probabilities of 0, 1, 2, 3, or 4 with minor discolorations in the sample. Find the expected number in the sample having minor discolorations.
8. Box 1 contains 4 nonconforming parts and 36 conforming parts. A part is selected from box 1, and placed into box 2 which has 8 nonconforming and 32 conforming. A part is selected from box 2. What is the probability the part selected from the second box is nonconforming?
9. Five parts are selected from a production line over an 8-hour period. The selections are independent of one another. If the probability is 0.15 on each draw that a defective part will be drawn, find the probability of the following events.
   (a) None of the five will be defective.
   (b) One of the five will be defective.
   (c) Two of the five will be defective.
   (d) Three of the five will be defective.
   (e) Four of the five will be defective.
   (f) Five of the five will be defective.
10. A batch contains 25 parts of which three are discolored. Three are drawn without replacement. Because of the small batch size, the second and the third draws are dependent on what happened before. Find the following probabilities of:
   (a) No discolored in the three.
   (b) One discolored in the three.
   (c) Two discolored in the three.
   (d) Three discolored in the three.

Summary

1. An **experiment** is any operation or procedure whose outcome cannot be predicted with certainty.
2. The **sample space** associated with an experiment consists of all possible outcomes associated with the experiment.
3. An **event** is some subset of the sample space.
4. In a **Venn diagram**, the sample space is represented as a rectangle, and the events are rectangles inside the sample space.
5. The **intersection** of *A* and *B* is the set of outcomes they have in common.
6. The **union** of two events, *A* and *B*, is the set of outcomes in *A* or *B*, or both.
7. The **complement** of *A* is the set of all outcomes in *S* that are not in *A*.
8. A **tree diagram** may be used to list the number of different outcomes in a sample space.
9. Suppose there are *k* sets of elements—n_1 in the first set, n_2 in the second set, … , and n_k in the *k*th set. A sample of *k* elements is formed by taking one element from each of the *k* sets. The number of different samples that can be formed is the product $n_1 n_2 \cdots n_k$. This is called the **multiplicative rule**.
10. Generally, when *r* objects are selected from a set of *n* distinct objects, there are $P_r^n = \frac{n!}{(n-r)!}$ **permutations** possible, and $\binom{n}{r} = \frac{n!}{r!(n-r)!}$ **combinations** possible. The order of selection is unimportant in combinations, but it is important when dealing with permutations.
    EXCEL **=PERMUT(NUMBER, NUMBER_CHOSEN)**
    =COMBIN(NUMBER, NUMBER_CHOSEN)
11. The **classical definition** might also be called the equally likely outcome definition. Suppose, an experiment has *n* equally likely outcomes, and an event *E* occurs if any one of *k* of these outcomes is an outcome of the experiment. The classical definition of probability defines the probability of event *E* as $P(E) = k/n$.
12. The **frequency probability definition** states that if an experiment is repeated many times, the probability of an event is the proportion of times the event occurs in *n* repetitions.
13. **Personal or subjective probability evaluations** are used when an experiment is not repeatable and the outcomes are not equally likely.
14. The **axioms of probability** for a finite sample space are as follows. Suppose you have a finite sample space *S* and an event *A* in *S*. Define *P*(*A*), the

probability of A, to be a value of a set function that satisfies the following three conditions:

Axiom 1: $0 \le P(A) \le 1$ for each event A in S.

Axiom 2: $P(S) = 1$.

Axiom 3: If A and B are mutually exclusive events in S, then $P(A \cup B) = P(A) + P(B)$.

Mutually exclusive events have empty intersections.

15. $P(A \cup B) = P(A) + P(B) - P(A \cap B)$. This rule is called the **general addition rule of probability for two events**.
16. If A and B are mutually exclusive, $P(A \cup B) = P(A) + P(B)$. This is called the **special addition rule**.
17. The following is called the **general addition rule for three events**.

$$P(A \cup B \cup C) = P(A) + P(B) + P(C) - P(A \cap B) - P(A \cap C) \\ - P(B \cap C) + P(A \cap B \cap C)$$

18. The **rule of complements** states that $P(A^c) = 1 - P(A)$.
19. The **conditional probability** of event A, given that event B is known to have occurred, is $P(A|B) = \frac{P(A \cap B)}{P(B)}$.
20. The **rule of total probability** states that if $S_1, S_2, \ldots, S_n$ are mutually exclusive events, and $D = (D \cap S_1) \cup (D \cap S_2) \cup \cdots (D \cap S_n)$ and one of the n events must occur, then

$$P(D) = P(S_1)\,P(D\,|\,S_1) + P(S_2)\,P(D\,|\,S_2) + \cdots + P(S_n)\,P(D\,|\,S_n)$$

21. The following is called **Bayes' rule**.

$$P(S_k\,|\,D) = \frac{P(S_k)P(D|S_k)}{P(S_1)P(D\,|\,S_1) + P(S_2)P(D\,|\,S_2)\cdots + P(S_n)P(D\,|\,S_n)}$$

22. The definition of **mathematical expectation** is as follows. If you win amount a_1 with probability p_1, amount a_2 with probability p_2, ..., and amount a_n with probability p_n your mathematical expectation is Exp = $\Sigma a_i p_i$. The amounts are real numbers, and the probabilities are between 0 and 1, and sum to 1.

CHAPTER 3

Probability Distributions for Discrete Random Variables

3.1 Random Variables

The concept of a **random variable** may be new to the reader of this book. A random variable assigns values to outcomes in the sample space. A random variable is best described by its probability distribution. The **probability distribution function** gives the values that X may assume and their probabilities. Random variables are best introduced by examples.

Experiment of rolling a pair of dice.

EXAMPLE

An experiment consists of rolling a pair of dice and letting the random variable X be the number of times the number 1 comes up when the pair of dice comes to rest. X can take on the values 0, 1, and 2 since when a pair of dice is rolled, the number 1 can turn up on none or one or both of the dice.

Figure 3-1 gives the EXCEL output for this experiment. The 36 outcomes are shown in the figure. The expression =COUNTIF(A2:B2,"1") is entered and a click-and-drag is performed from C2 to C19. This is also performed in G2:G19.

A	B	C	D	E	F	G	H	I
Die1	Die2	x		Die1	Die2	x	x	$p(x)$
1	1	2		4	1	1		
1	2	1		4	2	0	0	0.694
1	3	1		4	3	0	1	0.278
1	4	1		4	4	0	2	0.028
1	5	1		4	5	0		
1	6	1		4	6	0		
2	1	1		5	1	1		
2	2	0		5	2	0		
2	3	0		5	3	0		
2	4	0		5	4	0		
2	5	0		5	5	0		
2	6	0		5	6	0		
3	1	1		6	1	1		
3	2	0		6	2	0		
3	3	0		6	3	0		
3	4	0		6	4	0		
3	5	0		6	5	0		
3	6	0		6	6	0		

Figure 3-1 Sample space for rolling a pair of dice, X = number of times 1 occurs on a roll of the dice, and the probability distribution of X in columns H and I.

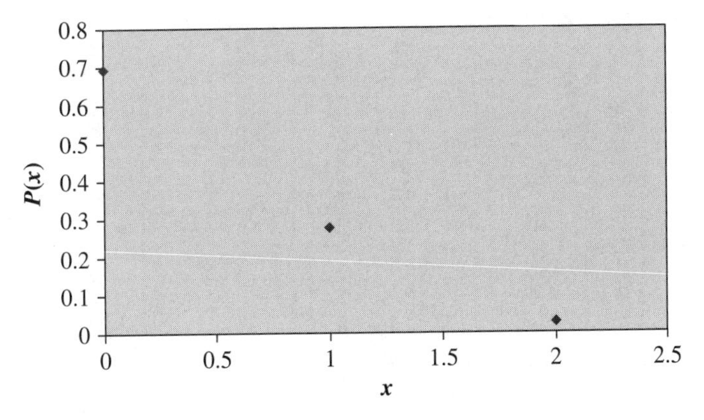

Figure 3-2 Plot of $P(X)$.

The number 0 occurs 25 times, the number 1 occurs 10 times, and 2 occurs once. The probability of 0 is 25/36, the probability of 1 is 10/36, and the probability of 2 is 1/36. The probability distribution for X is shown in the upper right corner of Figure 3-1. **Note that the probability distribution function for X has the properties $P(X) \geq 0$ and $\Sigma P(X) = 1$, where the sum is over the X values**. The interpretation of the probability distribution is as follows. When the dice are tossed a large number of times, the number 1 will turn up 0 times 69.4% of the time, the number 1 will turn up once about 27.8% of the time, and the number 1 will turn up on both dice 2.8% of the time. A plot of the probability distribution function is given in Figure 3-2. It was created using the chart wizard in EXCEL. It is said to be **skewed to the right** or it has a tail to the right.

Many random variables may be defined on the sample space of an experiment. They will have different probability distributions.

EXAMPLE

Consider the same experiment of rolling a pair of dice. Let random variable Y equal the sum on the faces that are turned up.

Figure 3-3 gives the EXCEL output for this experiment. The 36 outcomes are shown.

The expression =SUM(A2:B2) is entered in C2 and a click-and-drag is performed from C2 to C19. This is also performed in G2:G19. The probability distribution for Y is shown in the upper right corner of Figure 3-3. Note that the probability distribution for Y has the property $P(Y) \geq 0$ and $\Sigma P(Y) = 1$, where the sum is over the y values. The interpretation of the probability distribution is as follows. When the dice are tossed a large number of times, the sum 2 occurs 2.8% of the time, the sum 3 occurs 5.6 % of the time, and so forth. Figure 3-4 is a plot of $p(y)$ created by the chart wizard of EXCEL. It is said to be **symmetrical about 7**.

A	B	C	D	E	F	G	H	I
Die1	Die2	y		Die1	Die2	y	y	$p(y)$
1	1	2		4	1	5		
1	2	3		4	2	6	2	0.028
1	3	4		4	3	7	3	0.056
1	4	5		4	4	8	4	0.083
1	5	6		4	5	9	5	0.111
1	6	7		4	6	10	6	0.139
2	1	3		5	1	6	7	0.167
2	2	4		5	2	7	8	0.139
2	3	5		5	3	8	9	0.111
2	4	6		5	4	9	10	0.083
2	5	7		5	5	10	11	0.056
2	6	8		5	6	11	12	0.028
3	1	4		6	1	7		
3	2	5		6	2	8		
3	3	6		6	3	9		
3	4	7		6	4	10		
3	5	8		6	5	11		
3	6	9		6	6	12		

Figure 3-3 EXCEL output for the sample space of rolling a pair of dice, where Y = sum on the dice, and the probability distribution of Y is given.

Random variables are classified according to the number of values they can assume. The random variables encountered in this chapter can assume a finite number or a countably infinite number of values, and are called **discrete random variables. Continuous random variables** are considered in the next chapter. Their probabilities are areas under curves.

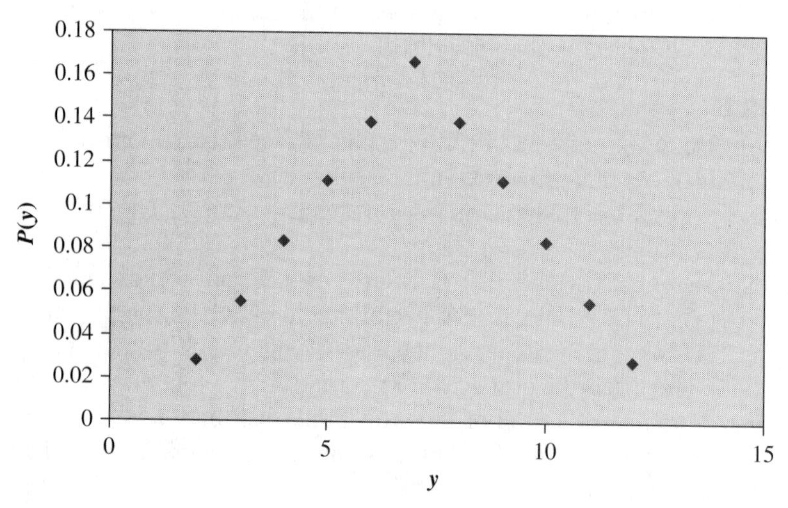

Figure 3-4 Plot of $P(y)$.

Another important function associated with discrete random variables is the **cumulative distribution function**. The cumulative distribution function is defined as

$$F(x) = P(X \leq x)$$

EXAMPLE

The cumulative distribution function associated with X in the first example of this section is

x	0	1	2
$F(x)$	0.694	0.972	1

and the cumulative distribution function associated with Y in the second example of this section is

Y	2	3	4	5	6	7	8	9	10	11	12
$F(y)$	0.028	0.084	0.167	0.278	0.417	0.584	0.723	0.834	0.917	0.973	1

The graph of $F(x)$ is given in Figure 3-5 and the graph of $F(y)$ is given in Figure 3-6.

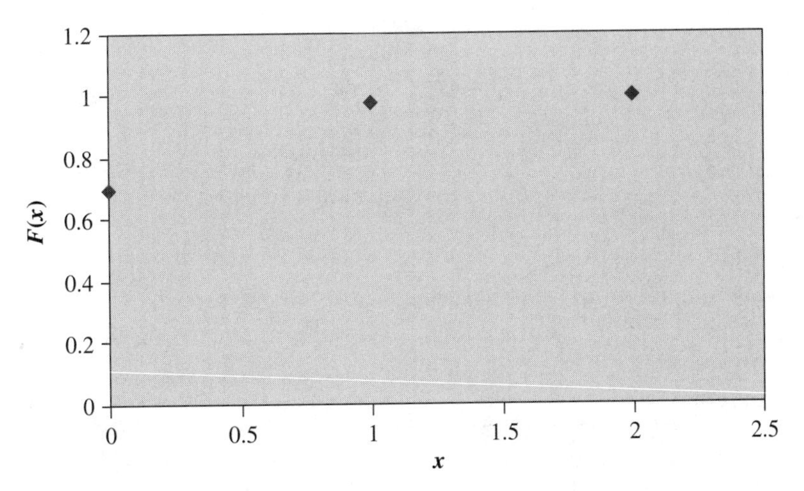

Figure 3-5 Plot of $F(x)$.

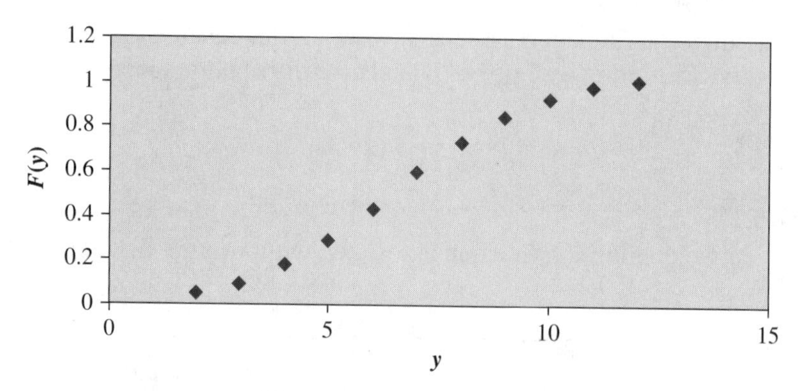

Figure 3-6 Plot of $F(y)$.

Cell phones may have surface flaws.

EXAMPLE

A cell phone manufacturer has been notified by an industrial engineer that a cell phone may have up to four surface flaws. If $X =$ the number of surface flaws per cell phone, then the probability distribution function has been found to have the following distribution:

x	0	1	2	3	4
$p(x)$	0.75	0.15	0.05	0.04	0.01

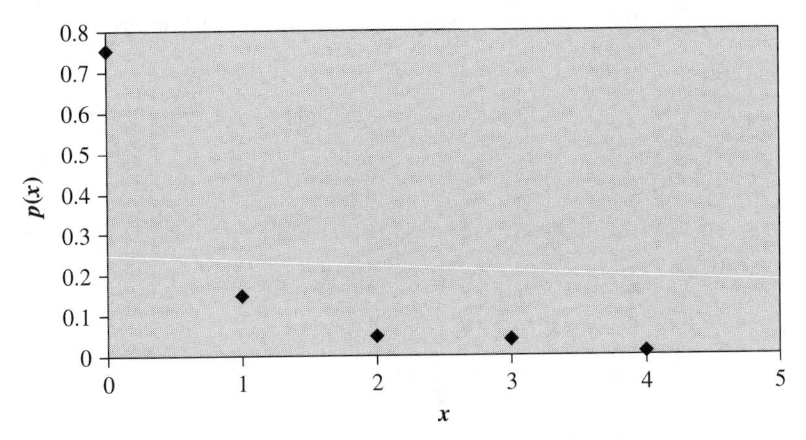

Figure 3-7 Probability distribution function of the number of surface flaws per cell phone.

The graph of $p(x)$ is shown in Figure 3-7.
The cumulative distribution function is shown in Figure 3-8.

x	0	1	2	3	4
$F(x)$	0.75	0.90	0.95	0.99	1

The probability of at most 2 defects is $F(2) = 0.95$.

The probability of at least 3 defects is $1 - F(2) = 0.05$.

The probability of exactly 2 defects is $F(2) - F(1) = 0.95 - 0.90 = 0.05$ or $p(2) = 0.05$.

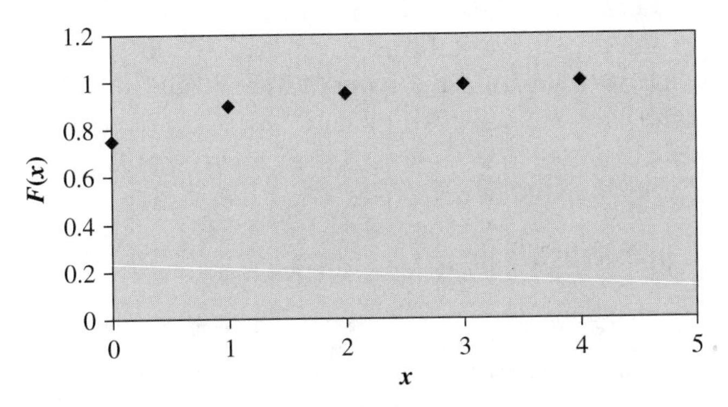

Figure 3-8 Cumulative distribution function of the number of surface flaws per cell phone.

Remember, a probability distribution function has two basic properties:

1. $p(x) \geq 0$
2. $\Sigma\, p(x) = 1$.

EXAMPLE

Check to see if the following can serve as probability distribution functions:

a. $p(x) = \frac{24}{50}\frac{1}{x}$ for $x = 1, 2, 3, 4$

b. $p(x) = x/55$ for $x = 1, 2, 3, 4, 5, 6, 7, 8, 9, 10$

c. $p(x) = \frac{x-4}{2}$ for $x = 3, 4, 5, 6$

a. 1. All the probabilities are positive.
   2. $\Sigma\, p(x) = \frac{24}{50}\left\{1 + \frac{1}{2} + \frac{1}{3} + \frac{1}{4}\right\} = 0.48 + 0.24 + 0.16 + 0.12 = 1$
   This function is a probability distribution function.

b. 1. All the probabilities are positive.
   2. $\Sigma\, p(x) = \frac{1+2+3+4+5+6+7+8+9+10}{55} = \frac{55}{55} = 1$
   This function is a probability distribution function.

c. $P(3) = -0.5$. Since this probability is negative, this function is not a probability distribution function.

EXAMPLE

A quality control engineer is in charge of receiving parts by the truck load. Each truck contains several thousand parts. Generally 1% of the parts are defective. The engineer can live with 1% but not 3% or 4%. She uses the following plan. Randomly sample 5 parts from the truck. If there are no defectives in the sample, accept the truck load as suitable. If 1% of the parts are defective, there is a 95% chance that she will accept the truck load, since P(5 out of 5 are nondefective) is $(.99)^5 = 0.951$. Now, if 4% of the parts are defective rather than 1%, find the probability that the sample of 5 will contain 0, 1, 2, 3, 4, or 5 defectives. That is, find the probability distribution of X, the number of defectives in a 4% defective truck load.

$$P(\text{0 defectives in a 4\% defective truck load}) = (.96)^5 = 0.8154$$

$$P(\text{1 defective in a 4\% defective truck load}) = (.04)(.96)^4 + (.96)(.04)(.96)^3 + (.96)^2(.04)(.96)^2 + (.96)^3(.04)(.96) + (.96)^4(.04) = 5(.04)(.96)^4 = 0.1699$$

$$P(\text{2 defectives and 3 nondefectives in a 4\% defective truck load}) = 10(.04)^2(.96)^3 = 0.0142$$

P(3 defectives and 2 nondefectives in a 4% defective truck load)
$$= 10(.04)^3(.96)^2 = 0.0006$$

P(4 defectives and 1 nondefective in a 4% defective truck load)
$$= 5(.04)^4(.96) = 0.0000$$

$$P(\text{5 defectives in a 4\% defective truck load}) = (.04)^5 = 0.0000$$

The probability distribution of X, where X is the number of defectives in a sample of 5 from a 4% defective truck load is as follows:

x	0	1	2	3	4	5
$p(x)$	0.8154	0.1699	0.0142	0.0006	0.0000	0.0000

EXAMPLE
Consider the following sampling plan. One item is inspected and if it passes inspection, the industrial process continues. If it does not pass inspection, the industrial process is stopped and inspected. The industrial process continues until some item does not pass inspection. An item is selected from the line every hour. Let X be the hour in which the first item does not pass inspection. If the process is producing 90% acceptable items, give the probability distribution for X. The possible values of X are 1, 2, 3, ... The probabilities are $p(1) = 0.1$, $p(2) = 0.9(.1)$, $p(3) = (0.9)^2(.1)$, and so forth. Summarizing, we find that $p(x) = (0.9)^{x-1}(.1)$ for $x = 1, 2, 3, \ldots$. Note that this is the first infinite discrete random variable we have discussed. Its possible outcomes are in a 1-to-1 correspondence with the positive integers.

First of all, note that the probability distribution function satisfies the two properties $p(x) \geq 0$ and $\Sigma\, p(x) = 1$. Certainly, all the probabilities are positive. If we add all the probabilities, we obtain $0.1 + (0.9)(.1) + (0.9)^2(.1) + (0.9)^3(0.1) + \cdots$. This is an infinite geometric series with the first term, $a = 0.1$ and the common ratio, $r = 0.9$. In algebra, it is shown that such a series has sum equal to

$$\frac{a}{1-r} = \frac{0.1}{1-0.9} = 1$$

When Figure 3-9 is consulted, it is seen that the above sum for the first 60 terms gives 0.9982. Furthermore, note that $F(29) = 0.9529$. This can be interpreted as follows. The probability that the first defective will occur on the first inspection or on the second or on the third or … or on the 29th is 0.9529. The cumulative distribution function evaluated at 29 is equal to 0.9529.

x	$p(x)$	$F(x)$	x	$p(x)$	$F(x)$	x	$p(x)$	$F(x)$
1	0.1	0.1	21	0.0122	0.8906	41	0.0015	0.9867
2	0.09	0.19	22	0.0109	0.9015	42	0.0013	0.988
3	0.081	0.271	23	0.0098	0.9114	43	0.0012	0.9892
4	0.0729	0.3439	24	0.0089	0.9202	44	0.0011	0.9903
5	0.0656	0.4095	25	0.008	0.9282	45	0.001	0.9913
6	0.059	0.4686	26	0.0072	0.9354	46	0.0009	0.9921
7	0.0531	0.5217	27	0.0065	0.9419	47	0.0008	0.9929
8	0.0478	0.5695	28	0.0058	0.9477	48	0.0007	0.9936
9	0.043	0.6126	29	0.0052	0.9529	49	0.0006	0.9943
10	0.0387	0.6513	30	0.0047	0.9576	50	0.0006	0.9948
11	0.0349	0.6862	31	0.0042	0.9618	51	0.0005	0.9954
12	0.0314	0.7176	32	0.0038	0.9657	52	0.0005	0.9958
13	0.0282	0.7458	33	0.0034	0.9691	53	0.0004	0.9962
14	0.0254	0.7712	34	0.0031	0.9722	54	0.0004	0.9966
15	0.0229	0.7941	35	0.0028	0.975	55	0.0003	0.997
16	0.0206	0.8147	36	0.0025	0.9775	56	0.0003	0.9973
17	0.0185	0.8332	37	0.0023	0.9797	57	0.0003	0.9975
18	0.0167	0.8499	38	0.002	0.9818	58	0.0002	0.9978
19	0.015	0.8649	39	0.0018	0.9836	59	0.0002	0.998
20	0.0135	0.8784	40	0.0016	0.9852	60	0.0002	0.9982

Figure 3-9 EXCEL output for $p(x) = (0.9)^{x-1}(.1)$ and $F(x)$ for $x = 1$ to 60.

The graph in Figure 3-10 continues on from 60 to infinity, and the $p(x)$ values get closer and closer to 0 but never touch 0. The graph in Figure 3-11 continues on from 60 to infinity, and the $F(x)$ values get closer and closer to 1 but never equal 1. EXCEL is used to construct Figure 3-9 in the following manner. The numbers, 1 through 60, are entered into A2:A61. The expression = 0.9^(A2-1)*0.1 is entered into B2, and a click-and-drag is performed from B2 to B61. $F(x)$ is formed in the following manner: B2 is entered into C2. Then = C2 + B3 is entered into C3 and a click-and-drag is performed from C3 to C61.

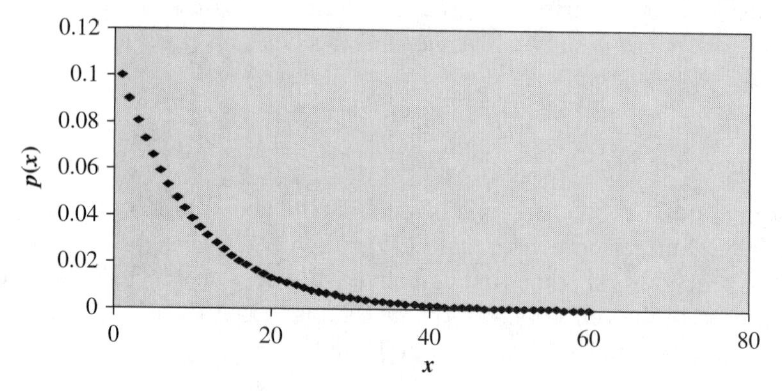

Figure 3-10 Plot of $p(x) = (0.9)^{x-1}(0.1)$ from Figure 3-9.

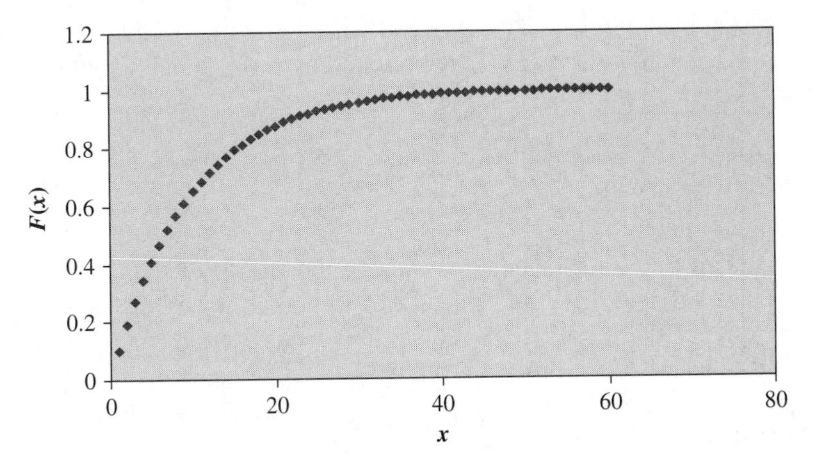

Figure 3-11 Plot of $F(x)$ from Figure 3-9.

3.2 The Mean and Standard Deviation of a Probability Distribution

The mean of a random variable is its mathematical expectation. If a discrete random variable can take on the values $x_1, x_2, \ldots, x_n$ with probabilities $p(x_1), p(x_2), \ldots,$ $p(x_n)$, the **mean** is defined to be

$$\mu = \sum x_i\, p(x_i) \quad \text{where } i \text{ goes from 1 to } n.$$

The sum may contain an infinite number of terms. The mean locates the center or the balance point of a distribution. Another important measure associated with random variables is the **variance**. The variance of a random variable is defined to be

$$\sigma^2 = \sum (x_i - \mu)^2\, p(x_i) \quad \text{where } i \text{ goes from 1 to } n.$$

The square root of the variance is the **standard deviation**.

$$\sigma = \sqrt{\sum (x_i - \mu)^2\, p(x_i)}$$

The standard deviation is used more often than the variance. However, it is necessary to find the variance before finding the standard deviation. Also, the variance is in square units, and the standard deviation is in the units of measurement.

There is an equivalent formula that is usually used to compute the variance and the standard deviation. The variance may be found using the algebraically equivalent formula

$$\sigma^2 = \sum x_i^2 p(x_i) - \mu^2$$

The standard deviation gives a measure of the spread of the distribution. The parameters μ and σ are important measures to know for any distribution. Let us consider some of the distributions we studied in Section 3.1. We shall find their means and standard deviations, and comment on their meanings.

EXAMPLE

The number of times 1 turns up when rolling a pair of dice has the following probability distribution:

x	0	1	2
$p(x)$	0.694	0.278	0.028

The following EXCEL output in Figure 3-12 shows that $\mu = 0.334$ (in C5) and $\sigma = 0.528$ (in F3).

If the dice were rolled 100 times and the sample mean of the 100 rolls were found, it would be near 0.334; and if the standard deviation of the 100 outcomes were found using the formula for S given in the first chapter, it would be near 0.528. The probability distribution gives the population description of the random variable; and when the probability distribution is known, μ and σ can be found. If $p(x)$ is not known, then the sample mean and the sample standard deviation provide estimates of μ and σ.

A	B	C	D	E	F
x	$p(x)$	$xp(x)$	$x^{2}p(x)$	Variance	Standard deviation
0	0.694	0	0	0.278444	
1	0.278	0.278	0.278		**0.527678**
2	0.028	0.056	0.112		
		0.334	0.39		

Figure 3-12 EXCEL worksheet for computing mean and standard deviation.

EXAMPLE

The sum of the two faces turned up when a pair of dice are rolled has the following probability distribution:

x	2	3	4	5	6	7	8	9	10	11	12
$p(x)$	0.028	0.056	0.083	0.111	0.139	0.167	0.139	0.111	0.083	0.056	0.028

The following EXCEL computation in Figure 3-13 shows that $\mu = 7$ and $\sigma = 2.4$.

The **empirical rule** applied to probability distributions states that for mound-shaped distributions such as the one above (see Figure 3-4), approximately 68% of the probability will be between $\mu - \sigma$ and $\mu + \sigma$, approximately 95% of the probability will be between $\mu - 2\sigma$ and $\mu + 2\sigma$, and 99.7% will be between $\mu - 3\sigma$ and $\mu + 3\sigma$. Between $\mu - \sigma = 4.6$ and $\mu + \sigma = 9.4$, there is $(0.111 + 0.139 + 0.167 + 0.139 + 0.111) = 0.67$ or 67%. Between $\mu - 2\sigma = 2.2$ and $\mu + 2\sigma = 11.8$, there is 0.944 or 94.4%. All of the probability is contained between $\mu - 3\sigma$ and $\mu + 3\sigma$. We see that the empirical rule very closely predicts the probability distribution for this mound-shaped distribution.

EXAMPLE

Consider the random variable X, where X has the probability distribution $0.9^{(x-1)}(0.1)$ for $x = 1, 2, 3, \ldots$. This is a discrete random variable with an infinite domain.

y	$p(y)$	$yp(y)$	$y^{2}p(y)$	Variance	Standard deviation
2	0.028	0.056	0.112		
3	0.056	0.168	0.504	5.802951	**2.408932**
4	0.083	0.332	1.328		
5	0.111	0.555	2.775		
6	0.139	0.834	5.004		
7	0.167	1.169	8.183		
8	0.139	1.112	8.896		
9	0.111	0.999	8.991		
10	0.083	0.83	8.3		
11	0.056	0.616	6.776		
12	0.028	0.336	4.032		
		7.007	54.901		

Figure 3-13 EXCEL worksheet for computing mean and standard deviation.

A	B
x	$xp(x)$
1	0.1
2	0.18
3	0.243
4	0.2916
5	0.32805
6	0.354294
7	0.372009
8	0.382638
9	0.38742
10	0.38742
	3.026431

Figure 3-14 Approximate mean of X using only the first ten terms of the sum ($\mu = \Sigma\, x\, 0.9^{(x-1)}\,(0.1)$) and EXCEL

The mean is given by $\mu = \Sigma x\ 0.9^{(x-1)}\,(0.1)$, where the sum goes from 1 to infinity. The standard deviation is given by $\sigma = \sqrt{\Sigma x^2\, 0.9^{x-1}\,(0.1) - \mu^2}$. EXCEL is used to evaluate the mean and the standard deviation. Suppose, only ten terms in the infinite sum that represent the mean are used to start. The expression =A2*0.9^(A2–1)*0.1 is entered into the cell B2 and a click-and-drag produces the output shown in cells B2: B11 in Figure 3-14. The sum is obtained in cell B12 and is 3.0264. Now, Figure 3-15 shows the mean obtained using 50, 100, 150, and 200 terms in the sum.

It is clear from Figure 3-15 that the sum is converging to 10.

This is a numerical technique that may be used for discrete random variables with infinite domains. We don't really need to add an infinite number of terms to see that $\mu = 10$. Once the mean is known, the standard deviation can be found as follows.

Figure 3-16 shows the sum of $x^2p(x)$ for 50, 100, 150, and 200 terms in the sum for the standard deviation.

The variance is $190 - 10^2 = 90$ and the standard deviation is 9.4868.

X	Sum
50	9.690733
100	9.997078
150	9.999978
200	10

Figure 3-15 Approximate mean of X using the first 50, 100, 150, and 200 terms of the sum ($\mu = \Sigma\, x\, 0.9^{(x-1)}\,(0.1)$) and EXCEL

X	Sum
50	170.9826
100	189.6762
150	189.9965
200	190

Figure 3-16 The sum of $x^2p(x)$ for 50, 100, 150, and 200 terms in the sum for the standard deviation. $\sigma = \sqrt{\Sigma x^2 p(x) - \mu^2}$

3.3 Chebyshev's Theorem

The empirical rule tells us what to expect concerning the probability distribution and its relationship to μ and σ for a probability distribution that is mound shaped (also called normally distributed). But, what about the probability distributions that are not mound shaped. A result proved by Chebyshev answers the question for the probability distribution of any shape. Chebyshev's theorem states that if we go at least k standard deviations from the mean, then we will find at most $1/k^2$ of the probability distribution. Thus if we go at least 2 standard deviations from the mean, there will be at most $1/2^2 = 0.25$ or 25% of the distribution. If we go at least 3 standard deviations from the mean, there will be at most 1/9 or 11% of the distribution. We can also state the theorem as follows. Within 2 standard deviations of the mean, there will be at least 75% of the distribution; and within 3 standard deviations of the mean, there will be at least 89% of the distribution.

$\mu - 2\sigma$	μ	$\mu + 2\sigma$	x
|	At least 75% of the probability	|	

EXAMPLE

Let X be the sum on two dice that are rolled. In Section 3.1, we derived the probability distribution function and the cumulative probability distribution functions. They are given in Figure 3-17.

We found that $\mu = 7$ and $\sigma = 2.4$. Show that there is at least 75% of the probability distribution within 2σ of μ and at least 89% within 3σ of μ. Now, $\mu - 2\sigma = 2.2$ and $\mu + 2\sigma = 11.8$. The amount of probability within 2σ of μ is $F(11) - F(2) = 0.973 - 0.028 = 0.945$, and this is at least 75%. The amount of probability could also be found by adding the following:

$$p(3) + p(4) + p(5) + p(6) + p(7) + p(8) + p(9) + p(10) + p(11) = 0.945$$

There is 100% between $\mu - 3\sigma = -0.2$ and $\mu + 3\sigma = 14.2$. The lower bound of 89% is seen to hold here.

x	$p(x)$	$F(x)$
2	0.028	0.028
3	0.056	0.084
4	0.083	0.167
5	0.111	0.278
6	0.139	0.417
7	0.167	0.584
8	0.139	0.723
9	0.111	0.834
10	0.083	0.917
11	0.056	0.973
12	0.028	1.000

Figure 3-17 Distributions of $p(x)$ and $F(x)$ for sum of dice.

EXAMPLE

Consider the random variable X, where X has the probability distribution $0.9^{(x-1)}(0.1)$, for $x = 1, 2, 3, \ldots$. We found the mean to be $\mu = 10$ and the standard deviation to be $\sigma = 9.49$. Figure 3-18 gives $p(x)$ and $F(x)$ for $x = 1$ through $x = 60$.

Find the percentage of probability within 2σ and 3σ of μ and compare with Chebyshev's theorem. Now, $\mu - 2\sigma = -8.98$ and $\mu + 2\sigma = 28.98$. The amount of probability between -8.98 and 28.98 is given by $F(28) = 0.9477$, which is at least 75%.

x	$p(x)$	$F(x)$	x	$p(x)$	$F(x)$	x	$p(x)$	$F(x)$
1	0.1	0.1	21	0.0122	0.8906	41	0.0015	0.9867
2	0.09	0.19	22	0.0109	0.9015	42	0.0013	0.988
3	0.081	0.271	23	0.0098	0.9114	43	0.0012	0.9892
4	0.0729	0.3439	24	0.0089	0.9202	44	0.0011	0.9903
5	0.0656	0.4095	25	0.008	0.9282	45	0.001	0.9913
6	0.059	0.4686	26	0.0072	0.9354	46	0.0009	0.9921
7	0.0531	0.5217	27	0.0065	0.9419	47	0.0008	0.9929
8	0.0478	0.5695	28	0.0058	0.9477	48	0.0007	0.9936
9	0.043	0.6126	29	0.0052	0.9529	49	0.0006	0.9943
10	0.0387	0.6513	30	0.0047	0.9576	50	0.0006	0.9948
11	0.0349	0.6862	31	0.0042	0.9618	51	0.0005	0.9954
12	0.0314	0.7176	32	0.0038	0.9657	52	0.0005	0.9958
13	0.0282	0.7458	33	0.0034	0.9691	53	0.0004	0.9962
14	0.0254	0.7712	34	0.0031	0.9722	54	0.0004	0.9966
15	0.0229	0.7941	35	0.0028	0.975	55	0.0003	0.997
16	0.0206	0.8147	36	0.0025	0.9775	56	0.0003	0.9973
17	0.0185	0.8332	37	0.0023	0.9797	57	0.0003	0.9975
18	0.0167	0.8499	38	0.002	0.9818	58	0.0002	0.9978
19	0.015	0.8649	39	0.0018	0.9836	59	0.0002	0.998
20	0.0135	0.8784	40	0.0016	0.9852	60	0.0002	0.9982

Figure 3-18 EXCEL output for $p(x) = (0.9)^{x-1}(.1)$ and $F(x)$ for $x = 1$ to 60.

Also $\mu - 3\sigma = -18.47$ and $\mu + 3\sigma = 38.47$. The amount of probability between -18.47 and 38.47 is given by $F(38) = 0.9818$, which is at least 89%.

EXAMPLE

The temperature associated with an industrial process has an unknown distribution. However, based on sampling, the approximate mean is known to be 25.3 degrees centigrade and the standard deviation is known to be 2.1 degrees centigrade. Applying Chebyshev's theorem, at least 75% of the time, the temperature will be between 21.1 and 29.5 degrees centigrade. At least 96% of the time, the temperature will be between 14.8 and 35.8 degrees centigrade. (Let $k = 5$ in Chebyshev's theorem.)

EXAMPLE

The pressure associated with an industrial process has a mean of 50.5 psi and a standard deviation of 0.2 psi. What can be said about the percent of time that the pressure is between 49.5 and 51.5 psi? Applying Chebyshev's theorem, and noting that the two limits are 5 standard deviations on either side of the mean, the probability is at least $1 - \frac{1}{5^2} = \frac{24}{25}$ or 96% that the pressure will be between 49.5 and 51.5 psi.

3.4 The Binomial Distribution

In the first three sections of this chapter, general properties of discrete random variables were discussed. Some types of discrete random variables occur so often that they have been singled out and studied in detail. Probably, the most frequently applied discrete random variable and distribution is the binomial. We will now study this random variable, its distribution, and many of its properties.

This penny is tossed four times and the number of heads that occur in the four tosses is counted. The number of heads that occur is a random variable.

Let us start by considering an example of a **binomial random variable**. Suppose X represents the number of heads to occur in 4 tosses of a fair coin. The random variable X can take one of the five values—0, 1, 2, 3, or 4. The probability distribution of X has a binomial distribution. Each flip of the coin is called a **trial**.

We say that X is based on 4 trials. Each trial has 2 outcomes possible—head or tail. Since the coin is fair, the probability of a head is 0.5 and the probability of a tail is 0.5. We say that a head occurring is a **success** and a tail is a **failure**. The outcome of interest is the number of successes that occur in the 4 trials. The trials are **independent** of one another. What happens on any trial is not influenced by what happens on any other trial. The probabilities of success and failure remain the same from trial to trial. The letter p represents the success probability and the letter q represents the failure probability. The probability of 0 successes (heads) in the 4 trials is p^4 or $(0.5)^4$. We multiply the probabilities because of independence. No successes means 4 tails were obtained. The probability of a tail followed by a tail followed by a tail followed by another tail is (0.5)(0.5)(0.5)(0.5) or 0.0625. The probability of 1 success (1 head) is the same as *HTTT* or *THTT* or *TTHT* or *TTTH*. In other words, the head could have occurred on the first or the second or the third or the fourth trial. We would add the four probabilities because the four events are mutually exclusive. The probability that $X = 1$ is $4(.5)^4 = 0.25$. The event $X = 2$ is equivalent to the events—*HHTT* or *HTHT* or *HTTH* or *TTHH* or *THTH* or *THHT*—occurring. Multiplying and adding probabilities as the rules of probability say, we should get $6(.5)^4 = 0.375$. Similarly, the probability of 3 successes is 0.25 and the probability of 4 successes is 0.0625. An EXCEL plot of this distribution is shown in Figure 3-19.

x	0	1	2	3	4
$p(x)$	0.0625	0.25	0.375	0.25	0.0625

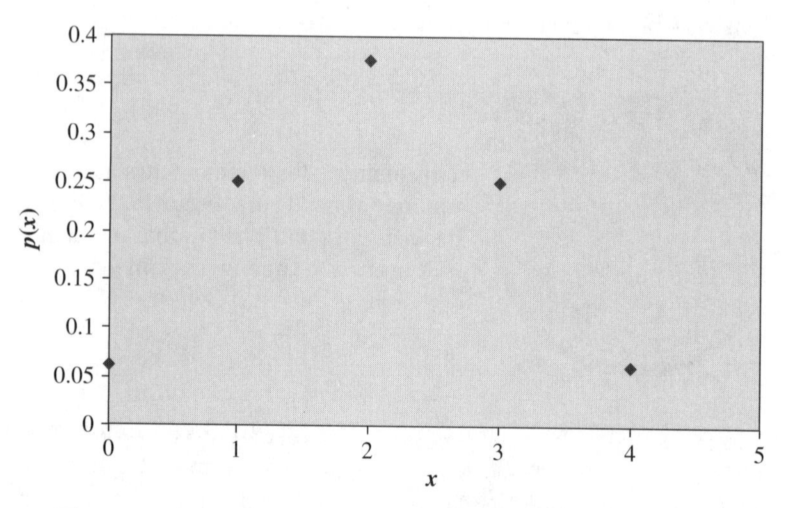

Figure 3-19 The binomial distribution with $n = 4$ and $p = 0.5$.

Having introduced the idea and terminology about a binomial distribution in the last paragraph, let us now generalize the ideas. Binomial random variables satisfy the following assumptions, called the **binomial assumptions**:

1. There are only two possible outcomes for each trial called success and failure.
2. The probabilities of a success and a failure remain the same from trial to trial.
3. There are n trials, and n is constant.
4. The n trials are independent of one another.

The binomial random **variable** is then defined as the number of successes in the n trials. A binomial random variable has a probability distribution given by the following:

$$b(x) = \binom{n}{x} p^x q^{n-x} \quad \text{for} \quad x = 0, 1, 2, \ldots, n$$

and

$$\binom{n}{x} = \frac{n!}{x!(n-x)!}$$ are called **binomial coefficients.**

Note that the binomial coefficients and the combination of n things taken x at a time are the same. The binomial coefficients count the number of outcomes that give x successes. In the above discussion concerning 4 flips of a coin, there were 6 outcomes that gave 2 heads. They were *HHTT* or *HTHT* or *HTTH* or *TTHH* or *THTH* or *THHT*. If we evaluate

$$\binom{4}{2} = \frac{4!}{2!(4-2)!} = 6$$

we see that the binomial coefficient accounts for the 6 outcomes that give 2 heads.

EXAMPLE

Use EXCEL to find the binomial coefficients for $x = 0, 1, \ldots, 10$ and $n = 10$. The numbers 0 through 10 are entered into A3:A13. The function =COMBIN(10,A3) is entered into B3 and a click-and-drag is performed to give Figure 3-20.

A plot of the data in Figure 3-20 is shown in Figure 3-21.

A	B
x	Binomial coefficients
0	1
1	10
2	45
3	120
4	210
5	252
6	210
7	120
8	45
9	10
10	1

Figure 3-20 Binomial coefficients for $n = 10$.

EXAMPLE

A sampling plan works as follows. A sample of 10 items is selected from a large lot. X represents the number of defectives found in the sample. If $X = 0$, the lot is accepted. If X equals 1 or more, the complete lot is inspected. If the lot is 0.5% defective, find the probability that the lot is accepted and the probability that the lot will be completely inspected. First, note that since the lot is large, taking a sample of 10 will not affect the probability of obtaining a defective item and the binomial assumptions may be assumed to hold. Now use EXCEL to calculate binomial probabilities as follows.

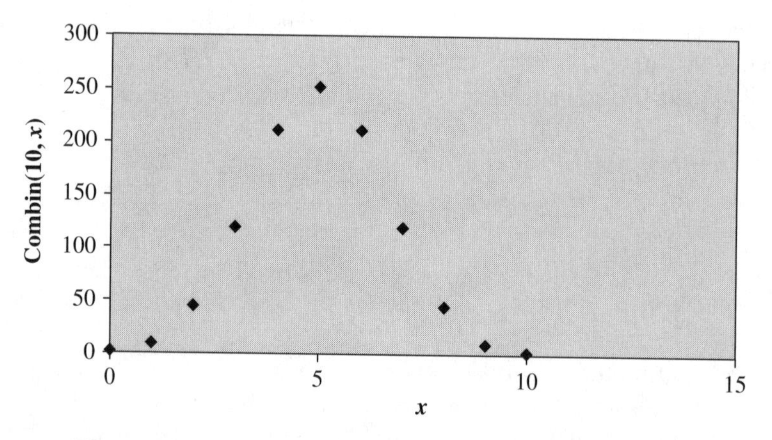

Figure 3-21 A plot of the binomial coefficients for $n = 10$.

The labels x, $b(x)$, and $B(x)$ are entered in A1:C1. The numbers 0 through 10 are entered in A2:A12. To calculate the binomial distribution, enter =BINOMDIST (A2,10,0.005,0) into cell B2 and perform a click-and-drag. The first parameter in the function is the number of successes, the second is the number of trials, and the third is the success probability; and if the fourth is a 0, the individual binomial probabilities are computed. If a 1 is entered in the fourth position, the cumulative binomial probabilities are computed.

A	B	C
X	***b(x)***	***B(x)***
0	0.9511101	0.95111
1	0.0477945	0.998905
2	0.0010808	0.999985
3	1.448E-05	1
4	1.274E-07	1
5	7.68E-10	1
6	3.216E-12	1
7	9.235E-15	1
8	1.74E-17	1
9	1.943E-20	1
10	9.766E-24	1

The probability that a 0.5% defective lot will be accepted is $B(0) = 0.95111$. The probability that a 0.5% defective lot will be rejected is $1 - B(0) = 0.04889$. If p is small, the binomial distribution is skewed to the right like this one.

EXAMPLE

Suppose, the lot in the previous example is 5% defective, rather than 0.5% defective. Compute the probabilities of accepting the lot as well as a complete inspection. If MINITAB is used to compute the probabilities, the pull-down **Calc ⇒ Probability Distributions ⇒ Binomial** will give the binomial distribution dialog box which is filled as shown in Figure 3-22. The numbers 0 through 10 are entered into column 1 of the MINITAB worksheet. In the dialog box shown in Figure 3-22, the probability is checked and the cumulative probability is also checked. The below output will be the result.

Row	x	$b(x)$	Cumulative
1	0	0.598737	0.59874
2	1	0.315125	0.91386
3	2	0.074635	0.98850
4	3	0.010475	0.99897
5	4	0.000965	0.99994
6	5	0.000061	1.00000
7	6	0.000003	1.00000
8	7	0.000000	1.00000
9	8	0.000000	1.00000
10	9	0.000000	1.00000
11	10	0.000000	1.00000

The probability that a 5% defective lot will be accepted is 0.598 and the probability that a 5% defective lot will be completely inspected is $1 - 0.598 = 0.402$.

A shortcut formula for the mean and standard deviation may be derived from the formula for the mean and standard deviation and some algebraic manipulation. It may be shown that the mean is found by multiplying the number of trials times the success probability, i.e., $\mu = np$. The standard deviation likewise can be found by use of the shortcut formula $\sigma = \sqrt{npq}$.

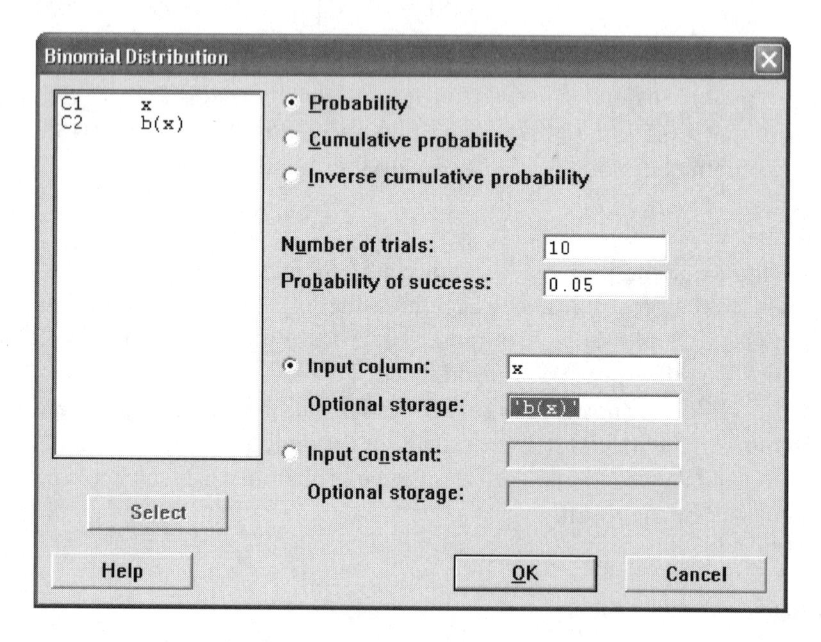

Figure 3-22 MINITAB binomial distribution dialog box.

EXAMPLE

Consider the following binomial distribution: Find the mean and standard deviation using the basic definition and the short cut formula, and show that you get the same answer.

x	0	1	2	3	4
$p(x)$	0.0625	0.25	0.375	0.25	0.0625

$$\mu = \sum xp(x) = 0(0.0625) + 1(0.25) + 2(0.375) + 3(0.25) + 4(0.0625) = 2$$

$$\mu = np = 4(.5) = 2$$

$$\sigma^2 = \sum x_i^2 p(x_i) - \mu^2 = 0(0.0625) + 1(0.25) + 4(0.375) + 9(0.25) + 16(0.0625) - 4 = 1$$

$$\sigma^2 = npq = 4(0.5)(0.5) = 1$$

Finally, consider the effect of p on the binomial distribution. This may be illustrated by plotting three probability distributions. The three distributions are:

1. Binomial with $n = 50$ and $p = 0.25$
2. Binomial with $n = 50$ and $p = 0.5$
3. Binomial with $n = 50$ and $p = 0.75$

These three plots are shown in Figure 3-23.

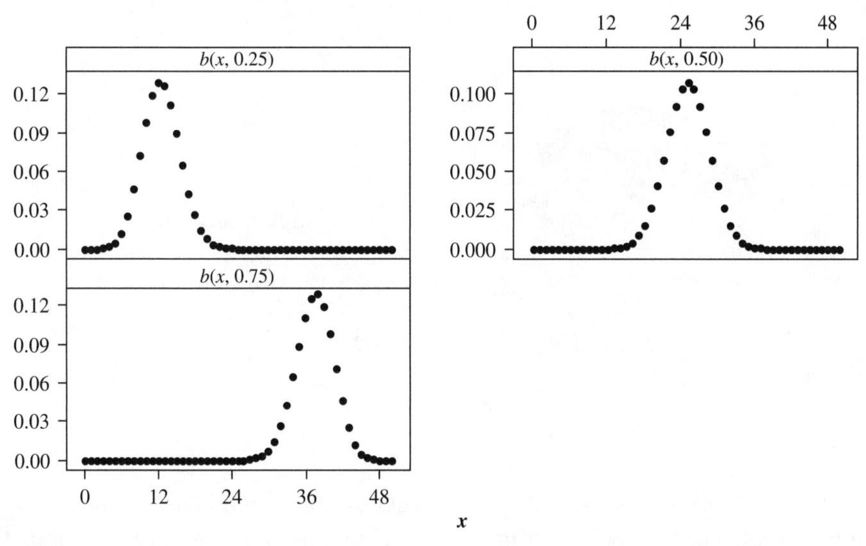

Figure 3-23 Binomial distributions with $n = 50$ and $p = 0.25$, $p = 0.50$, and $p = 0.75$.

When $p = 0.25$, the distribution is skewed to the right; when $p = 0.50$, the distribution is symmetrical; and when $p = 0.75$, the distribution is skewed to the left.

Note (concerning binomial probabilities): Most statistics books contain tables of binomial probabilities for various values of n and p. The author believes that such tables are an anachronism. With the wide spread availability of statistical software packages, many of which have the ability to compute binomial probabilities as well as many other probability distributions, the need for such tables no longer exists. In surveys of students in the author's classes, almost all students have EXCEL at home as well as available on all university computers. In addition, all engineers have EXCEL.

3.5 The Hypergeometric Distribution

EXAMPLE

Suppose, personal digital assistants (PDAs) are packaged 25 per box and a box contains 5 defective PDAs. A sample of 3 is selected. It is of interest to compute the probability that the sample contains 1 defective. The 1 defective can come from the 5 defectives in $\binom{5}{1}$ ways. The other 2 come from the 20 nondefectives in $\binom{20}{2}$ ways. The multiplicative rule states that the 1 defective and the 2 nondefectives can be selected in $\binom{5}{1}\binom{20}{2}$ ways. There are 5(190) = 950 samples with 1 defective and 2 nondefectives. The total space is the number of ways that 3 may be selected from 25 which is $\binom{25}{3}$ or 2300. The probability of selecting 3 that contain 1 defective is 950/2300 = 0.413. This is called a hypergeometric probability.

A personal digital assistant is mass produced and is subject to quality control techniques.

Problems often arise in which the hypergeometric is applicable. There is a dichotomy of elements involved. That is, the elements may be classified into two

types: defective/nondefective, male/female, hearts/nonhearts in a deck of cards, and so on. A sample from the population that has been dichotomized into the two groups is selected. The generic terms: success and failure are used to describe the two types of elements. The number of successes in the sample is the random variable of interest. With this introductory paragraph, we are ready to describe the random variable generally.

A population consists of N items of which s are classified as successes and $N - s$ are failures. A sample of size n is selected from the N. The hypergeometric random variable X is defined to be the number of successes in the sample. Paralleling the opening example of this section, the following probability distribution for a hypergeometric is obtained:

$$P(X = x) = \frac{\binom{s}{x}\binom{N-s}{n-x}}{\binom{N}{n}} \quad \text{for} \quad x = 0, 1, \ldots, n \text{ and } x \le s \text{ and } n - x \le N - a$$

The term $\binom{s}{x}$ counts the number of ways x successes can be selected from the s successes. The term $\binom{N-s}{n-x}$ counts the number of ways the $(n - x)$ failures can be selected from the $(N - s)$ failures. The $\binom{N}{n}$ counts the number of ways n can be selected from N. The $x \le s$ and $n - x \le N - a$ parts are needed because you cannot select more items than are available. (For example, you cannot select five items from four.) The computational difficulty is made easier by computer software.

EXAMPLE

Revisiting the lead off example of this section, find the probabilities of 0, 1, 2, or 3 defectives in a sample of three from a box of 25 PDAs containing 5 defectives using EXCEL.

The numbers 0, 1, 2, and 3 are entered into A2:A5, and the EXCEL function =HYPGEOMDIST(A2,3,5,25) is entered into B2 and a click-and-drag is performed. The first parameter is the x value, the second parameter is the sample size, the third parameter is the number of successes, and the fourth is the total population. Note that $h(1) = 0.413043$, the same as obtained earlier in this section on the hypergeometric.

x	$h(x)$
0	0.495652
1	0.413043
2	0.086957
3	0.004348

Figure 3-24 EXCEL output for finding the probabilities of 0, 1, 2, or 3 defectives in a box of 25 PDA's containing 5 defectives.

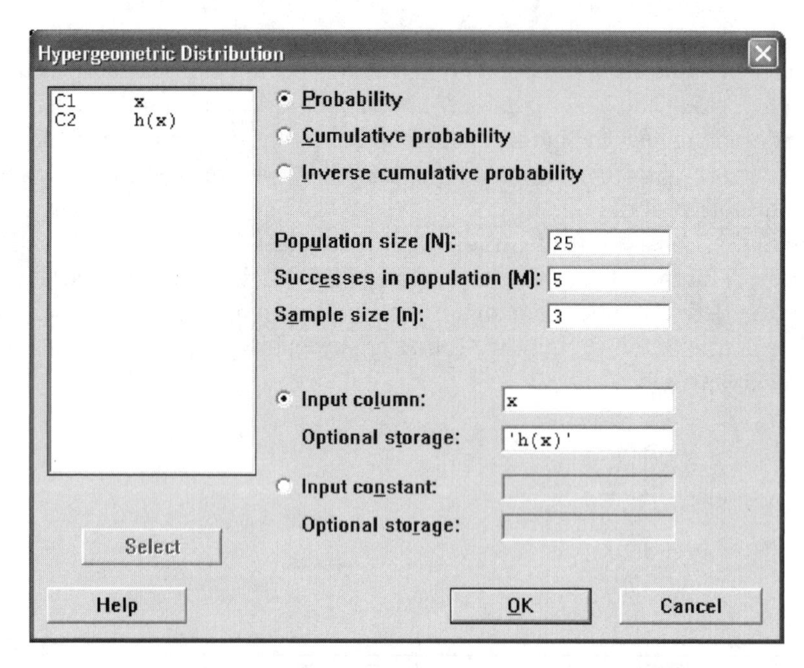

Figure 3-25 MINITAB hypergeometric distribution dialog box.

EXAMPLE

Give the MINITAB solution to the previous example. The pull-down menu **Calc ⇒ Probability Distributions ⇒ Hypergeometric** gives the dialog box in Figure 3-25.

Row	*x*	*h*(*x*)	Cumulative
1	0	0.495652	0.49565
2	1	0.413043	0.90870
3	2	0.086957	0.99565
4	3	0.004348	1.00000

The distribution given under $h(x)$ by MINITAB is the same as that obtained by EXCEL.

A shortcut formula for the mean and standard deviation may be derived from the formula for the mean and standard deviation and some algebraic manipulation. It may be shown that the mean of a hypergeometric distribution is $\mu = \frac{ns}{N}$. The standard deviation of a hypergeometric distribution likewise can be found by use of the shortcut formula

$$\sigma = \sqrt{\frac{ns(N-s)(N-n)}{N^2(N-1)}}$$

EXAMPLE

Consider the following hypergeometric distribution. Find the mean and the standard deviation using the basic definition and the shortcut formula and show that you get the same answer for $N = 50$, $s = 10$, and $n = 5$. The distribution of X = the number of successes in the sample of size 5 is as follows.

x	0	1	2	3	4	5
$h(x)$	0.310563	0.431337	0.20984	0.044177	0.003965	0.000119

$\mu = \Sigma xp(x) = 0(0.310563) + 1(0.431337) + 2(0.20984) + 3(0.0.44177) + 4(0.003965) + 5(0.000119) = 1$. Using the shortcut formula for the mean, $\mu = \frac{ns}{N} = \frac{5(10)}{50} = 1$, $\sigma^2 = \Sigma x_i^2 p(x_i) - \mu^2 = 0(0.310563) + 1(0.431337) + 4(0.20984) + 9(0.0.44177) + 16(0.003965) + 25(0.000119) - 1 = 1.734694 - 1 = 0.734694$, $\sigma = 0.857143$.

Using the shortcut formula for the standard deviation,

$$\sigma = \sqrt{\frac{ns(N-s)(N-n)}{N^2(N-1)}} = \sqrt{\frac{5(10)(40)(45)}{2500(49)}} = 0.857143$$

The hypergeometric and the binomial are very similar distributions when $n < 0.05N$. This is illustrated in the following example.

MP3s are packaged for shipment. A sample is taken to help insure quality. The number of defectives in a sample has a hypergeometric distribution.

EXAMPLE

A box of MP3s are packaged for shipment. The box contains 500 MP3s of which 25 have minor defects. A sample of 10 is selected. The number in the sample with a defect has a hypergeometric distribution with $N = 500$, $s = 25$, and $n = 10$. The number in the sample with a defect may also be approximated by a binomial with $n = 10$, $p = 0.05$, and $q = 0.95$. Since $n < 0.05N$, it may be assumed that as items are taken into the sample the 5% defects do not change significantly. The hypergeometric probabilities in row 2 and the binomial probabilities in row 3 in Figure 3-26 were generated using EXCEL.

Figure 3-27 gives a graphical comparison of the distributions in Figure 3-26.

x	0	1	2	3	4	5	6	7	8	9	10
$h(x)$	0.5959	0.3197	0.0739	0.0097	0.0008	4E-05	2E-06	3E-08	5E-10	4E-12	1E-14
$b(x)$	0.5987	0.3151	0.0746	0.0105	0.001	6E-05	3E-06	8E-08	2E-09	2E-11	1E-13

Figure 3-26 Comparison of hypergeometric and binomial probability distributions.

3.6 The Geometric Distribution

The **geometric random variable** has the same assumptions as the binomial, except the number of trials is not fixed. Random variable X is the trial number on which the first success occurs. The trials are independent of one another, and the probability is p that a success occurs on a given trial and q that a failure occurs on a given trial.

EXAMPLE

Suppose, a fair die is rolled until a 6 occurs. The probability that a 6 occurs on the first roll is $\frac{1}{6}$. The probability that a 6 occurs on the second roll is $P(2) = \frac{5}{6}\frac{1}{6} = \frac{5}{36}$. The trials are independent, and so, the probability of "not a 6" on the first roll followed by a 6 on the second roll is just the product of the two probabilities. The probability that a 6 occurs for the first time on the third roll is the same as "not a 6" on the first roll and "not a 6" on the second roll and a 6 on the third roll. The product of the

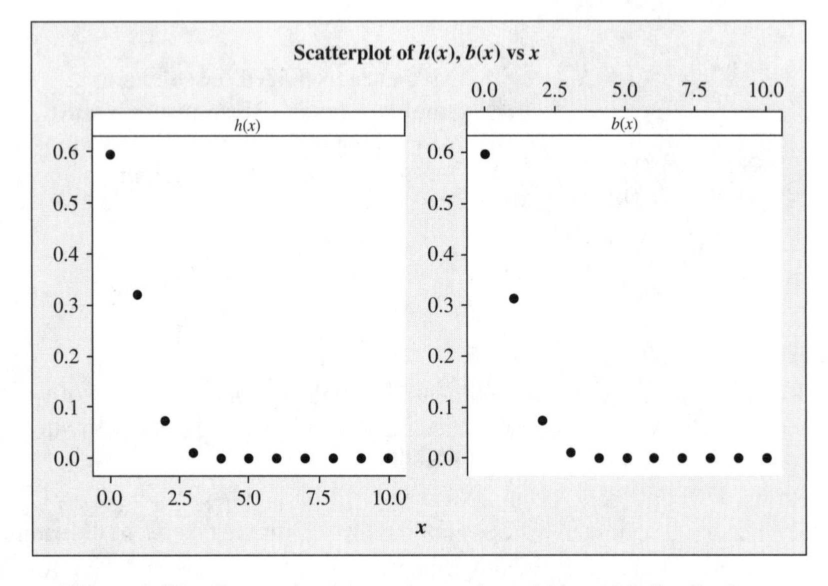

Figure 3-27 Comparing hypergeometric and binomial distributions.

x	1	2	3	4	5	6	7	8	9	10
$p(x)$	0.167	0.139	0.116	0.096	0.08	0.067	0.056	0.047	0.039	0.032
x	11	12	13	14	15	16	17	18	19	20
$p(x)$	0.027	0.022	0.019	0.016	0.013	0.011	0.009	0.008	0.006	0.005

Figure 3-28 First 20 probabilities for the geometric distribution $p(x)=\left(\frac{5}{6}\right)^{x-1}\frac{1}{6}$.

probabilities is $P(3)=\frac{5}{6}\frac{5}{6}\frac{1}{6}=\frac{25}{216}$. If this is continued, the geometric probability distribution is found to be

$$P(x)=\left(\frac{5}{6}\right)^{x-1}\frac{1}{6} \quad \text{for} \quad x=1, 2, 3, \ldots$$

EXCEL was used to generate Figure 3-28. It shows the probabilities associated with the first 20 values of the geometric variable x. The first 20 values account for over 97% of the probability associated with the distribution. An EXCEL plot of the computations in Figure 3-28 is shown in Figure 3-29.

Generalizing, the geometric distribution is found to be the following, where p is the success probability and $q = (1 - p)$ is the failure probability:

$$g(x)=q^{x-1}p \quad \text{for} \quad x=1, 2, 3, \ldots$$

The shortcut formulas for the mean and variance for a geometric random variable may be shown to be the following:

$$\mu=\frac{1}{p} \quad \text{and} \quad \sigma^2=\frac{1-p}{p^2}$$

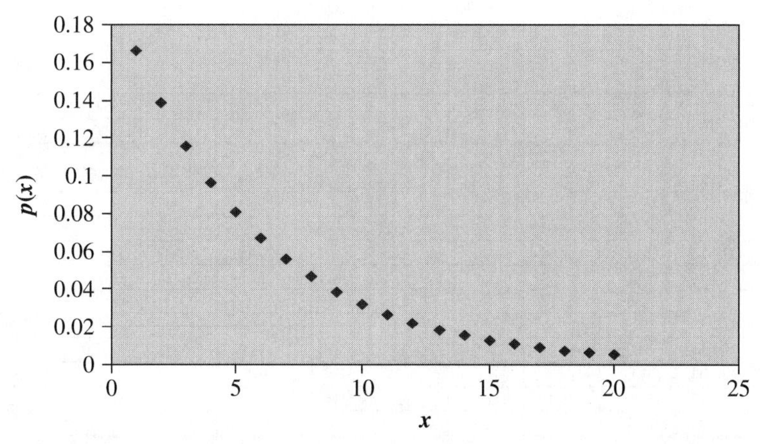

Figure 3-29 Plot of the geometric distribution $P(x)=\left(\frac{5}{6}\right)^{x-1}\frac{1}{6}$.

EXAMPLE

Traffic accident at 72nd and Dodge.

A traffic engineer has determined that the probability of at least one accident at 72nd and Dodge per day is 0.25. The location is observed until at least one accident occurs during a day. Success is at least one accident occurs during a day. Failure is no accident occurs during a day. The success probability is $p = 0.25$ and the failure probability is $q = 0.75$. The probability distribution function is $g(x) = 0.75^{x-1}0.25$ for $x = 1, 2, 3, \ldots$. Find the mean and variance using both the fundamental definitions and the shortcut formulas.

Even though the sums involved in finding the mean and the variance go from 1 to infinity, if the sums are truncated at 30, the mean and variance will be very close. Refer to Figure 3-30 in the following discussion. First of all, note that the sum of the probabilities, $\Sigma g(x) = 0.944 + 0.053 + 0.003 = 1$ to three decimal places. The mean is approximated by $3.12 + 0.712 + 0.070 = 3.994$ and is very close to the shortcut true value $\mu = \frac{1}{p} = \frac{1}{0.25} = 4$. The variance is $\sigma^2 = \Sigma x_i^2 p(x_i) - \mu^2$ and is approximated by $16.29 + 9.849 + 1.656 - 4^2 = 11.795$ which is very close to the true shortcut true value

$$\sigma^2 = \frac{1-p}{p^2} = \frac{1-0.25}{0.25^2} = 12.$$

x	1	2	3	4	5	6	7	8	9	10	Sum
$g(x)$	0.25	0.188	0.141	0.105	0.079	0.059	0.044	0.033	0.025	0.019	0.944
$xg(x)$	0.25	0.375	0.422	0.422	0.396	0.356	0.311	0.267	0.225	0.188	3.212
$x^{2}g(x)$	0.25	0.75	1.266	1.688	1.978	2.136	2.18	2.136	2.027	1.877	16.29
x	11	12	13	14	15	16	17	18	19	20	Sum
$g(x)$	0.014	0.011	0.008	0.006	0.004	0.003	0.003	0.002	0.001	0.001	0.053
$xg(x)$	0.155	0.127	0.103	0.083	0.067	0.053	0.043	0.034	0.027	0.021	0.712
$x^{2}g(x)$	1.703	1.52	1.338	1.164	1.002	0.855	0.724	0.609	0.509	0.423	9.849
x	21	22	23	24	25	26	27	28	29	30	Sum
$g(x)$	8E-04	6E-04	4E-04	3E-04	3E-04	2E-04	1E-04	1E-04	8E-05	6E-05	0.003
$xg(x)$	0.017	0.013	0.01	0.008	0.006	0.005	0.004	0.003	0.002	0.002	0.07
$x^{2}g(x)$	0.35	0.288	0.236	0.193	0.157	0.127	0.103	0.083	0.067	0.054	1.656

Figure 3-30 Using EXCEL to find the mean and variance of $g(x) = 0.75^{(x-1)}0.25$.

The geometric distribution has a **countably infinite sample space**. However, if a package such as EXCEL is available, it is seen that a finite number of terms are sufficient to handle any computations involving the distribution.

3.7 The Negative Binomial Distribution

The negative binomial distribution is the distribution of *X*, where *X* is the number of trials needed to obtain *r* successes. Consider an example before generalizing the formula for *X*.

EXAMPLE

Suppose we roll a die until a 6 occurs for the third time. Note that the third occurrence of the 6 might be on the third or fourth or fifth or ... roll of the die. If the third occurrence of a 6 is on the *x*th roll, then there must have been two 6s in the proceeding $(x - 1)$ rolls. The probability of two 6s in $(x - 1)$ rolls of a die is a binomial probability. Therefore the probability of two 6s in $(x - 1)$ rolls and then a 6 on the *x*th roll is

$$P(X=x)=\binom{x-1}{2}\left(\frac{1}{6}\right)^2\left(\frac{5}{6}\right)^{(x-1)-2}\left(\frac{1}{6}\right)=\binom{x-1}{2}\left(\frac{1}{6}\right)^3\left(\frac{5}{6}\right)^{(x-3)} \quad \text{for} \quad x=3, 4, \ldots$$

Using EXCEL, the numbers 3 through 12 are entered into B1:K1, and the expression =COMBIN(B1-1,2)*(1/6)^3*(5/6)^(B1-3) is entered into B2 and a click-and-drag is performed from B2 through K2 to form the first two rows of Figure 3-31. This same procedure is used to form the remaining rows of the table in Figure 3-31.

x	3	4	5	6	7	8	9	10	11	12	Sum
P(*x*)	0.005	0.012	0.019	0.027	0.033	0.039	0.043	0.047	0.048	0.049	0.323
x	13	14	15	16	17	18	19	20	21	22	Sum
P(*x*)	0.049	0.049	0.047	0.045	0.043	0.041	0.038	0.036	0.033	0.03	0.412
x	23	24	25	26	27	28	29	30	31	32	Sum
P(*x*)	0.028	0.025	0.023	0.021	0.019	0.017	0.015	0.014	0.012	0.011	0.185
x	33	34	35	36	37	38	39	40	41	42	Sum
P(*x*)	0.01	0.009	0.008	0.007	0.006	0.005	0.005	0.004	0.004	0.003	0.059
x	43	44	45	46	47	48	49	50	51	52	Sum
P(*x*)	0.003	0.002	0.002	0.002	0.002	0.001	0.001	0.001	9E-04	8E-04	0.016

Figure 3-31 The probabilities for $X = 3$ through $X = 52$ for the negative binomial distribution $P(X=x)=\binom{x-1}{2}\left(\frac{1}{6}\right)^3\left(\frac{5}{6}\right)^{(x-3)}$.

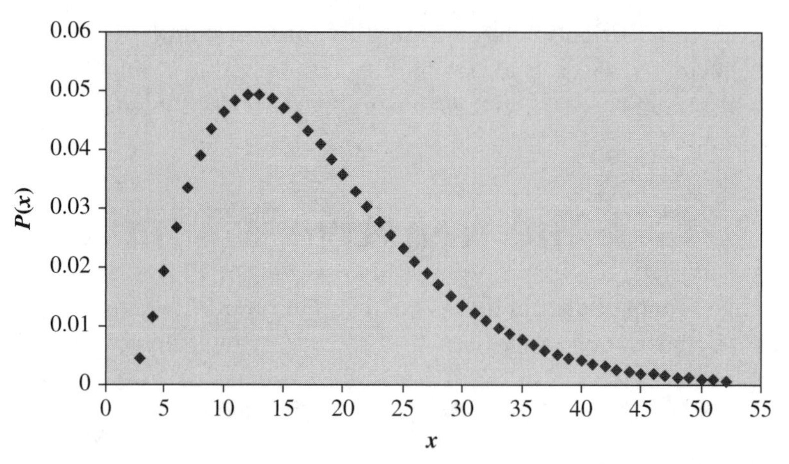

Figure 3-32 A plot of the negative binomial distribution $P(X=x)=\binom{x-1}{2}\left(\frac{1}{6}\right)^3\left(\frac{5}{6}\right)^{(x-3)}$.

The cumulative probability at 52 is (0.323 + 0.412 + 0.185 + 0.059 + 0.016) = 0.995 in Figure 3-31, that is $F(52) = 0.995$. A plot of the approximate negative binomial distribution given in Figure 3-31 is shown in Figure 3-32. The plot corresponds to x values from 3 to 52. The y value at $x = 52$ is 0.0008. Even though this distribution continues on forever, the amount of added probability from 53 onward is negligible (less than 0.005).

Generalizing from the last example, we have the following: Suppose the probability of a success is p and the probability of a failure is q on any given trial. The trials are independent of one another. The probability that the rth success occurs on the xth trial where $x = r, (r + 1), \ldots$ is

$$NB(x)=\binom{x-1}{r-1}(p)^r(q)^{(x-r)} \quad \text{for} \quad x = r, (r+1), \ldots$$

The shortcut formula for the mean of a negative binomial is $\mu=\frac{r}{p}$ and the variance is $\sigma^2=\frac{r(1-p)}{p^2}$. Sometimes, algebraic manipulations are used to derive the shortcut formulas for the mean and variance for the various distributions. Sometimes, techniques utilizing moment generating functions are utilized in the proofs. (Moment generating functions are discussed in mathematical statistics courses.) As engineering students, we are mainly interested in the results, not in the proofs. In a course in mathematical statistics, the derivations of the formulas would be the focus of the course and the applications would be of secondary interest.

EXAMPLE

This example compares the mean found by using the defining formula for the mean and the shortcut formula for the mean of the negative binomial distribution.

x	3	4	5	6	7	8	9	10	11	12	Sum
$P(x)$	0.005	0.012	0.019	0.027	0.033	0.039	0.043	0.047	0.048	0.049	0.323
$xP(x)$	0.014	0.046	0.096	0.161	0.234	0.313	0.391	0.465	0.533	0.592	2.845
x	13	14	15	16	17	18	19	20	21	22	Sum
$P(x)$	0.049	0.049	0.047	0.045	0.043	0.041	0.038	0.036	0.033	0.03	0.412
$xP(x)$	0.642	0.68	0.709	0.727	0.736	0.736	0.728	0.714	0.694	0.669	7.034
x	23	24	25	26	27	28	29	30	31	32	Sum
$P(x)$	0.028	0.025	0.023	0.021	0.019	0.017	0.015	0.014	0.012	0.011	0.185
$xP(x)$	0.642	0.611	0.579	0.545	0.511	0.477	0.443	0.41	0.379	0.348	4.945
x	33	34	35	36	37	38	39	40	41	42	Sum
$P(x)$	0.01	0.009	0.008	0.007	0.006	0.005	0.005	0.004	0.004	0.003	0.059
$xP(x)$	0.319	0.292	0.266	0.242	0.219	0.198	0.179	0.161	0.145	0.13	2.152
x	43	44	45	46	47	48	49	50	51	52	Sum
$P(x)$	0.003	0.002	0.002	0.002	0.002	0.001	0.001	0.001	9 04	8 04	0.016
$xP(x)$	0.117	0.104	0.093	0.083	0.074	0.066	0.058	0.052	0.046	0.04	0.733
x	3	4	5	6	7	8	9	10	11	12	

Figure 3-33 The mean of the negative binomial is 18.

Figure 3-33 is easily produced using EXCEL. The mean has an approximate value of 2.845 + 7.034 + 4.945 + 2.152 + 0.733 = 17.71. This approximate value of the mean is obtained by using the terms corresponding to $X = 3$ through $X = 52$. Keep in mind that the sum defining the mean actually has an infinite number of terms in it. The true mean using the shortcut formula is $\mu = \frac{r}{p} = \frac{3}{1/6} = 18$.

EXAMPLE

Quality control is applied to play stations to insure quality.

A quality control engineer has decided to inspect play stations until 4 defectives have been found. Suppose 1% of the play stations are defective. Let X stand for the number of inspections needed to find 4 defectives.

a. What is the mean number of inspections needed to find 4 defectives?

b. What is the standard deviation of X?

c. What is the probability that X will be greater than $\mu + \sigma$?

Note that X has a negative binomial distribution with $r = 4$, $p = 0.01$, and $q = 0.99$.

a. $\mu = \frac{r}{p} = \frac{4}{0.01} = 400.$

b. $\sigma^2 = \frac{r(1-p)}{p^2} = \frac{4(0.99)}{.0001} = 39600$ or $\sigma = 199$

c. Using EXCEL, the numbers 0 through 595 are placed in A1:A596 by using a click-and-drag, and =NEGBINOMDIST(A1,4,0.01) is entered into B1 and a click-and-drag is performed from B1 to B596. The function =SUM(B1: B596) is entered in B597. This gives the probability $P(X \leq \mu + \sigma)$, which is 0.84925. The probability we are seeking is 1 – 0.84925 or 0.15075.

3.8 The Poisson Distribution

A distribution that arises frequently in engineering is the Poisson distribution. In particular it arises when there are a large number of independent trials and a small success probability. The random variable X counts the number of successes to occur. It has a probability distribution represented by $Po(x)$ and given by the following:

$$Po(x) = e^{-\lambda} \frac{\lambda^x}{x!}, x = 0, 1, 2, \ldots$$

It can be shown that $\mu = \lambda$ and $\sigma^2 = \lambda$. The Poisson distribution and the binomial distribution give very similar probabilities when n is large and p is small as the next example will illustrate.

EXAMPLE

A traffic survey counts the number of accidents to occur at 72nd and Dodge during a week. A large number of cars pass through the intersection, and the probability that a given car is involved in an accident is $p = 0.0001$. During the week the traffic survey is conducted, 100,000 cars pass through the intersection. The mean number of accidents during such a week is $\mu = np = 100000(0.0001) = 10$. Figure 3-34 gives the binomial probabilities and the Poisson probabilities for $X = 0$ to 20. Note how close the binomial and the Poisson probabilities are to one another.

The EXCEL expression =BINOMDIST(A2,100000,0.0001,0) is entered into B2 and a click-and-drag is executed from B2 to B21. The EXCEL expression =POISSON(A2,10,0) is entered into C2 and a click-and-drag is executed from C2 to C21. The 10 in the Poisson expression is the mean of the Poisson. Remember $\lambda = np = 100{,}000(0.0001) = 10$. The 0 in the Poisson function causes the individual Poisson probabilities rather than cumulative Poisson probabilities to be printed. Figure 3-35 gives a plot of the two probability distributions. The distributions are

x	Binomial	Poisson
0	4.53772E-05	4.53999E-05
1	0.000453818	0.000453999
2	0.002269293	0.002269996
3	0.007564915	0.007566655
4	0.018913611	0.018916637
5	0.037829491	0.037833275
6	0.063052305	0.063055458
7	0.090078325	0.090079226
8	0.112601284	0.112599032
9	0.12511504	0.125110036
10	0.125116292	0.125110036
11	0.113742083	0.113736396
12	0.094784122	0.09478033
13	0.072909404	0.072907946
14	0.052076583	0.052077104
15	0.034716333	0.03471807
16	0.021696623	0.021698794
17	0.012761954	0.012763996
18	0.007089478	0.007091109
19	0.003731006	0.003732163
20	0.001865335	0.001866081
Sum	0.998413	0.998412

Figure 3-34 Comparing the binomial and poisson distributions ($n = 100000$ and $p = 0.0001$).

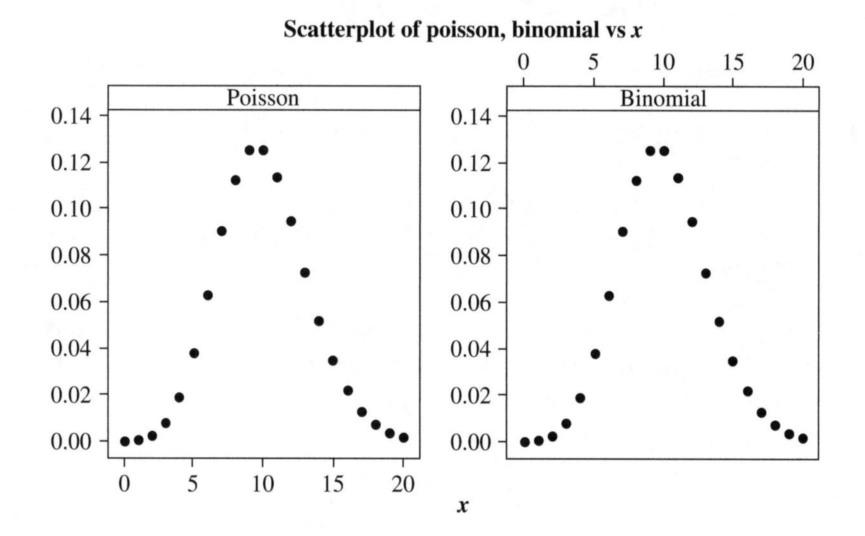

Figure 3-35 Comparing the binomial and Poisson distributions ($n = 100000$ and $p = 0.0001$).

very similar to the normal one for large n and small p. The normal distribution is continuous and will be discussed in the next chapter.

Most of the probabilities are associated with the values from 0 to 20. The sum of the Poisson probabilities in Figure 3-34 is 0.998412 and the sum of the binomial probabilities is 0.998413. Even though the variable takes on values from 0 to 100,000, there is very little probability beyond $X = 21$. Figure 3-35 is a MINITAB comparison of the graphs of the two probability functions.

EXAMPLE

The number of hits that a website experiences is a random variable.

The number of hits on a website during an 8-hour period follows a Poisson distribution. The mean number per 8-hour period is 10.

a. What is the probability of 15 or fewer hits during an 8-hour period using MINITAB? b. What is the probability of 25 or more hits during a 16-hour period using EXCEL?

a. Using MINITAB, the pull-down **Calc ⇒ Probability Distributions ⇒ Poisson Distribution** gives the dialog box shown in Figure 3-36. The output for the dialog box is as follows:

Cumulative Distribution Function

```
Poisson with mean = 10

 x    P( X <= x )
15      0.951260
```

The probability of 15 or fewer hits is 0.951260.

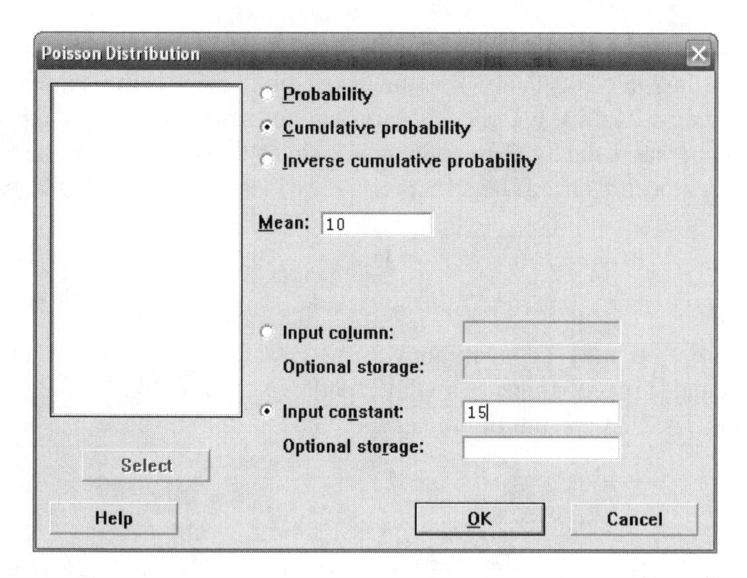

Figure 3-36 MINITAB dialog box for calculating $P(X \leq 15)$ for the Poisson distribution.

b. The mean number of hits during a 16-hour period is Poisson with $\lambda = 20$. The probability of 24 or fewer hits during a 16-hour period is given by the EXCEL command =Poisson(24,20,1) = 0.843227. The probability of 25 or more hits during the 16-hour period is 1 – 0.843227 = 0.156773. Note that 25 or more is the complement of 24 or less. The EXCEL command =Poisson(24,20,1) gives the cumulative probability of 24 or fewer for a Poisson distribution with mean 20.

3.9 The Multinomial Distribution

The multinomial distribution is an extension of the binomial distribution. There are n independent trials with not 2 but k possible outcomes per trial. Define discrete random variables $X_1, \ldots, X_k$ and probabilities $p_1, \ldots, p_k$ as follows:

X_i = the number of times that outcome i occurs in the n trials and p_i = the probability that outcome i occurs. The **joint probability mass function** that $X_1 = x_1, \ldots, X_k = x_k$ is

$$P(X_1 = x_x, \ldots, X_k = x_k) = \frac{n!}{x_1! \cdots x_k!} p_1^{x_1} p_2^{x_2} \cdots p_k^{x_k}, \; x_i = 0, 1, \ldots, n \text{ and } \sum x_i = n$$

EXAMPLE

A fair die is thrown 5 times. On any one throw, outcome 1 is that an even number appears, outcome 2 is that a 1 or 3 appears, and outcome 3 is that a 5 appears. Find the probability that in the 5 throws, outcome 1 occurs twice, outcome 2 occurs twice, and outcome 3 occurs once. The joint probability mass function is as follows:

$$P(X_1 = 2, X_2 = 2, X_3 = 1) = \frac{5!}{2!2!1!} 0.5^2 0.33^2 0.17^1$$

The expression may be evaluated using EXCEL. The EXCEL expression used to evaluate the probability is =MULTINOMIAL(2,2,1)*0.5^2*0.33^2*0.17^1 = 0.138848. The EXCEL multinomial function is used to evaluate

$$\frac{n!}{x_1! \cdots x_k!}, \ \Sigma x_i = n$$

EXAMPLE

Based on past experience, the make up of Dr. Stephens' statistics for engineers class has consisted of 20% civil engineers, 30% chemical engineers, 40% industrial engineers, and 10% electrical engineers. Given that these probabilities still hold, what are the probabilities that her 2007 classes will consist of 25 civil engineers, 30 chemical engineers, 40 industrial engineers, and 10 electrical engineers?

The expression for the probability is

$$P(X_1 = 25, X_2 = 30, X_3 = 40, X_4 = 10) = \frac{105!}{25!30!40!10!} 0.2^{25} 0.3^{30} 0.4^{40} 0.1^{10}$$

The EXCEL expression yields 0.00074 and is given as follows:

=MULTINOMIAL(25,30,40,10)*0.2^25*0.3^30*0.4^40*0.1^10.

The EXCEL function MULTINOMIAL(25, 30, 40, 10) combined with the exponentials 0.2^25*0.3^30*0.4^40*0.1^10 save the student considerable work.

EXAMPLE

A traffic engineer knows that at a certain intersection over a 24-hour period, no accidents occur with probability 0.25, one accident will occur with probability 0.60, and two or more accidents occur with probability 0.15. What is the probability that over ten 24-hour periods, no accidents occur 3 times, one accident occurs 6 times, and two or more accidents occur once?

The expression for the probability is

$$P(X_1 = 3, X_2 = 6, X_3 = 1) = \frac{10!}{3!6!1!} 0.25^3 0.6^6 0.15^1 = 840(0.015625)(0.046656)(0.15) = .0918$$

The same answer is obtained using the EXCEL function MULTINOMIAL and the exponents (=MULTINOMIAL(3,6,1)*0.25^3*0.6^6*0.15).

3.10 Simulation

Often we are interested in obtaining some sample values from a random variable but it may be expensive or difficult to obtain the values. In such cases we may try to **simulate** the values that we would likely obtain without actually observing the variable.

EXAMPLE

Consider the sums that we would obtain from observing the outcomes when a pair of dice is rolled 50 times. A random number generator is used to simulate what we might obtain. We found earlier in Section 3.1 that the probability distribution for the sum on the dice had the distribution given in columns 1, 2, and 3 of Figure 3-37. Column 4 gives ranges of three-digit random numbers determined by column 3.

In column 4 are shown ranges of three-digit random numbers. EXCEL produces three-digit numbers as follows. The expression =RANDBETWEEN(0,999) is placed in a cell and a click-and-drag is performed over 5 columns and 10 rows. The random numbers generated by EXCEL are shown in Figure 3-38.

x	$P(x)$	Cumulative	Range
2	0.028	0.028	000–027
3	0.056	0.084	028–083
4	0.083	0.167	084–166
5	0.111	0.278	167–277
6	0.139	0.417	278–416
7	0.166	0.583	417–582
8	0.139	0.722	583–721
9	0.111	0.833	722–832
10	0.083	0.916	833–915
11	0.056	0.972	916–971
12	0.028	1.000	972–999

Figure 3-37 Simulating the sum on a pair of dice.

160	183	539	69	251
719	587	343	825	942
48	575	729	447	132
843	985	662	663	643
378	360	755	845	859
606	626	144	683	284
642	363	57	772	703
809	827	702	232	172
344	864	16	457	222
638	36	912	240	406

Figure 3-38 Fifty random numbers between 000 and 999 generated by EXCEL.

The probabilities that a three-digit random number falls in the range in Figure 3-37 are equal to the probabilities given in the column labeled $p(x)$ in Figure 3-37. The probabilities for the sums on the dice are therefore given by the $p(x)$ column of Figure 3-37.

When the random numbers in Figure 3-38 are translated to outcomes when rolling the dice, the results are shown in Figure 3-39.

When a dot plot of the sample is generated by MINITAB, the results are shown in Figure 3-40. Suppose another simulated sample is desired. Figure 3-41 shows a random sample of 50 random numbers between 0 and 999 generated by MINITAB. These random numbers are converted to simulated outcomes when rolling the dice by using Figure 3-37. This gives a second simulated sample shown in Figure 3-42. A histogram of this second sample is shown in Figure 3-43. Random samples between 000 and 999 may be obtained using EXCEL, MINITAB, SAS, SPSS, or STATISTIX.

4	5	7	3	5
8	8	6	9	11
3	7	9	6	4
10	12	8	8	7
6	6	9	10	10
8	8	4	8	6
8	6	3	9	8
9	9	8	5	5
6	10	2	7	5
8	3	10	5	6

Figure 3-39 Outcomes on the dice determined by the random numbers.

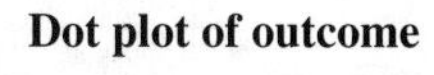

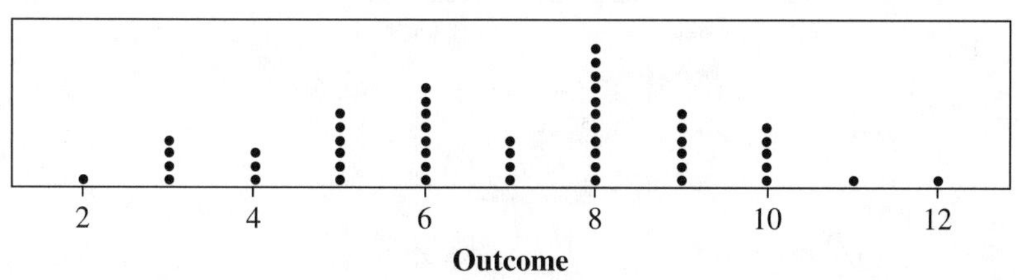

Figure 3-40 Dot plot of simulated sample of outcomes from Figure 3-39.

85	389	174	700	609
215	351	604	622	824
601	202	292	52	82
110	938	574	555	67
131	935	111	448	986
836	631	448	570	90
3	727	597	78	800
410	347	111	979	123
92	415	531	131	963
146	615	770	179	682

Figure 3-41 Fifty random numbers between 000 and 999 generated by MINITAB.

4	6	5	8	8
5	6	8	8	9
8	5	6	3	3
4	11	7	7	3
4	11	4	7	12
10	8	7	7	4
2	9	8	3	9
6	6	4	12	4
4	6	7	4	11
4	8	9	5	8

Figure 3-42 Second sample of simulated outcomes on the dice determined by the random numbers in Figure 3-41.

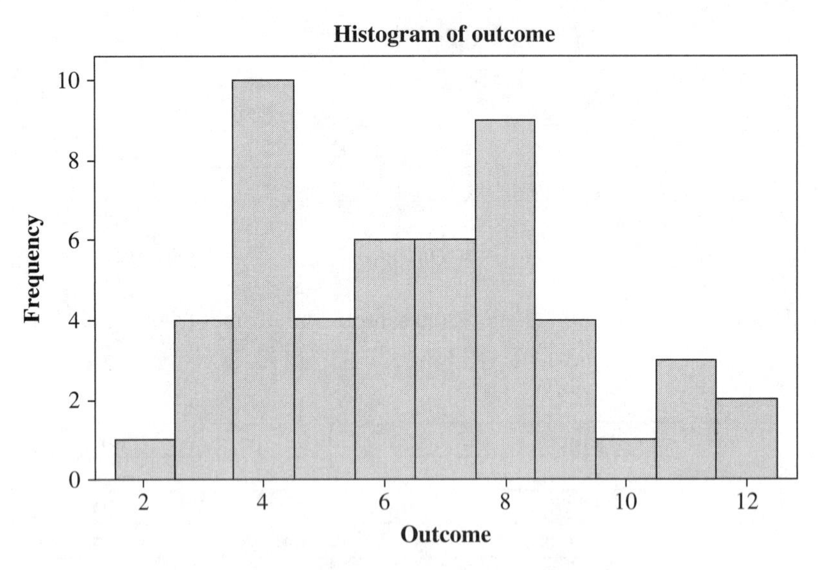

Figure 3-43 Histogram of second simulated sample.

EXAMPLE

In order to simulate the number of defective items in the samples from a production line an engineer uses random numbers. The probability distribution for X, the number of defectives found in samples, is known to be the following:

X	0	1	2	3
$p(x)$	0.50	0.25	0.20	0.05

The engineer wishes to simulate what his results would be for a 30-day period. Give results that would simulate what she might experience over a 30-day period. First, build a table like that shown in Figure 3-44.

x	$p(x)$	Cumulative	Range
0	0.50	0.50	00–49
1	0.25	0.75	50–74
2	0.20	0.95	75–94
3	0.05	1.00	95–99

Figure 3-44 Simulation setup.

Then select thirty 2-digit random numbers between 00 and 99.

45	50	64	18	91	98	89	83	78	81
99	96	44	92	23	49	91	30	29	80
15	34	46	46	72	4	4	47	31	65

These random numbers are replaced by simulated number of defectives using Figure 3-44 to obtain the following simulated results:

0	1	1	0	2	3	2	2	2	2
3	3	0	0	0	0	2	0	0	2
0	0	0	0	1	0	0	0	0	1

Figure 3-45 shows a time series plot of what the engineer might expect over a 30-day period. The index numbers represent days and the defective numbers represent the number of defectives found on a given day.

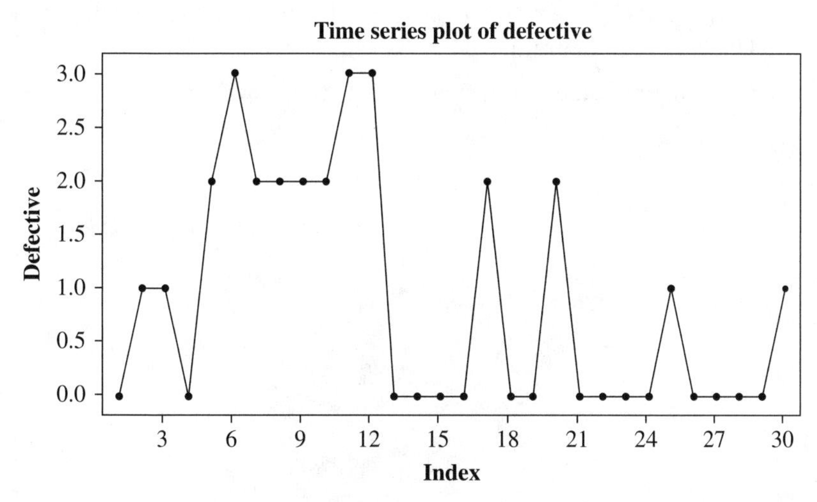

Figure 3-45 Simulated time series plot over a 30-day period for the number of defectives found.

Exercises

1. A regular tetrahedron is a solid structure that has four faces that are identical. The numbers 1, 2, 3, and 4 are stamped on the four faces. When the tetrahedron is tossed upon a table, one of the faces will be down and the other three will be facing up. The down face will be 1, 2, 3, or 4. Suppose three tetrahedra are tossed and that X represents the sum on the three faces that are facing down. X can take on the values 3, 4, … , 12. See Figure 3-46.
   (a) Use EXCEL or some other software package to generate the $4 \times 4 \times 4 = 64$ possible outcomes when the above experiment is performed.
   (b) Find the probability distribution for X, the sum of the three down-turned faces when the three tetrahedra are tossed.
   (c) Find the mean and standard deviation of X.
2. Answer the following questions concerning the random variable in problem 1:
   (a) Graph a scatter plot for the probability distribution of X.
   (b) Find the amount of probability within two and three standard deviations of the mean and compare this with the results given by Chebyshev's theorem.
   (c) Give the cumulative distribution function for X.
   (d) Simulate a sample of size 20 for the variable X.

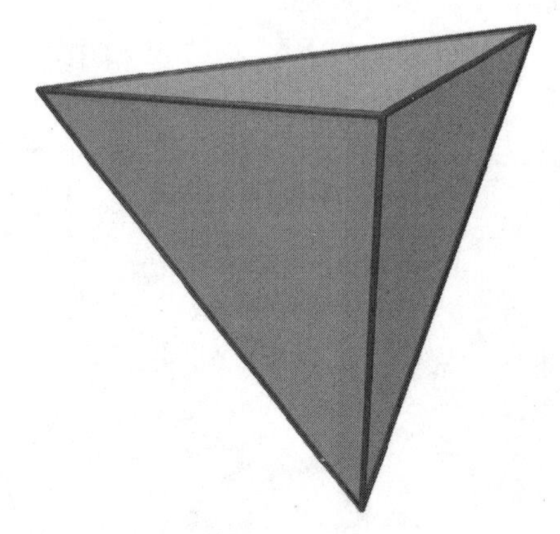

Figure 3-46 Regular tetrahedron found at web address *http://en.wikipedia.org/wiki/Tetrahedron*.

3. Engineers investigating the decibel levels at which people play their iPods have found the following. (There has been some fear of hearing loss from the high levels at which people play their MP3 players.) The level X has the following distribution:

X	**Range in decibels**	$p(x)$
1	70–79	0.05
2	80–89	0.10
3	90–99	0.20
4	100–109	0.25
5	110–120	0.40

   (a) Show that $p(x)$ is a probability distribution function.

   (b) Construct a scatter diagram for $p(x)$.

   (c) Find the mean and standard deviation for the random variable X.

   (d) Give the cumulative distribution for the random variable X.

   (e) Construct a graph for $F(x)$.

   (f) Simulate a random sample of size 50.

   (g) Find the mean and standard deviation of the simulated sample and compare them with the mean and standard deviation of the random variable.

4. Chemical engineers claim that 70% of all Americans have perfluorooctanoic acid (PFOA) in their blood stream. This chemical is a likely human carcinogen. Given that the 70% figure is correct, find the following for blood tests for PFOA in human blood: (Thirty Americans were randomly tested for PFOA.)

   (a) What is the mean or expected number that you would find in a sample of 30 who test positive for PFOA?

   (b) What is the standard deviation of the number found in samples of 30 who test positive for PFOA?

   (c) What is the probability of finding 15 or fewer in a sample of 30 who test positive for PFOA?

   (d) What is the probability of finding 25 or more in 30 who test positive for PFOA?

5. An engineer has 50 of her favorite websites book-marked. Ten of them are technical in nature and she often uses them to get information pertinent

to her job. The other 40 are nontechnical sites such as Amazon.com from which she orders books for her reading pleasure. Suppose 5 of the sites are selected at random.

(a) Use the basic definition of the hypergeometric distribution and the combination function of EXCEL to find the probabilities that in the 5 selected sites, 0, 1, 2, 3, 4, or 5 are technical sites.

(b) Do part a using the built-in function in MINITAB.

6. The regular tetrahedron in Figure 3-46 is tossed until the face 1 is the face that is down and 2, 3, and 4 are up. X represents the toss on which face 1 turns down first.

(a) Plot the probability distribution function and the cumulative probability distribution function in separate panels on the same graph.

(b) Find the mean and variance of the random variable X.

(c) What shape does the graph of $p(x)$ have?

(d) What is the probability that face 1 of the tetrahedron faces down for the 1st time on or before the 5th, 10th, or 15th toss?

(e) What would you think if $X > 15$?

7. A petroleum engineer knows that within a certain region, the probability that a drilling will lead to discovering oil is 0.45. Let X represent the drilling on which oil is discovered for the third time. The drillings are independent of one another.

(a) Give the probability distribution function for the random variable X.

(b) Plot the probability distribution function and the cumulative probability distribution function in separate panels on the same graph.

(c) Find the mean and variance of the random variable X.

(d) What shape does the graph of $p(x)$ have?

(e) What is the probability oil will be discovered for the 3rd time on or before the 5th, 10th, or 15th drilling?

8. The number of calls that an industrial engineer makes on her cell phone during a day has a Poisson distribution and the average is 15. Answer the following questions regarding the random variable $X =$ the number of calls made per day on the cell phone:

(a) Give the probability distribution function for the random variable X.

(b) Calculate the probabilities and the cumulative probabilities for $X = 0$ to $X = 30$.

(c) What is the probability that she makes more than 30 calls?

(d) What is the probability that she makes between 5 and 15 calls inclusive?

(e) Plot the probability distribution function and the cumulative probability distribution function in separate panels on the same graph.

9. A traffic engineer is studying accidents and finds that 15% are 1-car accidents, 65% are 2-car accidents, 15% are 3-car accidents, and 5% are 4-car accidents in the area he is studying. He randomly selects 25 accidents. Find the following probabilities:

(a) All 25 are 2-car accidents.

(b) All 25 involve the same number of cars.

(c) In the 25 selected, 6 are 1-car accidents, 6 are 2-car accidents, 6 are 3-car accidents, and 7 are 4-car accidents.

10. A group of six engineers got together for a night of poker and before the games began, they posed some probability problems concerning a deck of cards to sharpen their thinking. See, how many of the problems they posed, that you can work. Figure 3-47 gives the structure of a deck of cards in case you need a reminder.

(a) If five cards are dealt (with replacement), what is the probability all five are from the same suit? Work using the multinomial distribution.

(b) A card is dealt (with replacement) until a Heart first appears. Let X stand for the deal on which a Heart first appears. Give $p(x)$ and $F(x)$ for $x = 1$ through 20. Work using the Geometric distribution.

Hearts	Clubs	Diamonds	Spades
Ace	Ace	Ace	Ace
Two	Two	Two	Two
Three	Three	Three	Three
Four	Four	Four	Four
Five	Five	Five	Five
Six	Six	Six	Six
Seven	Seven	Seven	Seven
Eight	Eight	Eight	Eight
Nine	Nine	Nine	Nine
Ten	Ten	Ten	Ten
Jack	Jack	Jack	Jack
Queen	Queen	Queen	Queen
King	King	King	King

Figure 3-47 Standard deck of cards.

(c) Each of the six engineers is dealt a card, they look at it, and then replace it in the deck before the next card is dealt to the next engineer. Let X represent the number of face cards dealt to the six engineers. Find $p(x)$ and $F(x)$. Work using the binomial distribution.

(d) A hand of five cards is dealt to one of the engineers without replacement. The random variable X is defined as the number of Aces in the five cards. Find the probability as well as the cumulative probability functions for X. Use the Hypergeometric distribution to work this problem.

(e) A card is dealt, put back into the deck, and another card is dealt and this is continued. Let X represent the round on which a club occurs for the third time. Graph $p(x)$ and $F(x)$ for $x = 3, 4, \ldots, 25$. Use the negative binomial distribution to solve this problem.

Summary

1. A **random variable** assigns values to outcomes in the sample space.
2. The **probability distribution function** gives the values that X may assume and their probabilities.
3. **A probability distribution function for X has the properties $p(x) \geq 0$ and $\Sigma\, p(x) = 1$, where the sum is over the x values.**
4. A distribution function is **skewed to the right** if its graph has a tail to the right. It is **skewed to the left** if its graph has a tail to the left. A distribution is **symmetrical** if the part to the left of the mean is the mirror image of the part to the right of the mean.
5. The **cumulative distribution function** is defined as $F(x) = P(X \leq x)$.
6. If a **discrete random variable** can take on the values $x_1, x_2, \ldots, x_n$ with probabilities $p(x_1), p(x_2), \ldots, p(x_n)$, the **mean** is defined to be $\mu = \Sigma x_i\, p(x_i)$ where i goes from 1 to n.
7. The **variance** of a random variable is defined to be $\sigma^2 = \Sigma\, (x_i - \mu)^2 p(x_i)$ where i goes from 1 to n. The square root of the variance is the **standard deviation.** $\sigma = \sqrt{\Sigma(x_i - \mu)^2\, p(x_i)}$.
8. The **variance** may be found using the equivalent formula $\sigma^2 = \Sigma\, x_i^2 p(x_i) - \mu^2$.
9. The **empirical rule** applied to probability distributions states that for mound-shaped distributions, approximately 68% of the probability will be between $\mu - \sigma$ and $\mu + \sigma$, approximately 95% of the probability will be between $\mu - 2\sigma$ and $\mu + 2\sigma$, and 99.7% will be between $\mu - 3\sigma$ and $\mu + 3\sigma$.

10. **Chebyshev's theorem** states that if we go at least k standard deviations from the mean, then we will find at most $1/k^2$ of the probability distribution. Thus if we go at least 2 standard deviations from the mean, there will be at most 25% of the distribution. If we go at least 3 standard deviations from the mean, there will be at most 11% of the distribution.
11. **Binomial random variables** satisfy the following assumptions called the **binomial assumptions:**
    (a) **There are only two possible outcomes for each trial, called success and failure.**
    (b) **The probability of success and failure remains the same from trial to trial.**
    (c) **There are n trials and n is constant.**
    (d) **The n trials are independent of one another.**
12. The binomial random variable is then defined as the number of successes in the n trials. A binomial random variable has a probability distribution given by the following:

$$b(x) = \binom{n}{x} p^x q^{n-x} \quad \text{for} \quad x = 0, 1, 2, \ldots, n$$

and

$$\binom{n}{x} = \frac{n!}{x!(n-x)!}$$ are called **binomial coefficients**.

13. It may be shown that the **mean of the binomial distribution** is found by multiplying the number of trials times the success probability, i.e., $\mu = np$. The **standard deviation of the binomial distribution** likewise can be found by the use of the shortcut formula $\sigma = \sqrt{npq}$.
14. A population consists of N items of which s are classified as successes and $N - s$ are failures. A sample of size n is selected from the N. The **hypergeometric random variable** is defined to be the number of successes in the sample.

$$P(X = x) = \frac{\binom{s}{x}\binom{N-s}{n-x}}{\binom{N}{n}} \quad \text{for} \quad x = 0, 1, \ldots, n \text{ and } x \le s \text{ and } n - x \le N - a.$$

The term $\binom{s}{x}$ counts the number of ways x successes can be selected from the s successes. The term $\binom{N-s}{n-x}$ counts the number of ways the $(n-x)$ failures can be selected from the $(N-s)$ failures. The $\binom{N}{n}$ counts the number of ways n can be selected from N. The $x \le s$ and $n-x \le N-a$ parts are needed because you cannot select more items than are available.

15. It may be shown that the **mean of a hypergeometric distribution** is $\mu = \frac{ns}{N}$. The **standard deviation of a hypergeometric distribution** likewise can be found by the use of the shortcut formula $\sigma = \sqrt{\frac{ns(N-s)(N-n)}{N^2(N-1)}}$.
16. The **geometric random variable** has the same assumptions as the binomial, except the number of trials is not fixed. Random variable X is the trial number on which the first success occurs. The trials are independent of one another and the probability is p that a success occurs on a given trial and q that a failure occurs on a given trial.
17. The **geometric distribution** is found to be the following where p is the success probability and $q = (1-p)$ is the failure probability:

$$g(x) = q^{x-1}p \quad \text{for} \quad x = 1, 2, 3, \ldots$$

18. The shortcut formulas for the mean and variance for a geometric random variable may be shown to be the following:

$$\mu = \frac{1}{p} \quad \text{and} \quad \sigma^2 = \frac{1-p}{p^2}$$

19. The **negative binomial distribution** is the distribution of X, where X is the number of trials needed before the rth success. Suppose the probability of a success is p and the probability of a failure is q on any given trial. The trials are independent of one another. The probability that the rth success occurs on the xth trial where $x = r, (r+1), \ldots$ is

$$NB(x) = \binom{x-1}{r-1}(p)^r(q)^{(x-r)} \quad \text{for} \quad x = r, (r+1), \ldots$$

20. The shortcut formula for the mean of a negative binomial is $\mu = \frac{r}{p}$ and the variance of a negative binomial is $\sigma^2 = \frac{r(1-p)}{p^2}$.
21. The **Poisson distribution** arises when there are a large number of independent trials and a small success probability. The random variable X counts the

number of successes to occur. It has a probability distribution represented by $Po(x)$ and given by the following:

$$Po(x) = e^{-\lambda} \frac{\lambda^x}{x!} \quad x = 0, 1, 2, \ldots$$

It can be shown that $\mu = \lambda$ and $\sigma^2 = \lambda$.

22. The **multinomial distribution** is an extension of the binomial distribution. There are n independent trials with not 2 but k possible outcomes per trial. Define discrete random variables $X_1, \ldots, X_k$ and probabilities $p_1, \ldots, p_k$ as follows.

    X_i = the number of times that outcome i occurs in the n trials and p_i = the probability that outcome i occurs. The **joint probability mass function** that $X_1 = x_1, \ldots, X_k = x_k$ is

    $$P(X_1 = x_x, \ldots, X_k = x_k) = \frac{n!}{x_1! \cdots x_k!} p_1^{x_1} p_2^{x_2} \ldots p_k^{x_k}, \; x_i = 0, 1, \ldots, n \text{ and } \Sigma x_i = n$$

23. Often, we are interested in obtaining some sample values from a random variable but it may be expensive or difficult to obtain the values. In such cases we may try to **simulate** the values that we would likely obtain without actually observing the variable.

CHAPTER 4

Probability Densities for Continuous Random Variables and Introduction to MAPLE

4.1 Continuous Random Variables

Continuous random variables possibly take on an infinite number of values. Therefore, it is not possible to build a table having values and probabilities as was done in the discrete case. Characteristics such as pressure, temperature, density, and other engineering measurements take on values continuously in intervals. For example, adult female heights for a population take on values between 50 and 75 inches. We might represent their heights by the curve shown in Figure 4-1. This curve shows the probabilities associated with the variable height. The curve tells us that there are more females with heights near 62.5 inches than there are with heights close to 50 or 75 inches. There are a few very short and a few very tall but most are in the middle near 62.5 inches.

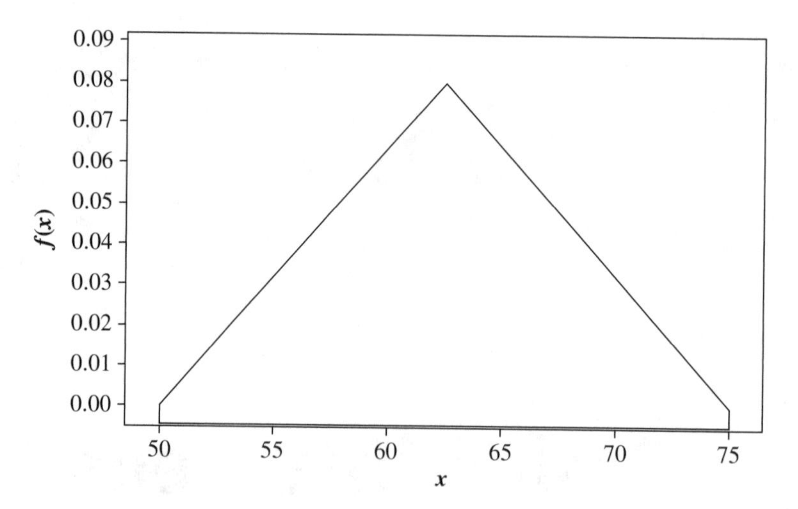

Figure 4-1 Density function for adult female heights.

A **probability density function (pdf)** shows how the characteristic is distributed. It has two properties:

1. The curve is always positive, $f(x) \geq 0$.
2. The total area under the curve is 1, that is, $\int_{50}^{75} f(x)\,dx = 1$.

In this case the equation of the pdf, shown in Figure 4-1, is as follows:

$$f(x) = \begin{cases} 0.0064x - 0.32, & 50 < x < 62.5 \\ -0.0064x + 0.48, & 62.5 \leq x \leq 75 \\ 0, & \text{otherwise} \end{cases}$$

If the graph is sketched, Figure 4-1 is obtained. The area under the curve is given by

$$\int_{50}^{62.5} (0.0064x - 0.32)\,dx + \int_{62.5}^{75} (-0.0064x + 0.48)\,dx$$

MAPLE is a software package that performs many different mathematical functions, including integration. The software package MAPLE may be used to find the area under the graph as follows:

```
> evalf(int(.0064*x-0.32,x=50..62.5)
+ int(-.0064*x+0.48,x=62.5..75));          1.000000000
```

The student should study the command structure for performing the integration since it will be used quite often in this chapter. The number 1 is shown after the integration command, indicating that the area under the curve is equal to 1. Probabilities are interpreted as areas under the pdf curve. The probability that a female has a height over 67 inches is represented by $(X \geq 67)$ and is equal to the shaded area in the tail of Figure 4-2.

High school female basketball players have heights in the right tail of the density curve.

The probability or area is equal to $\int_{67}^{75} (-0.0064x + 0.48)\,dx$. The antiderivative of this integrand is $-0.0032x^2 + 0.48x$ and is evaluated at the upper limit minus the evaluation at the lower limit, or $18 - 17.7952 = 0.2048$. Or, if MAPLE is used to find the value of the integral, we find:

```
> evalf(int(-.0064*x+.48,x=67..75));        0.2048000000
```

As can be seen, we obtain the same answer using MAPLE as we do if we work the problem "by hand" as we would in a calculus course. We will use MAPLE to evaluate most of the integrals in this book. Using MAPLE to do the integration allows us to concentrate on the statistics and not to be constantly reviewing our calculus.

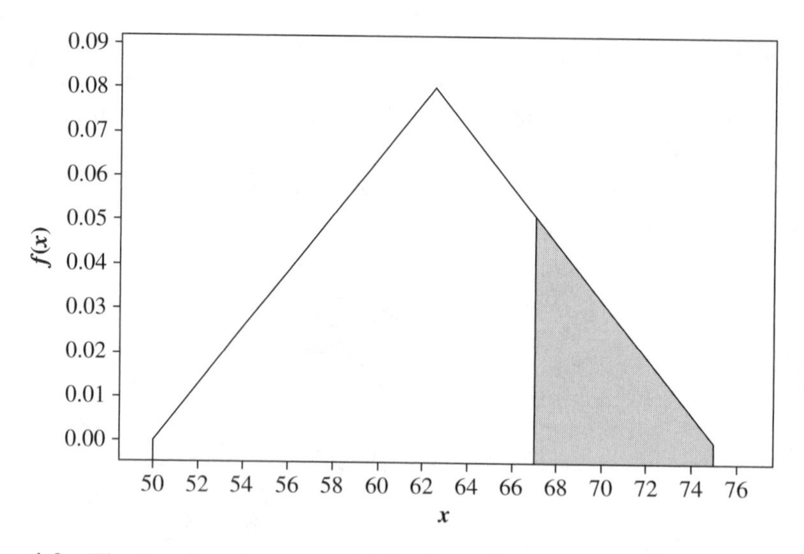

Figure 4-2 The females who are over 67 inches tall are represented by the grey area.

EXAMPLE

Engineers have developed a new tire with lifetimes (in thousands) given by the following density curve:

$$f(x) = \begin{cases} 0.02e^{-0.02x}, & x > 0 \\ 0, & x < 0 \end{cases}$$

What percent of the tires have mileages less than 60,000 miles? The answer is the integral from 0 to 60 of the density function, or using MAPLE,

```
> evalf(int(.02*exp(-.02*x),x=0..60));    0.6988057881
```

About 70% of the tires will have lifetimes less than 60,000 miles.

The lifetimes of tires have probability density functions that are of interest to engineers.

Let's suppose the student is at home rather than at school and has EXCEL, but not MAPLE on their home computer. The interval from 0 to 60 is divided into 30 equal parts of length 60/30 = 2 and the midpoints of the intervals are entered into column A. The expression =0.02*EXP(-0.02*A1) is entered into B1 and a click-and-drag is performed from B1 to B30. The expression =2*B1 is entered into C1 and a click-and-drag is performed from C1 to C30. Each cell from C1 to C30 contains the area of a rectangle and when the areas are added up, the sum (=SUM(C1:C30)) approximates the value of the integral. The value of the sum (0.69875) is very close to the MAPLE answer, 0.6988057881. The results are shown in Figure 4-3.

We are using the approximation

$$\int_0^{60} 0.02e^{-0.02x}dx \approx \sum_1^{30} f(x_i)\Delta x$$

The term $\Delta x = 2$ and $f(x_i) = 0.02e^{-0.02x_i}$.

Figure 4-4 shows the area that is approximated above.

Another property of continuous random variables is that the probability of a single point is zero, that is $P(X = a) = 0$. The area associated with a single point is zero. From this we have the following result:

$$\begin{aligned} P(a \le X \le b) &= P(X = a) + P(a < X < b) + P(X = b) \\ &= 0 + P(a < X < b) + 0 = P(a < X < b) \end{aligned}$$

x	$f(x)$	$f(x)\Delta x$	x	$f(x)$	$f(x)\Delta x$
1	0.019604	0.039208	31	0.010759	0.021518
3	0.018835	0.037671	33	0.010337	0.020674
5	0.018097	0.036193	35	0.009932	0.019863
7	0.017387	0.034774	37	0.009542	0.019085
9	0.016705	0.033411	39	0.009168	0.018336
11	0.01605	0.032101	41	0.008809	0.017617
13	0.015421	0.030842	43	0.008463	0.016926
15	0.014816	0.029633	45	0.008131	0.016263
17	0.014235	0.028471	47	0.007813	0.015625
19	0.013677	0.027354	49	0.007506	0.015012
21	0.013141	0.026282	51	0.007212	0.014424
23	0.012626	0.025251	53	0.006929	0.013858
25	0.012131	0.024261	55	0.006657	0.013315
27	0.011655	0.02331	57	0.006396	0.012793
29	0.011198	0.022396	59	0.006146	0.012291
				Sum=	0.69875

Figure 4-3 Approximating the value of an integral.

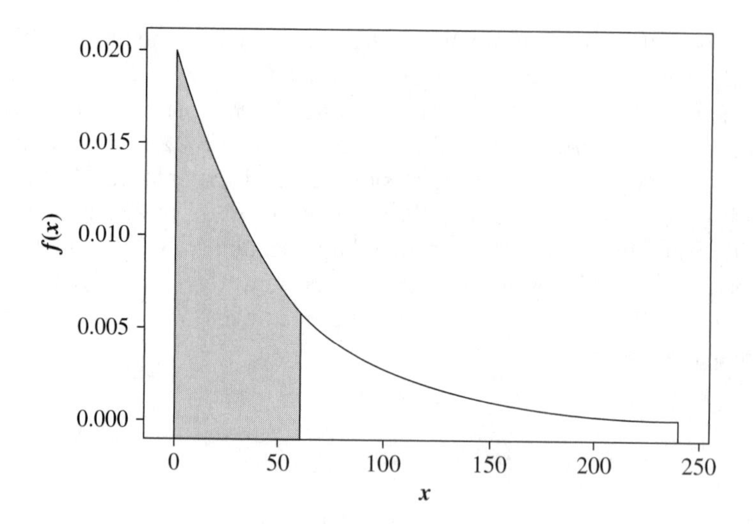

Figure 4-4 $P(X \le 60)$.

For a discrete variable, we found that $\mu = \Sigma x p(x)$. The corresponding result for a continuous random variable is $\mu = \int_{-\infty}^{\infty} x f(x)\,dx$. And, corresponding to the property for a discrete random variable that $\sigma^2 = \Sigma(x_i - \mu)^2 p(x_i)$ is the result

$$\sigma^2 = \int_{-\infty}^{\infty} (x-\mu)^2 f(x)\,dx$$

for a continuous variable. A shortcut formula that is equivalent for the variance is

$$\sigma^2 = \int_{-\infty}^{\infty} x^2 f(x)\,dx - \mu^2$$

Likewise, the cumulative distribution function for a continuous random variable is defined as follows:

$$F(x) = P(X \le x) = \int_{-\infty}^{x} f(x)\,dx$$

EXAMPLE

For the tire mileages pdf given above, find the mean, the variance, the standard deviation, and the cumulative distribution function.

$$f(x) = \begin{cases} 0.02e^{-0.02x}, & x > 0 \\ 0, & x < 0 \end{cases}$$

The mean mileage is given by

$$\mu = \int_{-\infty}^{\infty} xf(x)\,dx = \int_{0}^{\infty} 0.02e^{-0.02x}x\,dx$$

The MAPLE solution is as follows:

```
> evalf(int(0.02*x*exp(-0.02*x),x=0..infinity));   50.
```

We see that the mean mileage of these tires is 50,000 miles.

The variance, using the defining formula, is

$$\sigma^2 = \int_{-\infty}^{\infty} (x-50)^2 f(x)\,dx = \int_{0}^{\infty} (x-50)^2 0.02e^{-0.02x}dx$$

The MAPLE solution is as follows:

```
> evalf(int((x-50)^2*0.02*exp(-0.02*x),x=0..infinity));   2500.
```

Or, if we use the shortcut formula,

$$\sigma^2 = \int_{-\infty}^{\infty} x^2 f(x)\,dx - \mu^2 = \int_{0}^{\infty} x^2 0.02e^{-0.02x}dx - 50^2$$

The MAPLE solution is as follows:

```
> evalf(int(x^2*0.02*exp(-0.02*x),x=0..infinity))-2500;   2500.
```

The standard deviation is 50.

The cumulative distribution function is

$$F(x) = \begin{cases} 0, & x < 0 \\ \int_{0}^{x} 0.02 * e^{-0.02t}dt, & x > 0 \end{cases}$$

The MAPLE solution is as follows:

```
> evalf(int(0.02*exp(-0.02*t),t=0..x));
```
$1.-1.e^{(-0.02000000000\,x)}$

Note that the dummy integration variable is t and x is considered a fixed number. The function, $F(x)$ is

$$F(x) = \begin{cases} 0, & x < 0 \\ 1 - e^{-0.02x}, & x \geq 0 \end{cases}$$

Note the following values:

$$F(0) = 0,\ F(\infty) = 1 \quad \text{and} \quad F(60) = 1 - e^{-1.2} = 0.6988$$

$F(60) = P(X < 60)$ is the same value calculated when we first encountered this pdf.

EXAMPLE

Water usage is an important variable for cities.

Civil engineers have determined that the daily water usage in a particular city in hundreds of thousands of gallons is a random variable having the probability density

$$f(x) = \begin{cases} (1/9)x e^{-x/3}, & x > 0 \\ 0, & x < 0 \end{cases}$$

The graph for this water usage is shown in Figure 4-5.

```
> evalf(int((x/9)*exp(-x/3),x=0..infinity));
```
1.

The above MAPLE integration shows that the function is a pdf.

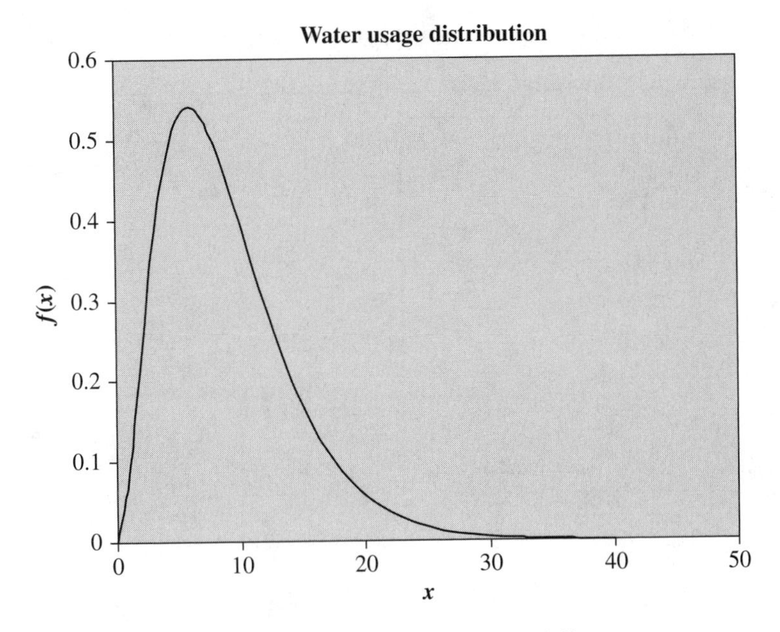

Figure 4-5 Probability density function for daily water usage.

The mean daily water usage in thousands of gallons is

$$\mu = \int_{-\infty}^{\infty} xf(x)\,dx = \int_{0}^{\infty} \frac{x^2}{9} e^{-\frac{x}{3}} dx$$

The MAPLE evaluation of the integral is

```
> evalf(int((x^2/9)*exp(-x/3),x=0..infinity)); 6.
```

We see that the mean is 600,000 gallons.

Consider the solution using EXCEL. By considering Figure 4-5, we see that even though the curve is defined for all x, that if we go to 50, we have all the values of practical significance. Suppose first of all that we wish to confirm that the total area under the curve is 1. Break the interval from 0 to 50 into 50 subintervals each one unit long. Enter the numbers at the middle of the subintervals into column A of the worksheet. Enter 0.5 into A1 and 1.5 into A2 and then perform a click-and-drag. The function values =(A1/9)*EXP(-A1/3) are entered into B1 and a click-and-drag is performed. Then =B1*1 is entered into C1. This is the area of the first rectangle from 0 to 1 and height of the function at 0.5. A click-and-drag from C1 to C50 gives the areas of the first 50 rectangles. The expression =SUM(C1:C50) gives 1.004584, the approximation to the area under the curve. If 100 or 1000 rectangles are used instead, the area approximation gets closer to 1.

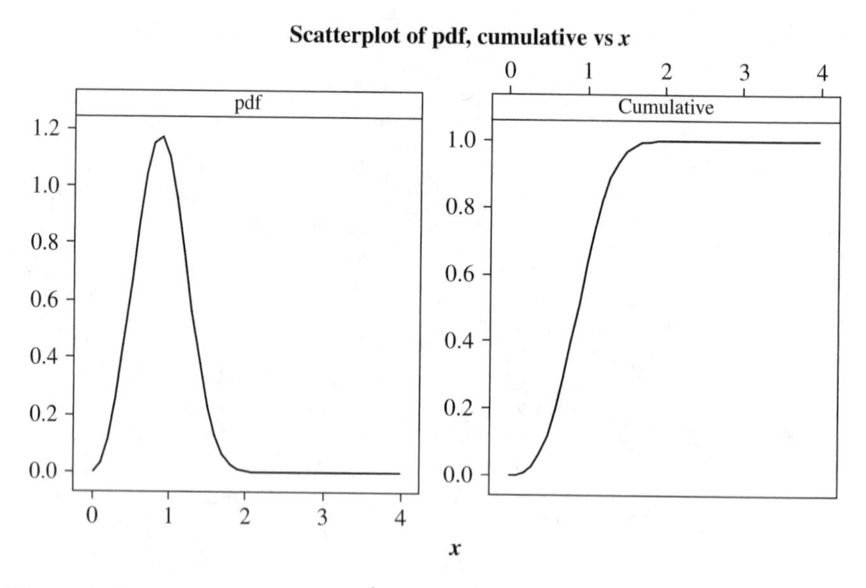

Figure 4-6 A plot of $f(x) = 3x^2e^{-x^3}$ and $F(x)$ in separate panels of the same graph.

EXAMPLE

A variable has the following cumulative probability function: Find the pdf for X.

$$F(x) = \begin{cases} 0, & x < 0 \\ 1 - e^{-x^3}, & x \geq 0 \end{cases} \qquad f(x) = \frac{dF(x)}{dx} = \begin{cases} 0, & x < 0 \\ 3x^2e^{-x^3}, & x \geq 0 \end{cases}$$

Or, we may have MAPLE find the derivative.

```
> diff(1-exp(-x^3),x);
```

$$3x^2 e^{(-x^3)}$$

Figure 4-6 looks at both $f(x)$ and $F(x)$ in separate panels of the same graph.

Sometimes, we know the shape of the probability density function from experience with the variable. All that is needed is to adjust the function by multiplying a constant to make the total area under the curve equal to 1.

EXAMPLE

Find the value of K, using MAPLE, that makes the following function a pdf. Then plot the probability density function and the cumulative distribution function using MINITAB.

$$f(x) = \frac{Kx^2}{e^x}, \; 0 < x < 2$$

```
> evalf(int(K*x^2/exp(x),x=0..2));
```

$$0.646647168K$$

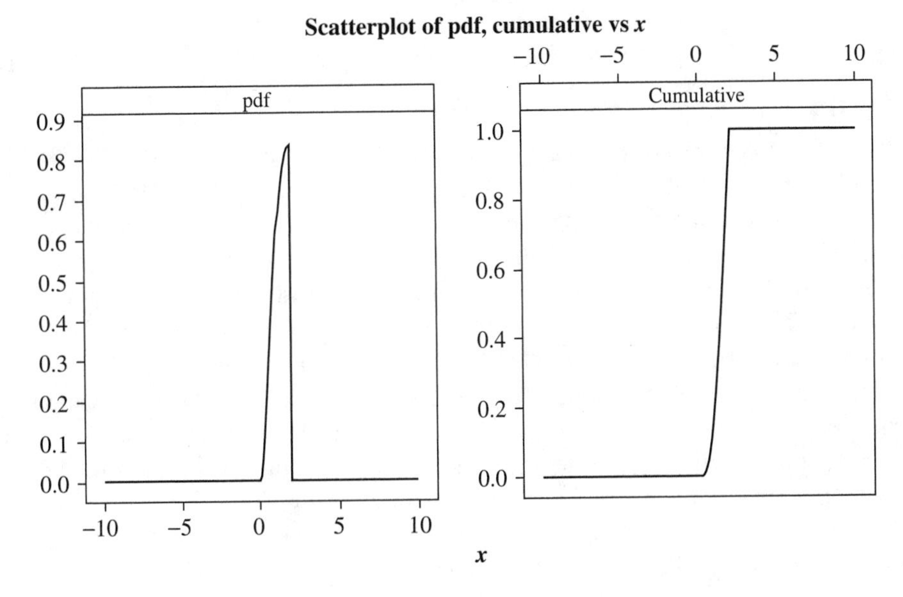

Figure 4-7 A plot of $f(x)=\dfrac{1.546x^2}{e^x}$ and $F(x)$ in separate panels of the same graph.

This must be equal 1 to be a pdf. Setting $0.647K$ equal to 1, we find that $K =$ 1.546. Therefore, $f(x)=\frac{1.546x^2}{e^x}, 0<x<2$ is a pdf. The cumulative distribution function for $0 < x < 2$ is found as follows:

```
> evalf(int(1.546*t^2/exp(t),t=0..x));
-1.546000000 (-2.e^x + 2 + 2.x + x^2) e^(-1.x)
```

$$F(x)=\begin{cases} 0, & x<0 \\ -1.546(-2e^x+2+2x+x^2)e^{-x}, & 0<x<2 \\ 1, & x>2 \end{cases}$$

A plot of $f(x)$ and $F(x)$ in separate panels of the same graph is shown in Figure 4-7.

4.2 The Normal Distribution

One of the most widely used continuous distributions is the normal distribution. The normal distribution has the following probability density function:

$$f(x)=\frac{1}{\sqrt{2\pi}\sigma}e^{-\frac{(x-\mu)^2}{2\sigma^2}}, -\infty < x < \infty$$

The graph of the normal curve may be constructed using any of the many software packages. We will illustrate how to construct it using EXCEL and MINITAB.

EXAMPLE

Electrical engineers find that Xbox 360s have lifetimes that are normally distribu ted with a mean equal to 48 months and a standard deviation equal to 6 months. First, we will describe how to plot the graph using EXCEL. Put the x values in the worksheet from 4 standard deviations below the mean to 4 above the mean. In this case, the numbers would extend from $48 - 4(6) = 24$ to $48 + 4(6) = 72$ months in the A column. The expression =NORMDIST(A1,48,6,0) is entered into cell B1 and a click-and-drag is performed from B1 to B49. The function =NORMDIST(A1,48,6,0) has four parameters. The first is a value on the x-axis, the second is the mean of the normal curve, the third is the standard deviation of the normal curve, and the fourth is 0 or 1. If a 0 is entered, the height up to the normal curve is computed. If a 1 is entered, the area to the left of the x-value entered is computed. Entering a 0 for the fourth parameter will create the points on the normal curve. All that remains is to plot the points using the chart wizard. The result is shown in Figure 4-8.

What are some of the characteristics of the normal curve?

1. The total area under the curve is 1.
2. The curve is asymptotic to the x-axis. That is, it gets closer and closer to the x-axis but never touches as you go to plus infinity and minus infinity.

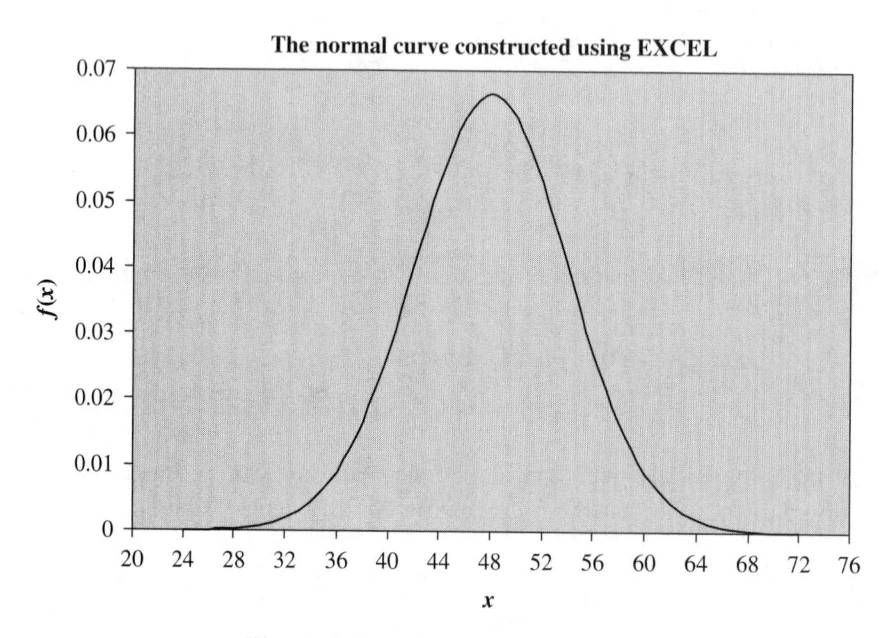

Figure 4-8 Lifetimes of Xbox 360s.

3. The curve is symmetrical about the mean. The mean, μ, locates the center of the curve.
4. The EXCEL function =NORMDIST(x,μ,σ,1) may be used to find the area to the left of x.
5. The standard deviation, σ, determines the shape of the curve. If σ is small, the curve is tall and skinny. If σ is large, the curve is more spread out.
6. There is 68% of the area under the curve within one standard deviation of the mean, 95% within two standard deviations of the mean, and 99.7% within three standard deviations of the mean.

If the values from columns A and B are copied from EXCEL into C1 and C2 of MINITAB then MINITAB may be used to graph the normal curve. This is shown in Figure 4-9. Use the pull-down **Graph ⇒ Scatterplot**.

Figure 4-10 shows the effects of μ and σ on shape and location of the normal curve. The mean locates the center of the curve and the standard deviation affects the shape of the curve.

MAPLE may be used to find the total area under the curve and show that it indeed equals to 1.

```
> evalf(int((1/(sqrt(2*Pi)*6))*exp(-(x-48)^2/72),
x=-infinity..infinity));     1.
```

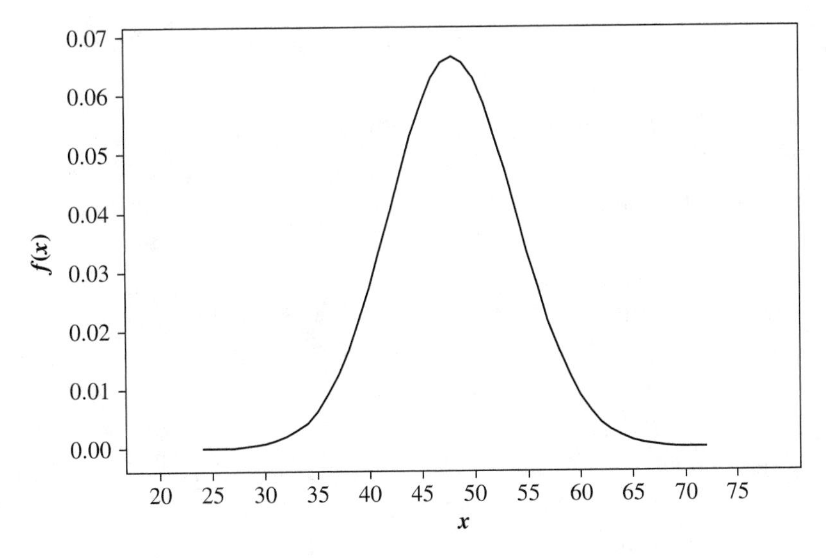

Figure 4-9 MINITAB plot of lifetimes of Xbox 360s.

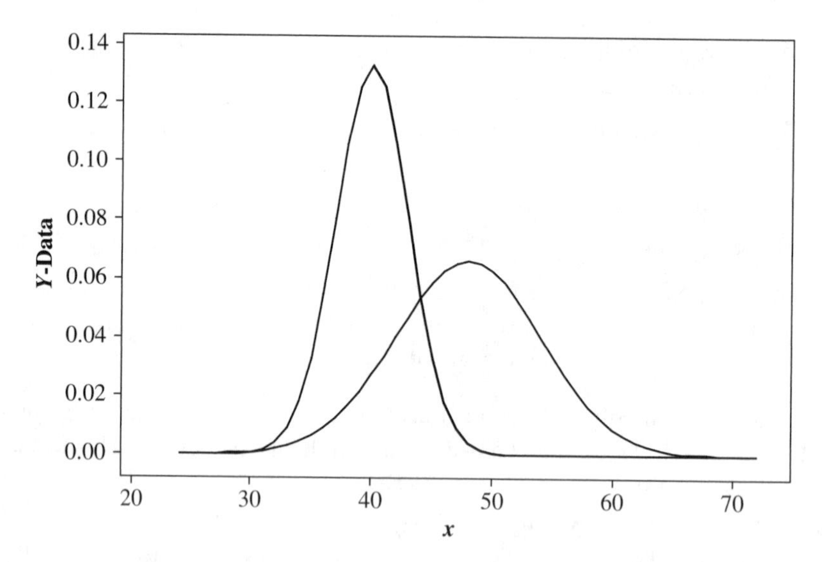

Figure 4-10 Normal curve with $\mu = 40$, $\sigma = 3$ and normal curve with $\mu = 48$, $\sigma = 6$.

EXAMPLE

Communications engineers have determined that lifetimes of Ace cell phones are normally distributed with a mean equal to 60 months and a standard deviation equal to 5 months. Determine the following:

a. What is the probability that a cell phone of this type has a lifetime less than 55 months?

b. What is the probability that a cell phone of this type has a lifetime greater than 70 months?

c. Find a lifetime such that only 5% of the cell phones last this long or longer.

a.

If MINITAB is used to find the shaded area, the pull-down **Calc ⇒ Probability Distributions ⇒ Normal** is used to produce the following output (see Figure 4-11):

Cumulative Distribution Function

```
Normal with mean = 60 and standard deviation = 5

x    P(X <= x)
55      0.158655
```

We see that the probability is 0.158655.

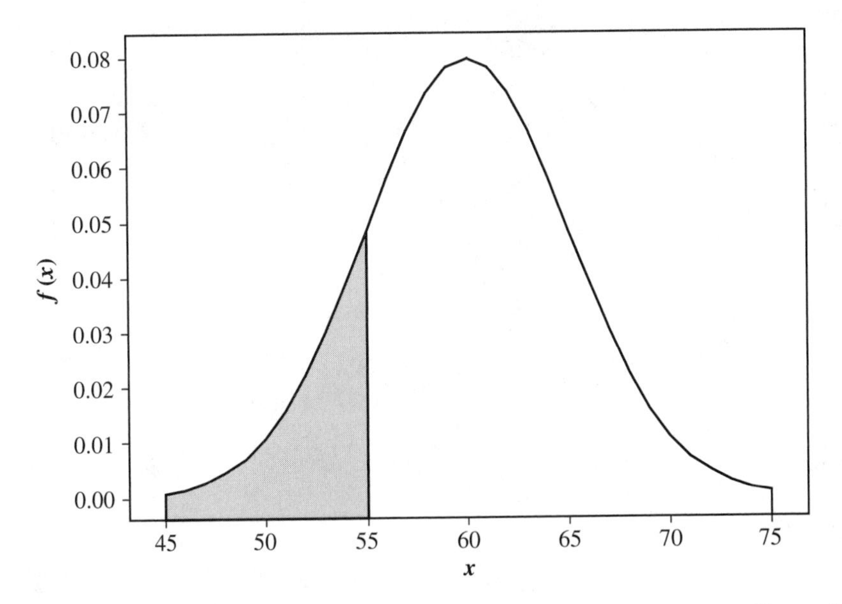

Figure 4-11 Shaded area represents cell phones with lifetimes less than 55 months.

If MAPLE is used, the following results are obtained:

```
> evalf(int((1/(sqrt(2*Pi)*5))*exp(-(x-60)^2/50),
x=-infinity..55)); 0.1586552540
```

If EXCEL is used, the following =NORMDIST(55,60,5,1) gives 0.158655.

b.

The MINITAB solution is

Cumulative Distribution Function

```
Normal with mean = 60 and standard deviation = 5

X   P(X <= x)
70       0.977250
```

The probability asked for is $P(X > 70)$ which equals $1 - P(X < 70) =$ 0.02275 (see Figure 4-12).

The MAPLE solution is

```
>evalf(int((1/(sqrt(2*Pi)*5))*exp(-(x-60)^2/50),
x=70..infinity)); 0.0227501320
```

The EXCEL solution is =1-NORMDIST(70,60,5,1) = 0.02275.

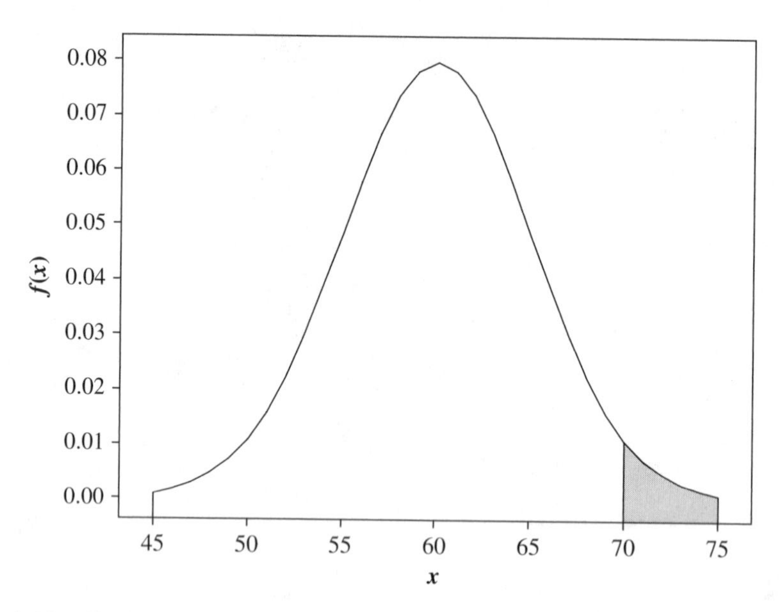

Figure 4-12 Shaded area represents cell phones with lifetimes greater than 70 months.

c. This part is worked by using the inverse function of the normal distribution. If MINITAB is used, the solution is

Inverse Cumulative Distribution Function

```
Normal with mean = 60 and standard deviation = 5
X  P(X <= x)
0.95  68.2243
```

The EXCEL solution is =NORMINV(0.95, 60, 5) = 68.2243.

4.3 The Normal Approximation to the Binomial Distribution

A coin is flipped 16 times and the random variable of interest is X = the number of heads to appear in the 16 flips. This distribution is a binomial distribution with $n = 16$ trials and $p = 0.5$. The binomial probabilities are shown in Figure 4-13. In this figure, the area of each rectangle is equal to the binomial probability of the number of heads shown on the x-axis. Each rectangle extends 0.5 units on either side of the x-value and the height of the rectangle is the probability that goes with

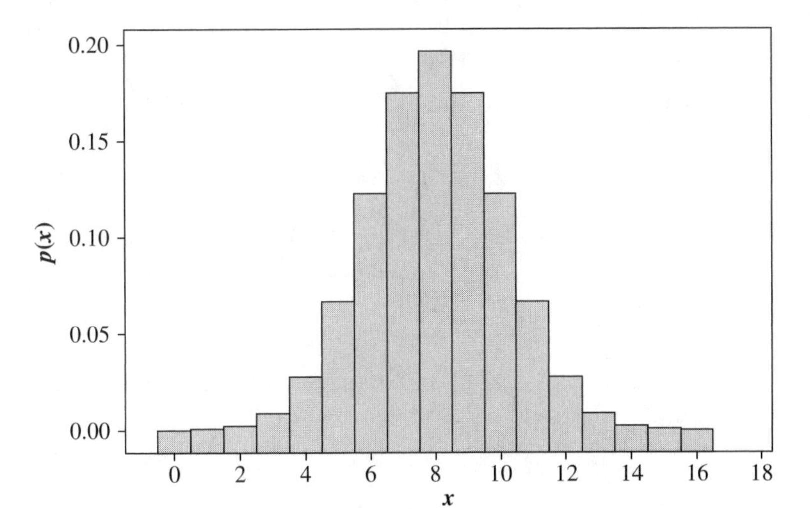

Figure 4-13 The binomial distribution for $n = 16$ and $p = 0.5$ illustrated as rectangles.

that value. For example, the rectangle at $x = 10$ extends from 9.5 to 10.5 and it has height equal to 0.122192. If the area of the rectangle is found, it is $1 \times 0.122192 = 0.122192$. The sum of the 17 rectangle areas is 1. The mean of the distribution is $\mu = np = 16(0.5) = 8$. The standard deviation of the distribution is

$$\sigma = \sqrt{npq} = \sqrt{16(0.5)(0.5)} = \sqrt{4} = 2$$

This is basically a review of what was learned in Section 3.4. From the histogram of the binomial distribution, it is clear that the most likely outcome is for 8 heads to appear, and the least likely occurrence is either 0 or 16 heads in the 16 flips.

Figure 4-14 shows a normal distribution with mean $\mu = 8$ and standard deviation $\sigma = 2$. The two distributions are strikingly similar. However, one is discrete and the other one is continuous. The normal distribution may be used to approximate the binomial distribution when $np > 5$ and $nq > 5$.

Figure 4-15 shows the two distributions together. Sometimes, the normal curve is lower than the rectangle and sometimes, the normal curve is higher than the rectangle.

The application of this connection is shown in Figure 4-16. Suppose our interest is in the probability of flipping 6, 7, 8, or 9 heads. That is, we are interested in finding $P(X = 6, 7, 8, \text{or } 9)$. This would be the area under the rectangles located at 6, 7, 8, and 9. By studying Figure 4-16, it is clear that this would be approximately the area under the normal curve from where $x = 5.5$ to where $x = 9.5$. Let us compare the area under the rectangles and the area under the normal curve. Using EXCEL, the binomial probabilities or the areas under the rectangles =BINOMDIST(6,16,0.5,0)+

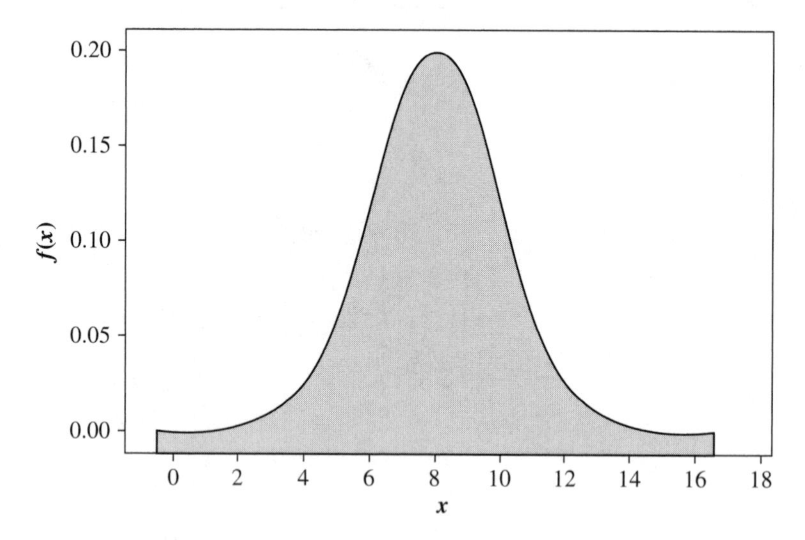

Figure 4-14 The normal distribution with mean = 8 and standard deviation = 2.

BINOMDIST(7,16,0.5,0)+BINOMDIST(8,16,0.5,0)+BINOMDIST(9,16,0.5,0) = 0.667694. The EXCEL area under the normal curve =NORMDIST(9.5,8,2,1)-NORMDIST(5.5,8,2,1) = 0.667723. Notice how amazingly close the two areas are to each other.

Before the modern software packages such as EXCEL, MINITAB, SAS, SPSS, and STATISTIX, this approximation was often used to work applied problems. It is still an important approximation.

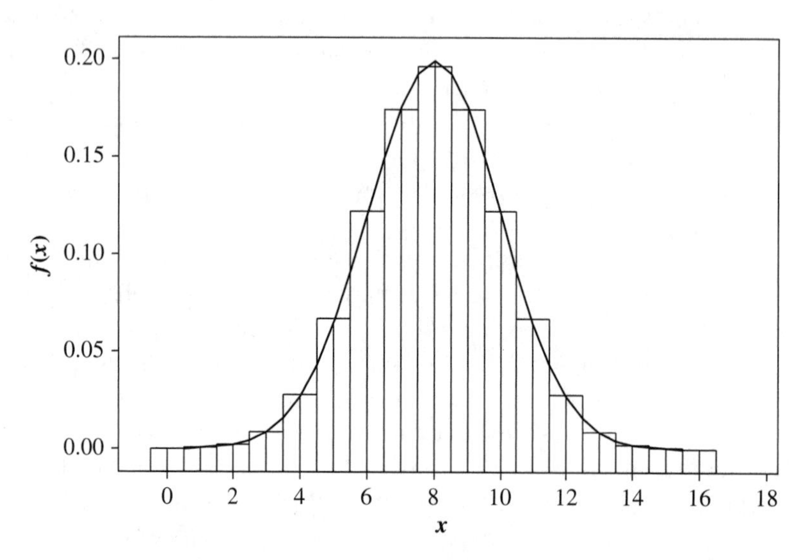

Figure 4-15 The binomial distribution and the normal distribution shown together.

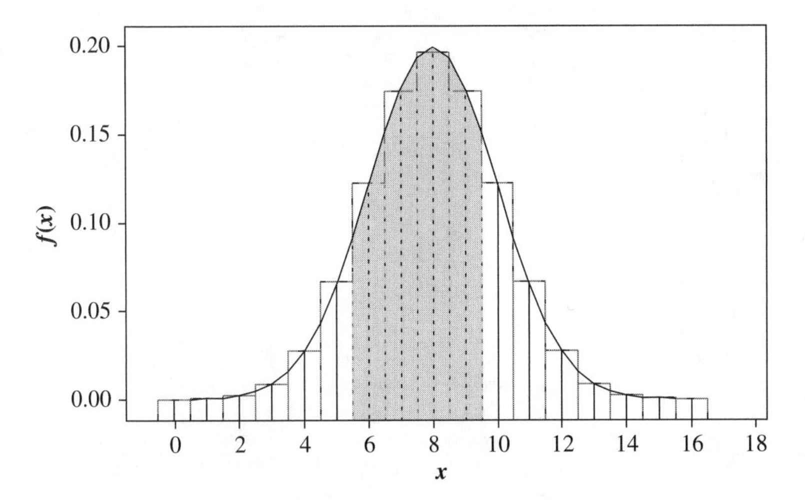

Figure 4-16 The area under the rectangles at 6, 7, 8, and 9 is approximated by the area under the normal curve from $x = 5.5$ to $x = 9.5$.

EXAMPLE

A manufacturer knows that 1% of their personal digital assistants (pda's) will need repair within a year. What is the probability that 80 or fewer in 10,000, that are shipped to Wall Market within a year, will need repair? The basic variable is a binomial where X = the number needing repair in the 10,000 pda's sent to Wall Market. First, make sure the normal approximation is appropriate: $n(p) = 10000(.01) = 100 > 5$ and $n(q) = 10000(.99) = 9900 > 5$. The approximation is appropriate. We are asked to find $P(X \le 80)$. A normal curve with the mean

$\mu = n(p) = 10000(0.01) = 100$ and the standard deviation

$$\sigma = \sqrt{npq} = \sqrt{10000(.01)(.99)} = \sqrt{99} = 9.95$$

is fitted to the binomial distribution. The EXCEL solution is given by the following: =NORMDIST(80.5,100,9.95,1) = 0.025.

The exact answer =BINOMDIST(80,10000,0.01,1) = 0.022. The binomial answer was not readily available just 25 years ago. The only choice was to do the normal approximation. Computers and their abilities have changed the nature of statistics.

EXAMPLE

A production line makes surgical masks. Five percent are defective. Approximate the probability that a sample of 125 will contain 10 or more defectives. The normal approximation can be made since $np > 5$ and $nq > 5$. We are asked to approximate

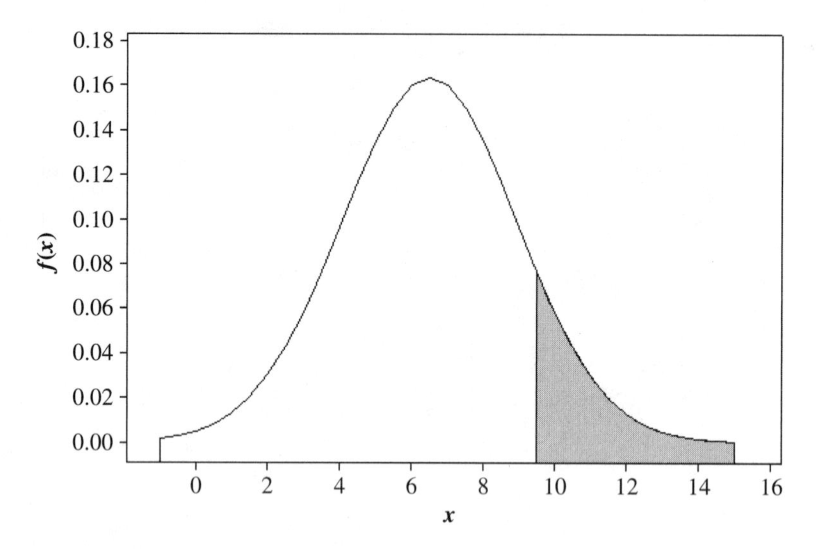

Figure 4-17 Normal curve area that approximates binomial probability $P(X \geq 10)$.

$P(X \geq 10)$. We fit a normal curve through the binomial distribution. The normal curve has mean = 6.25 and standard deviation = 2.44. We need to find the area to the right of 9.5 under the normal curve. Figure 4-17 shows the area under the normal curve that we need to find. The MINITAB area is

Cumulative Distribution Function

```
Normal with mean = 6.5 and standard deviation = 2.44
 x    P(X <= x)
9.5      0.890559
```

Now subtract this from 1 to get 0.11.

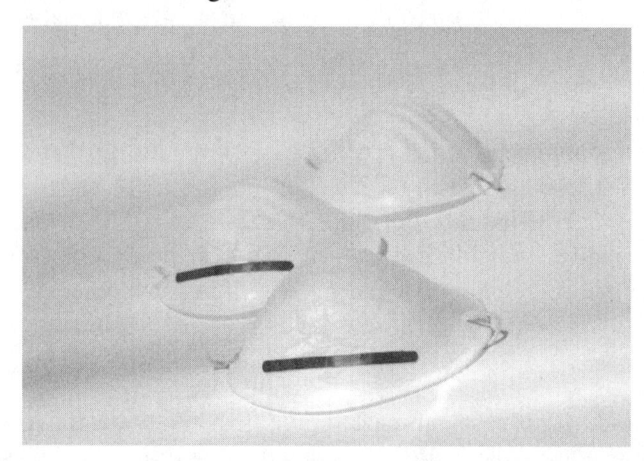

Surgical masks are manufactured by a production line. If the percent defective in a sample is small enough, the lot is acceptable.

4.4 The Uniform Distribution

A continuous model that is extremely simple and yet finds many applications in statistics is the uniform model. This model states basically that every interval of a fixed length has the same probability. Random variable X is uniformly distributed over the interval from a to b if it has the following pdf:

$$f(x)=\begin{cases} 0, & x<a \\ \dfrac{1}{b-a}, & a\le x\le b \\ 0, & x>b \end{cases}$$

The mean of a uniformly distributed random variable is $\mu=\int_a^b \frac{x}{b-a}\,dx$. If this integral is evaluated using MAPLE, the following is obtained:

```
> evalf(int(x/(b-a),x=a..b));
```

$$\frac{0.5000000000(b^2-1.a^2)}{b-1.a}$$

This may be simplified to $\mu = \frac{a+b}{2}$. The variance is given by

$$\sigma^2=\int_a^b\left(x-\frac{a+b}{2}\right)^2\frac{1}{(b-a)}dx$$

Again, letting MAPLE do the integration,

```
> evalf(int(x^2/(b-a),x=a..b)-(a+b)^2/4);
```

$$\frac{0.3333333333(b^3-1.a^3)}{b-1.a}-0.2500000000(a+b)^2$$

The student is asked to see how good their algebra is by showing that this expression simplifies to $\sigma^2=\frac{(b-a)^2}{12}$.

EXAMPLE

Filling machines vary in the amount of fill they administer.

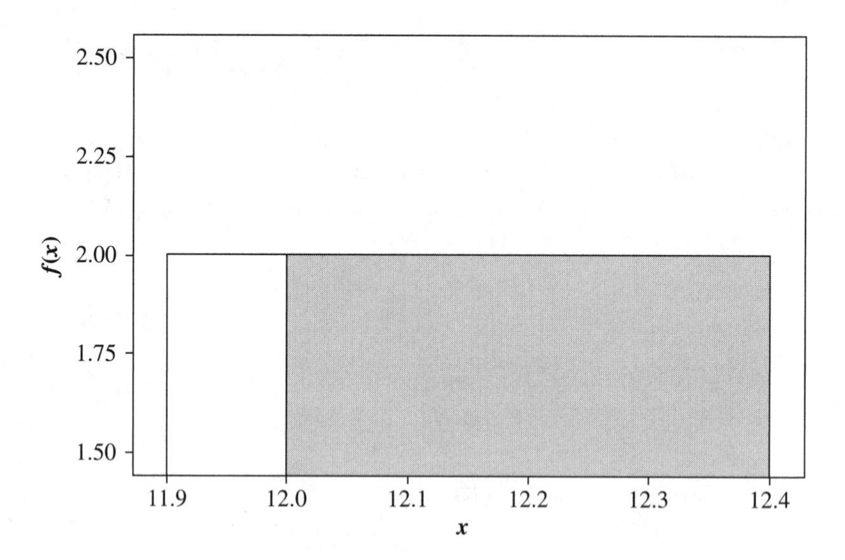

Figure 4-18 Uniform distribution between 11.9 and 12.4.

A quality control engineer knows that the amounts of fill that a machine puts into 12-ounce bottles are actually uniformly distributed between 11.90 and 12.40 ounces. Use MINITAB to draw $P(X > 12.0)$. This is shown in Figure 4-18.

The average fill per container is (11.9 + 12.4)/2 = 12.15 ounces. The percent of containers with more than 12.0 ounces is

$$\int_{12.0}^{12.4} \frac{1}{12.4 - 11.9}\, dx = \int_{12.0}^{12.4} 2\, dx = 2(.4) = 0.8, \text{ or } 80\%$$

The standard deviation of fills is $\sigma = \sqrt{\frac{(12.4 - 11.9)^2}{12}} = 0.14$ ounces.

4.5 The Log-Normal Distribution

The log-normal random variable, X, has a probability density function as follows:

$$f(x) = \begin{cases} \dfrac{1}{\sqrt{2\pi}\beta} x^{-1} e^{-\frac{(\ln x - \alpha)^2}{2\beta^2}}, & x > 0, \beta > 0 \\ 0, & \text{otherwise} \end{cases}$$

If this variable is transformed by taking its logarithm, i.e., $Y = \ln(X)$, the random variable Y has a normal distribution with mean α and standard deviation β. It can be shown that the density of Y is

$$f(y) = \frac{1}{\sqrt{2\pi}\beta} e^{-\frac{(y-\alpha)^2}{2\beta^2}}, -\infty < y < \infty$$

Going back to the log-normal variable, X, the mean of the log-normal may be shown to equal

$$\mu = e^{\alpha+\beta^2/2}$$

And the variance may be shown to equal

$$\sigma^2 = e^{2\alpha+\beta^2}(e^{\beta^2} - 1)$$

Remember, if you are in a mathematical statistics course, your interest is in deriving the equation for the mean and the variance of X. If you are in an engineering statistics course, your interest is in applying and using the results and understanding how you may apply the results to engineering. We shall now look at how to use the software to apply the above results.

To obtain some idea of what this log-normal density looks like, consider it when α is 0 and β is 1 and when α is 1 and β is 1. In Figure 4-19, C2 is the log-normal graph when $\alpha = 0$ and $\beta = 1$ and C3 is the log-normal graph when $\alpha = 1$ and $\beta = 1$.

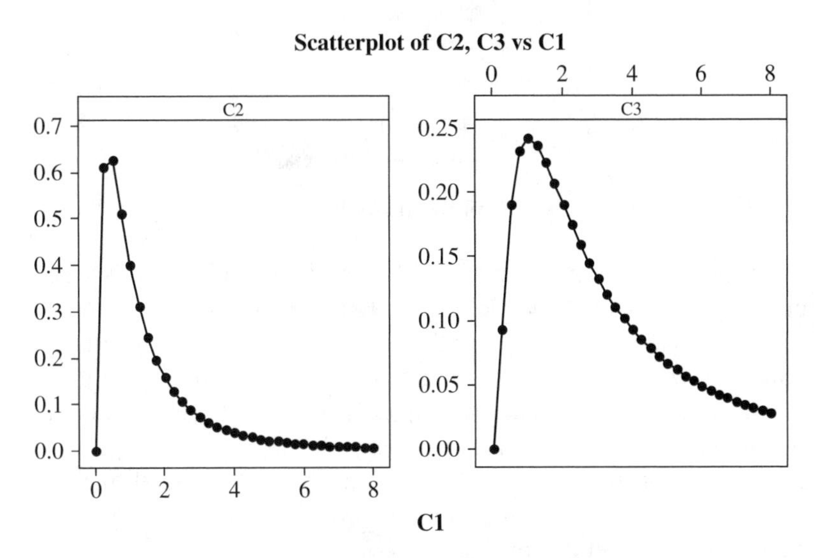

Figure 4-19 C2 is the log-normal graph when $\alpha = 0$ and $\beta = 1$ and C3 is the log-normal graph when $\alpha = 1$ and $\beta = 1$.

The mean of the graph, where $\alpha = 0$ and $\beta = 1$, is

$$\mu = e^{0+1^2/2} = e^{0.5} = 1.65$$

The variance is

$$\sigma^2 = e^{2(0)+1^2}\left(e^{1^2} - 1\right) = e(e-1) = 4.67$$

And, the standard deviation is 2.16.
The mean of the graph, where $\alpha = 1$ and $\beta = 1$, is

$$\mu = e^{1+1^2/2} = e^{1.5} = 4.48,$$

The variance is

$$\sigma^2 = e^{2(1)+1^2}\left(e^{1^2} - 1\right) = e^3(e-1) = 34.51$$

And, the standard deviation is 5.88.

EXAMPLE

Cities' water supplies have fluoride added.

Engineers for the EPA have found that fluoride in a large city's water supply have concentrations that follow a log-normal distribution with parameters $\alpha = 3.2$ and $\beta = 1$. What is the probability that the concentration exceeds 8 parts per million (ppm)? Suppose we look at several solutions to this problem.
EXCEL solution: The command =LOGNORMDIST(8,3.2,1) is entered into any cell. The parameters are 8, the value in the probability statement $P(X < 8)$; 3.2, the mean of ln(X); and 1, the standard deviation of the ln(X). Recall that ln(X) has a normal distribution with mean = 3.2 and standard deviation = 1. The value 0.131238 is subtracted from 1 to get the answer.

$$P(X > 8) = 1 - P(X < 8) = 0.868.$$

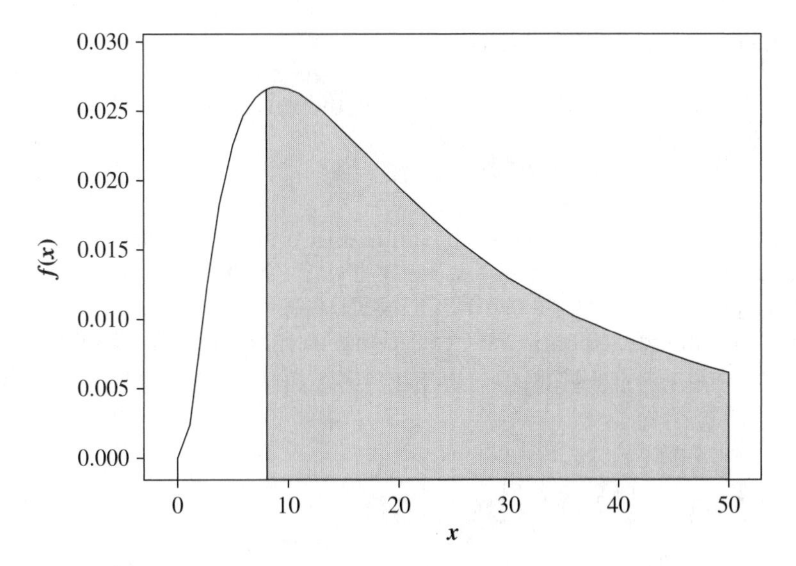

Figure 4-20 The probability that fluoride concentration exceeds 8 ppm.

MINITAB solution: Figure 4-20 is a MINITAB representation of $P(X > 8)$. The area is found as follows:

The pull-down **Calc ⇒ Probability Distributions ⇒ Log-normal** gives a dialog box that allows you to find the area as follows.

Cumulative Distribution Function

```
Log-normal with location = 3.2  and  scale = 1

x     P(X <= x)
8       0.131238
```

The probability is found as $1 - 0.131238 = 0.868$

Normal curve solution: Recall that $Y = \ln(X)$ has a normal distribution with $\mu = 3.2$ and $\sigma = 1$. We are seeking $P(X > 8)$, which is the same as $P(\ln(X) > \ln(8)) = P(\ln(X) > 2.0794)$. Therefore, if we find the area to the right of 2.0794 under a normal curve having a mean equal to 3.2 and a standard deviation equal to 1, we should find the same area as in Figure 4-20. Using EXCEL to find this area under the normal curve, we give the command =1-NORMDIST(2.0794,3.2,1,1), which gives the answer 0.868, the same as was obtained using the log-normal distribution.

EXAMPLE

Zonyl RP grease-resistant coating leaches out chemicals according to a log-normal distribution. The parameters $\alpha = 0.62$ and $\beta = 0.25$. Find the mean and standard deviation of the log-normal distribution.

The mean amount of leaching is

$\mu = e^{0.62+0.25^2/2} = e^{0.65} = 1.92$ and the variance is equal to

$$\sigma^2 = e^{2(0.62)+0.25^2}\left(e^{0.25^2} - 1\right) = e^{1.3025}\ (e^{0.0625} - 1) = 0.24$$

The standard deviation is 0.49. The amount of probability within 3 standard deviations of the mean is $P(0.45 < X < 3.39)$. Using EXCEL, the answer is given by =LOGNORMDIST(3.39,0.62,0.25)-LOGNORMDIST(0.45,0.62,0.25) = 0.991877.

The log-normal distribution has been found useful in describing failure times of equipment, risk analysis of nuclear power plants, and current gains in transistors.

4.6 The Gamma Distribution

The gamma function is an applied mathematics function that is defined as an integral with limits from 0 to infinity. Since the gamma distribution has a probability density function that contains the gamma function, it is necessary to study the gamma function before studying the gamma distribution. The gamma function is defined as follows:

$$\Gamma(\alpha) = \int_0^\infty x^{\alpha-1} e^{-x} dx, \alpha > 0$$

This improper integral has certain properties that we shall consider. An integration by parts where $u = x^{\alpha-1}$ and $dv = e^{-x}dx$ yields $du = (\alpha-1)x^{\alpha-2}\,dx$ and $v = -e^{-x}$. An application of the integration by parts technique yields the following recursive formula:

$$\Gamma(\alpha) = \int_0^\infty x^{\alpha-1} e^{-x}\,dx = (\alpha-1)\int_0^\infty x^{\alpha-2} e^{-x}\,dx = (\alpha-1)\Gamma(\alpha-1)$$

For $\alpha = 2$, $\Gamma(2) = (2 - 1)\Gamma(2 - 1) = 1\Gamma(1)$. The gamma function evaluated at $\alpha = 1$ is

$$\Gamma(1) = \int_0^\infty x^{1-1} e^{-x}\,dx = \int_0^\infty e^{-x}\,dx = 1$$

Therefore, $\Gamma(2) = 1$. For $\alpha = 3$, $\Gamma(3) = (3 - 1)\Gamma(3 - 1) = 2\,\Gamma(2) = 2(1) = 2!$. For $\alpha = 4$, $\Gamma(4) = (4 - 1)\Gamma(4 - 1) = 3\,\Gamma(3) = 3(2!) = 3!$. Continuing in this manner, we find for any positive integer n, $\Gamma(n) = (n - 1)!$. It may also be shown that $\Gamma(1/2) = \sqrt{\pi}$. The gamma function has many interesting properties. The actual numerical values of the gamma function are found in EXCEL.

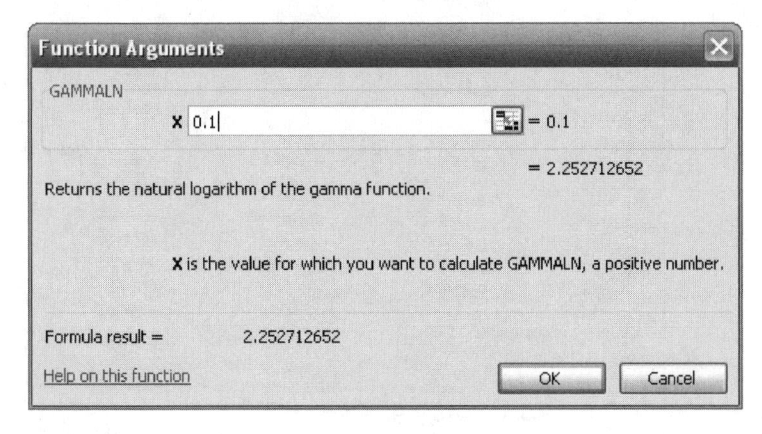

Figure 4-21 Finding $\Gamma(0.1)$ using the EXCEL function GAMMALN.

EXAMPLE

Evaluate $\Gamma(0.1)$, $\Gamma(0.5)$, $\Gamma(5)$, and $\Gamma(3.46465)$ using EXCEL.

The EXCEL function =GAMMALN(0.1) gives 2.2527 in Figure 4-21. Using the property that $y = \ln(x)$ is equivalent to $x = e^y$. We have $\ln(\Gamma(0.1)) = 2.2527$ is equivalent to $\Gamma(0.1) = e^{2.2527}$, or using EXCEL again, we obtain =EXP(GAMMALN(0.1)) = 9.5135. Similarly, =EXP(GAMMALN(0.5)) gives 1.7725, =EXP(GAMMALN(5)) gives 24, and EXP(GAMMALN(3.46465)) gives 3.1969. Figure 4-22 gives the EXCEL generated values for the gamma function. Figure 4-23 gives the MINITAB plot for this function.

Now that the gamma function has been discussed, the gamma distribution will be defined.

x	$f(x)$	x	$f(x)$	x	$f(x)$	x	$f(x)$
0.01	99.43259	0.6	1.489192	1.6	0.893515	3.5	3.323351
0.03	32.785	0.7	1.298055	1.7	0.908639	3.75	4.422988
0.05	19.47009	0.8	1.16423	1.8	0.931384	4	6
0.07	13.7736	0.9	1.068629	1.9	0.961766	4.25	8.285085
0.09	10.61622	1	1	2	1	4.5	11.63173
0.1	9.513508	1.1	0.951351	2.25	1.133003	4.75	16.58621
0.2	4.590844	1.2	0.918169	2.5	1.32934	5	24
0.3	2.991569	1.3	0.897471	2.75	1.608359	5.25	35.21161
0.4	2.21816	1.4	0.887264	3	2	5.75	78.78448
0.5	1.772454	1.5	0.886227	3.25	2.549257	6	120

Figure 4-22 Values of the gamma function, $\Gamma(x)$.

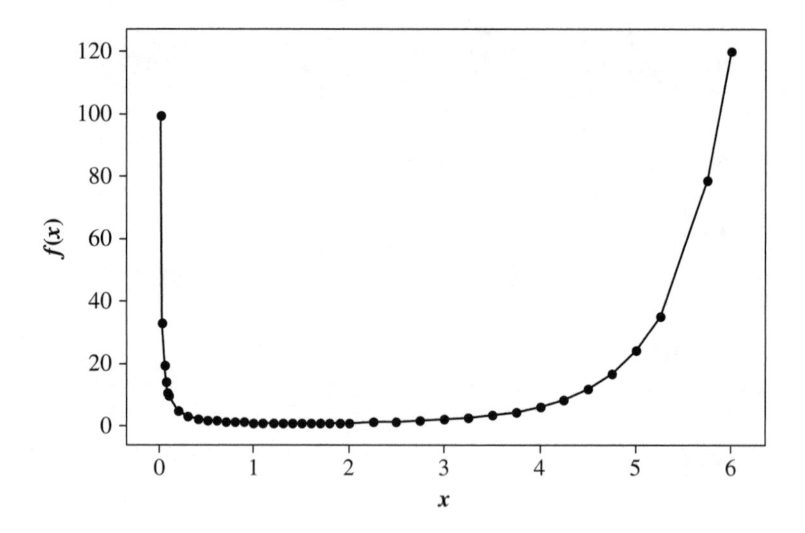

Figure 4-23 Plot of the gamma function.

The **gamma distribution** with parameters α and β is defined as follows:

$$f(x) = \frac{1}{\beta^{\alpha}\Gamma(\alpha)} x^{\alpha-1} e^{-x/\beta}, x > 0, \alpha > 0, \beta > 0$$

Figure 4-24 shows four different gamma distributions. For the variable C2, $\alpha = 1$ and $\beta = 1$; for the variable C3, $\alpha = 2$ and $\beta = 1$; for the variable C4, $\alpha = 3$

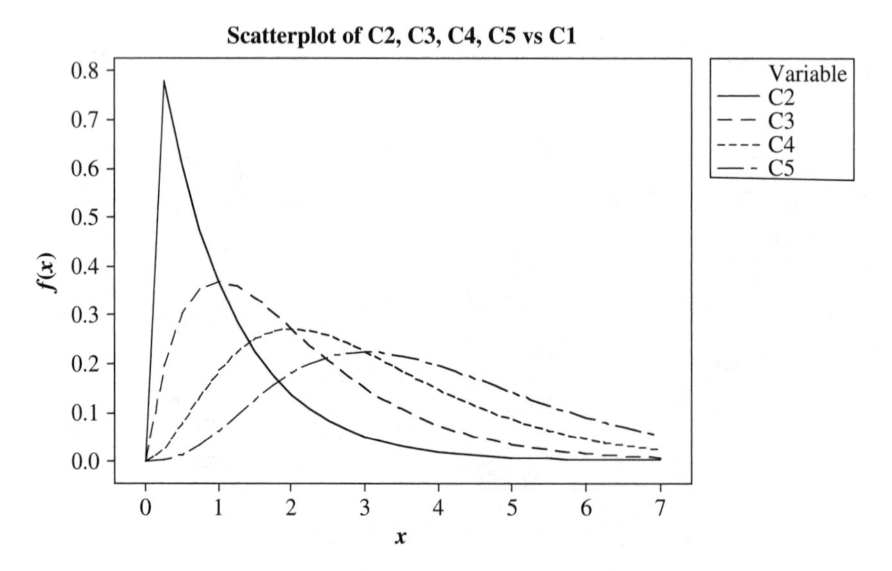

Figure 4-24 Different gamma distributions.

and $\beta = 1$; and for the variable C5, $\alpha = 4$ and $\beta = 1$. The mean of the gamma distribution is $\mu = \alpha\beta$ and the variance is $\sigma^2 = \alpha\beta^2$.

The **exponential distribution** is a special case of the gamma distribution with $\alpha = 1$. The density becomes

$$f(x) = \frac{1}{\beta^{\alpha}\Gamma(\alpha)} x^{\alpha-1} e^{-\frac{x}{\beta}}, x > 0, \alpha = 1, \beta > 0 = \frac{1}{\beta^{1}\Gamma(1)} x^{1-1} e^{-\frac{x}{\beta}}, x > 0 \qquad f(x) = \frac{1}{\beta} e^{-\frac{x}{\beta}}, x > 0$$

The mean of the exponential distribution is $\mu = \alpha\beta = (1)\beta = \beta$ and the variance is $\sigma^2 = \alpha\beta^2 = (1)\,\beta^2$ and $\sigma = \beta$.

Summarizing, the exponential has probability density function

$$f(x) = \frac{1}{\beta} e^{-\frac{x}{\beta}}, x > 0$$

and $\mu = \beta$ and $\sigma = \beta$.

EXAMPLE

The time to failure for mp3s follows an exponential distribution with mean time to failure equal to 3 years. Find the percent that have times to failure equal to 4 years or more. Solve using EXCEL, MINITAB, and MAPLE.

EXCEL

Sometimes the pdf for the exponential is expressed as $f(x) = \lambda e^{-\lambda x}$. The mean and standard deviation is $1/\lambda$. The dialog box for the exponential is shown in Figure 4-25.

Function Arguments

EXPONDIST

X 4 = 4

Lambda 0.3333 = 0.3333

Cumulative true = TRUE

= 0.736367713

Returns the exponential distribution.

Cumulative is a logical value for the function to return: the cumulative distribution function = TRUE; the probability density function = FALSE.

Formula result = 0.736367713

Help on this function OK Cancel

Figure 4-25 Dialog box for the exponential distribution in EXCEL.

The parameter $\lambda = 1/3 = 0.3333$. The **true** indicates cumulative distribution or $P(X < 4)$. Therefore $P(X > 4) = 1 - 0.737 = 0.263$.

MINITAB

The pull-down **Calc ⇒ Probability Distributions ⇒ Exponential** gives the dialog box shown in Figure 4-26.

Cumulative Distribution Function

```
Exponential with mean = 3

x     P(X <= x)
4      0.736403
```

$$P(X > 4) = 1 - 0.736 = 0.264$$

The MAPLE solution is as follows:

MAPLE

```
> evalf(int((1/3)*exp(-x/3),x=4..infinity));      0.2635971382
```

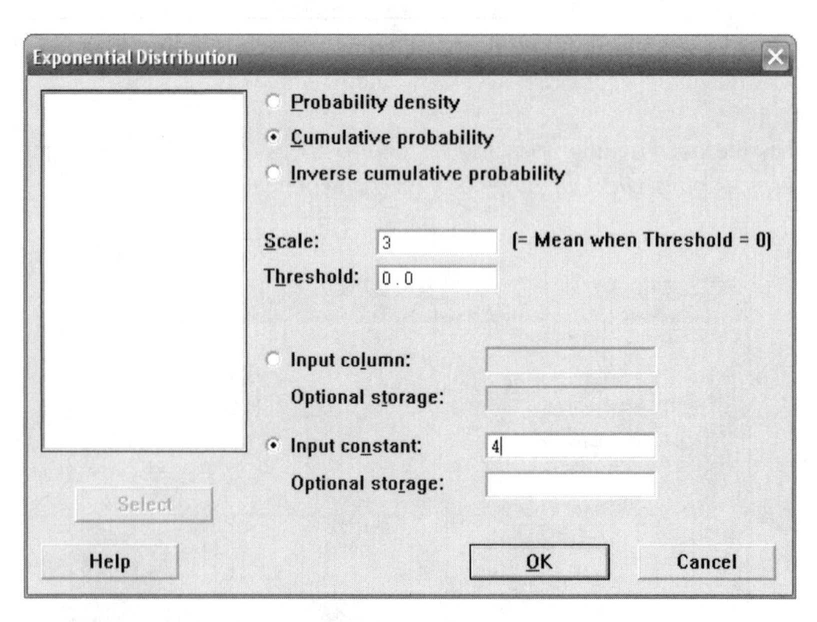

Figure 4-26 MINITAB dialog box for exponential distribution.

4.7 The Beta Distribution

The beta distribution is defined on the unit interval. It has a probability density function that is a function of the gamma function. The beta distribution has the following probability density function:

$$f(x)=\begin{cases}\dfrac{\Gamma(\alpha+\beta)}{\Gamma(\alpha)\Gamma(\beta)}x^{\alpha-1}(1-x)^{\beta-1}, & 0<x<1,\alpha>0,\beta>0\\ 0, & \text{otherwise}\end{cases}$$

The mean of the beta distribution is $\mu=\frac{\alpha}{\alpha+\beta}$ and the variance is

$$\sigma^2=\frac{\alpha\beta}{(\alpha+\beta)^2(\alpha+\beta+1)}$$

EXAMPLE

Consider the beta distribution when $\alpha=4$ and when $\beta=4$. Verify that the beta density is a density function by showing that it integrates to 1 using MAPLE. Find the mean and the variance using MAPLE and show that the mean and variance given are obtained by also using the above expression. Also graph the beta density.

$$f(x)=\frac{\Gamma(4+4)}{\Gamma(4)\Gamma(4)}x^{4-1}(1-x)^{4-1}=\frac{7!}{3!3!}x^3(1-x)^3=140x^3(1-x)^3$$

is the density function. (Note that we made use of $\Gamma(n)=(n-1)!$.)

Integrating by the use of MAPLE, we find that $f(x)$ is a density since the integral equals 1.

```
> evalf(int(140*x^3*(1-x)^3,x=0..1));     1.
```

Likewise the mean is found to be 0.5.

```
> evalf(int(140*x*x^3*(1-x)^3,x=0..1));   0.5000000000
```

Using the expression $\mu=\frac{4}{4+4}$ also yields 0.5 for the mean. The variance is found using MAPLE as follows:

```
> evalf(int(140*x^2*x^3*(1-x)^3,x=0..1))-0.5^2;     0.0277777778
```

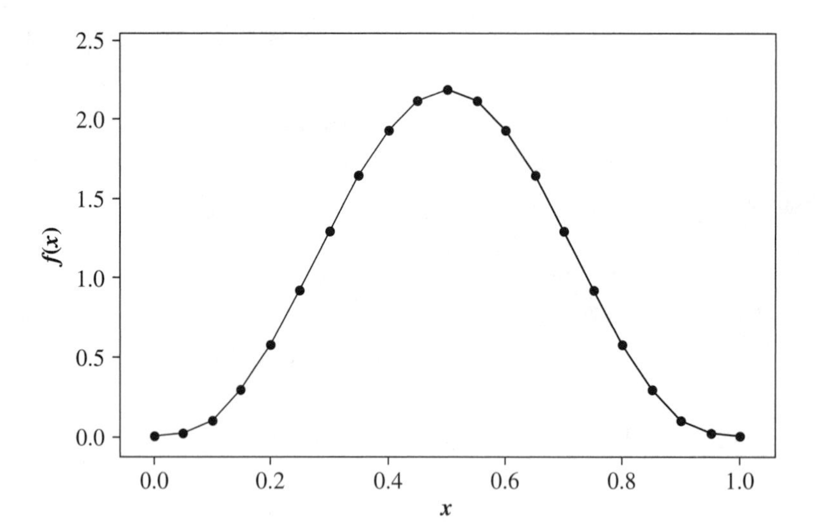

Figure 4-27 The beta density function $f(x) = 140x^3(1 - x)^3$.

Using the expression

$$\sigma^2 = \frac{\alpha\beta}{(\alpha+\beta)^2(\alpha+\beta+1)} = \frac{16}{64(9)} \text{ gives 0.02777 for the variance.}$$

The beta density is shown in Figure 4-27.

EXAMPLE

Highway construction requires engineering skills.

A civil engineer knows from experience that the proportion of highway sections requiring repairs in any given year in Douglas county is a random variable with a beta distribution having $\alpha = 1$ and $\beta = 3$. Use EXCEL, MINITAB, and MAPLE to

determine the probability that at most 40% of the highway sections will require repairs in any given year.

EXCEL

The filled in dialog box is shown in Figure 4-28. ($P(X < 0.4) = 0.784$)

MINITAB

The MINITAB output is as follows:

Cumulative Distribution Function

```
Beta with first shape parameter = 1 and second = 3

  x      P(X <= x)
0.4          0.784
```

A MINITAB graph of the area is given in Figure 4-29.

MAPLE

The density function is $f(x) = 3(1 - x)^2$, $0 < x < 1$. The MAPLE solution is

```
> evalf(int(3*(1-x)^2,x = 0..(0.4)));        0.7840000000
```

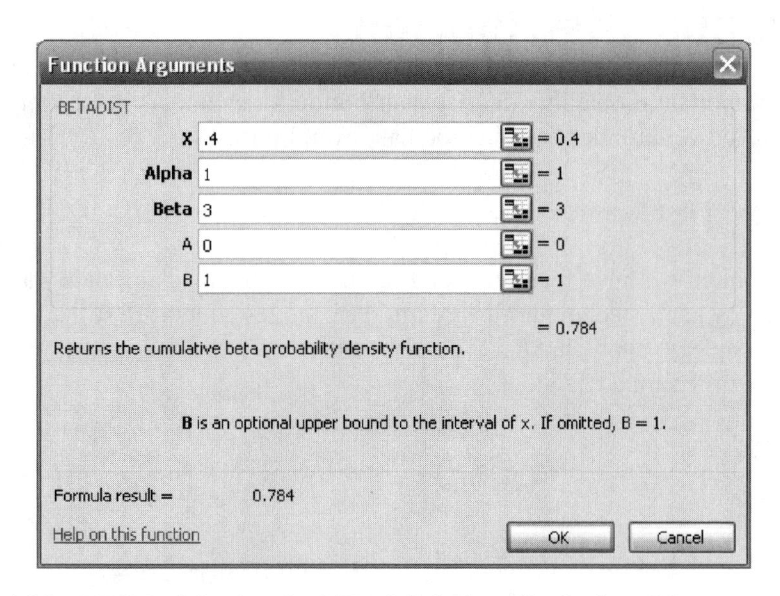

Figure 4-28 EXCEL dialog box for $P(X < 0.4)$ for beta distribution with $\alpha = 1$ and $\beta = 3$.

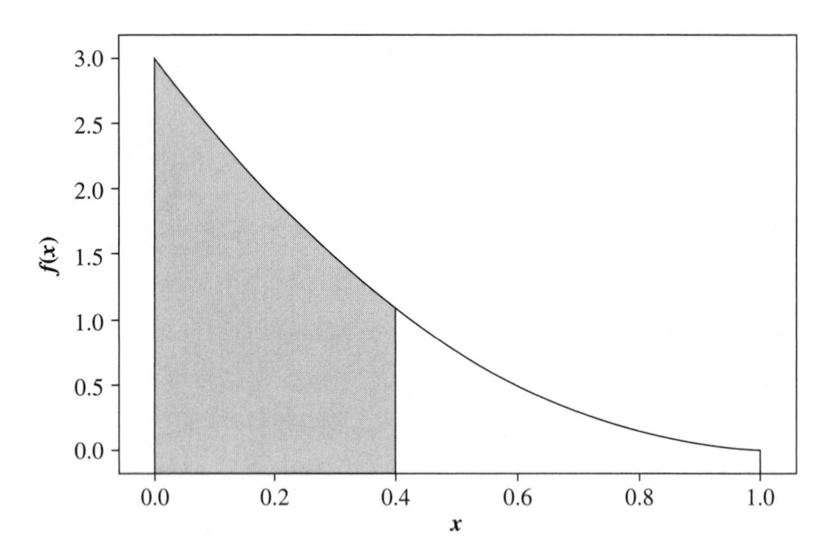

Figure 4-29 $P(X < 0.4)$ for beta distribution with $\alpha = 1$ and $\beta = 3$.

4.8 The Weibull Distribution

Another important probability density distribution for engineers is the Weibull distribution. The Weibull density function has several forms.

First form $\quad f(x) = \alpha\beta x^{(\beta-1)} e^{-\alpha x^{\beta}}, x > 0, \alpha > 0, \beta > 0$

Even though we have recommended the use of statistical software throughout the book, "by hand" techniques learned in the calculus course are still acceptable.

For example, to show that the Weibull is a density, make the change of variable $y = x^{\beta}$. The differential change is

$$dy = \beta x^{\beta-1}\, dx \quad \text{or} \quad dx = \frac{dy}{\beta x^{\beta-1}}$$

$$\int_0^{\infty} \alpha\beta x^{(\beta-1)} dx e^{-\alpha x^{\beta}} = \int_0^{\infty} \alpha\, dy e^{-\alpha y} = \int_0^{\infty} e^{-\alpha y} \alpha\, dy = 1$$

The mean of the Weibull is

$$\mu = \int_0^{\infty} x \alpha\beta x^{\beta-1} e^{-\alpha x^{\beta}} dx$$

Make the change of variable $y=\alpha x^{\beta}$, the limits remain 0 and ∞, $dy=\alpha\beta x^{\beta-1}dx$ or $dx=\frac{dy}{\alpha\beta x^{\beta-1}}$ and $x=\left(\frac{y}{\alpha}\right)^{\frac{1}{\beta}}$. The integral becomes

$$\mu=\alpha^{-\frac{1}{\beta}}\int_0^\infty y^{\frac{1}{\beta}}e^{-y}dy=\alpha^{-\frac{1}{\beta}}\Gamma\left(1+\frac{1}{\beta}\right)$$

Similarly, it can be shown that the variance is

$$\sigma^2=\alpha^{-\frac{2}{\beta}}\left\{\Gamma\left(1+\frac{2}{\beta}\right)-\left[\Gamma\left(1+\frac{1}{\beta}\right)\right]^2\right\}$$

EXAMPLE

This brand of cell phone battery has lifetimes that follow a Weibull distribution.

The length of life of a cell phone battery has the first form Weibull distribution with parameters $\alpha=1$ and $\beta=2$. The mean length of life for this phone battery is

$$\mu=1^{-\frac{1}{2}}\Gamma\left(1+\frac{1}{2}\right)=\Gamma(1.5)$$

The EXCEL function =EXP(GAMMALN(1.5)) gives 0.886 years for μ. The probability that it has a life over 2 years is

$$\int_2^\infty 1(2)x^{2-1}e^{-1x^2}dx$$

Evaluating this integral using MAPLE, we get

```
> evalf(int(2*x*exp(-x^2),x=2..infinity));          0.01831563889
```

Very few of these batteries have lifetimes exceeding 2 years.

A second form that you will sometimes see for the Weibull distribution is the following density function:

$$\textbf{Second form} \quad f(x)=\frac{\alpha}{\beta}x^{\alpha-1}e^{-\frac{x^{\alpha}}{\beta}},\ x>0,\alpha>0,\beta>0$$

For this functional form for the density, it can be shown that the mean and the variance are as follows:

$$\mu=\beta^{\frac{1}{\alpha}}\Gamma\left(1+\frac{1}{\alpha}\right) \quad \text{and} \quad \sigma^2=\beta^{\frac{2}{\alpha}}\left\{\Gamma\left(1+\frac{2}{\alpha}\right)-\left[\Gamma\left(1+\frac{1}{\alpha}\right)\right]^2\right\}$$

A third form that you will see for the Weibull distribution is the following density function.

$$\textbf{Third form} \quad f(x)=\frac{\alpha}{\beta^{\alpha}}x^{\alpha-1}e^{-\left(\frac{x}{\beta}\right)^{\alpha}},\ x>0,\ \alpha>0,\ \beta>0$$

For this functional form for the density, it can be shown that the mean and the variance are as follows:

$$\mu=\beta\Gamma\left(1+\frac{1}{\alpha}\right) \quad \text{and} \quad \sigma^2=\beta^2\left\{\Gamma\left(1+\frac{2}{\alpha}\right)-\left[\Gamma\left(1+\frac{1}{\alpha}\right)\right]^2\right\}$$

It is this third form that is utilized by MINITAB and EXCEL.

EXAMPLE

An industrial engineer finds that the life of a bearing (in hundred of hours) follows a Weibull distribution, the third form with $\alpha = 2$ and $\beta = 50$. Use the third form of the Weibull distribution and let X stand for the bearing life in hundred of hours. Find the probability X is less than 1000 hours using EXCEL, MINITAB, and MAPLE. Also give the mean and standard deviation of the lives of the bearings.

These bearings have lifetimes that are random variables with Weibull distributions.

EXCEL

Figure 4-30 gives the dialog box for finding the $P(X < 10)$ using the third form of the Weibull model with $\alpha = 2$ and $\beta = 50$. We see that the probability is very small that the life will be less than 1000 hours. The mean lifetime of such bearings is

$$\mu = \beta \Gamma\left(1 + \frac{1}{\alpha}\right) = 50\Gamma(1.5)$$

Function Arguments

WEIBULL

X	10	= 10
Alpha	2	= 2
Beta	50	= 50
Cumulative	true	= TRUE

= 0.039210561

Returns the Weibull distribution.

Cumulative is a logical value: for the cumulative distribution function, use TRUE; for the probability mass function, use FALSE.

Formula result = 0.039210561

Help on this function

OK Cancel

Figure 4-30 EXCEL dialog box for finding Weibull probabilities.

The gamma function of 1.5 is 0.886. The mean lifetime of such bearings is equal to 44.3 hundred or 4430 hours. The MAPLE verification of the mean is

```
> evalf(int(0.0008*x^2*exp(-x^2/2500),x=0..infinity));
44.31134627
```

The variance is

$$\sigma^2 = \beta^2 \left\{ \Gamma\left(1+\frac{2}{\alpha}\right) - \left[\Gamma\left(1+\frac{1}{\alpha}\right)\right]^2 \right\} = 2500(\Gamma(2) - \Gamma(1.5)^2) = 537.51$$

and the standard deviation is 23.18. The MAPLE verification of this variance is

```
> evalf(int(0.0008*x^3*exp(-x^2/2500),x=0..infinity))-44.311^2;
536.535279
```

MINITAB

Using the pull-down **Calc ⇒ Probability Distribution ⇒ Weibull** gives the dialog box shown in Figure 4-31 in which the shape parameter is the alpha value and the scale parameter is the beta value. The output is shown below.

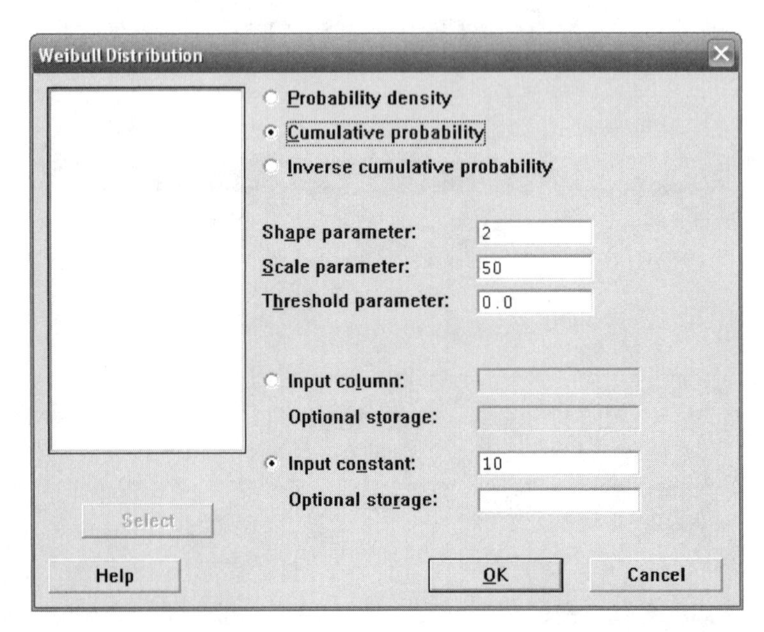

Figure 4-31 Dialog box for the Weibull distribution with $\alpha = 2$ and $\beta = 50$.

Cumulative Distribution Function

```
Weibull with shape = 2 and scale = 50

 x    P(X <= x)
10     0.0392106
```

MAPLE

The MAPLE solution is

```
> evalf(int(0.0008*x*exp(-x^2/2500),x=0..10)); 0.03921056085
```

4.9 Joint Distributions—Discrete and Continuous

A cell phone receives up to two calls per hour and makes up to two calls per hour. X represents the number of calls received and Y represents the number of calls made. An engineer developed the joint probability distribution shown in the table from historical data. This table gives the **bivariate probabilities** $p(x, y) = P(X = x, Y = y)$ for $X = 0$, 1, and 2 and $Y = 0$, 1, and 2.

	$X = 0$	$X = 1$	$X = 2$
$Y = 0$	0.02	0.04	0.03
$Y = 1$	0.04	0.15	0.17
$Y = 2$	0.05	0.23	0.27

A bivariate probability function satisfies two requirements:

1. $0 \le p(x, y) \le 1$ for all values of x and y.
2. $\sum_x \sum_y p(x,y) = 1$.

Note that $\sum_x \sum_y p(x,y) = 1$ for the above bivariate distribution.

$P(X = 0) = 0.02 + 0.04 + 0.05 = 0.11$, $P(X = 1) = 0.04 + 0.15 + 0.23 = 0.42$, and $P(X = 2) = 0.03 + 0.17 + 0.27 = 0.47$. This is called the **marginal probability distribution** of X and can be summarized as follows:

x	0	1	2
$p_1(x)$	0.11	0.42	0.47

Similarly, Y has a marginal probability distribution as follows:

y	0	1	2
$p_2(y)$	0.09	0.36	0.55

Recalling the conditional probability formula

$$P(A \mid B) = \frac{P(A \cap B)}{P(B)}$$

and applying it to this example,

$$p(X=0 \mid Y=0) = \frac{p(X=0, Y=0)}{p(Y=0)} = \frac{0.02}{0.09} = 0.222$$

Similarly, $p(1 \mid 0) = 0.444$ and $p(2 \mid 0) = 0.333$. The conditional distribution of X, given $Y = 0$, is

x	0	1	2
$p(x \mid 0)$	0.222	0.444	0.333

There are two other conditional distributions for X, given $Y = 1$ and $Y = 2$. Similarly there are three conditional distributions for Y, given $X = 0$, $X = 1$, and $X = 2$.

Two continuous random variables have a **bivariate joint probability density function** if they have the following properties:

1. $f(x, y) \geq 0$ for all x and y
2. $\int_{-\infty}^{\infty} \int_{-\infty}^{\infty} f(x,y)\, dx\, dy = 1$

EXAMPLE

Use MAPLE to determine the value of c that makes the following a joint pdf:

$$f(x, y) = \begin{Bmatrix} cxy^2, & 0 < x < 2, 0 < y < 2 \\ 0, & \text{elsewhere} \end{Bmatrix}$$

```
> evalf(int(int(c*x*y^2,x=0..2),y=0..2)); 5.333333333c
```

Since the integral must equal 1 to be a joint pdf, set $5.333c = 1$, or $c = 0.1875$

The marginal density functions are found by integrating the other variable out of the joint density function.

EXAMPLE

Find the marginal of X and the marginal of Y for the joint pdf:

$$f(x, y) = \begin{Bmatrix} 0.1875xy^2, & 0 < x < 2, 0 < y < 2 \\ 0, & \text{elsewhere} \end{Bmatrix}$$

The marginal of X is

$$f_1(x) = \int_0^2 0.1875xy^2 dy$$

The MAPLE solution is

```
> evalf(int(0.1875*x*y^2,y=0..2)); 0.5000000000x
```

$$f_1(x) = 0.5x,\ 0 < x < 2$$

The marginal of Y is

$$f_2(y) = \int_0^2 0.1875xy^2 dx$$

The MAPLE solution is

```
> evalf(int(0.1875*x*y^2,x=0..2)); 0.3750000000y^2
```

$$f_2(y) = 0.375y^2,\ 0 < y < 2$$

Note that for this example $f(x, y) = f_1(x) f_2(y)$. When $f(x, y) = f_1(x) f_2(y)$, we say that X and Y are **independent random variables**.

The **conditional density function for X, given Y**, is

$$f_1(x|y) = \frac{f(x,y)}{f_2(y)}$$

and the **conditional density function for Y given X** is

$$f_2(y|x) = \frac{f(x,y)}{f_1(x)}$$

We have considered bivariate distributions so far in this section. Any number of variables can be considered. The following example considers a **trivariate distribution**.

EXAMPLE

Random variables X, Y, and Z have the following joint probability density function. Determine the value that c would need to be for it to be a density.

$$f(x, y, z) = \begin{Bmatrix} cxy^2z^3, & 0 < x < 3, 0 < y < 2, 0 < z < 1 \\ 0, & \text{elsewhere} \end{Bmatrix}$$

We need to solve the following for c:

$$\int_0^1 \int_0^2 \int_0^3 cxy^2z^3 dx\,dy\,dz = 1$$

Once again, we turn to MAPLE for help. The integration is not hard but it takes a lot of time.

```
> evalf(int(int(int(c*x*y^2*z^3,x=0..3),y=0..2),z=0..1)); 3.c
```

Set $3c = 1$ and get $c = 1/3$.

$$f(x, y, z) = \begin{cases} 0.3333xy^2z^3, & 0 < x < 3, 0 < y < 2, 0 < z < 1 \\ 0, & \text{elsewhere} \end{cases}$$

Next, find the three marginal densities and see if the variables are independent. Three double integrations will be needed.

```
> evalf(int(int(0.333*x*y^2*z^3,y=0..2),z=0..1)); 0.2220000000x
```

$$f_1(x) = 0.222x,\ 0 < x < 3$$

```
> evalf(int(int(0.333*x*y^2*z^3,x=0..3),z=0..1)); 0.3746250000y^2
```

$$f_2(y) = 0.375y^2,\ 0 < y < 2$$

```
> evalf(int(int(0.333*x*y^2*z^3,x=0..3),y=0..2)); 3.996000000z^3
```

$$f_3(z) = 3.996z^3,\ 0 < z < 1$$

Since $f(x, y, z) = f_1(x)\,f_2(y)\,f_3(z)$, the three variables are independent.

4.10 Checking Data for Normality

When presented with a set of sample data, it is of interest to determine the form of the population distribution. In particular, it will be of interest in later chapters to test whether a set of data was taken from a population having a normal distribution. Several of the software packages have **normality tests**. These are tests that help answer the question: "Is it reasonable to assume that this data was taken from a population that is normally distributed?" The author prefers to use one of the following tests to answer this question: the Shapiro-Wilk normality test, the Anderson-Darling test of normality, the Ryan-Joiner test of normality, or the Kolmogorov-Smirnov test of normality. All the tests are basically conducted the same way. A test

statistics is calculated and it is assumed that the data is taken from a normally distributed population. The probability that the sample came from a normally distributed population is computed. It is accepted as reasonable that the data came from a normally distributed population unless the probability is unusually small (usually less than 0.05).

EXAMPLE

An engineer wishes to test that the following pressure readings (see Figure 4-32) were taken from a normal distribution:

Using MINITAB and the pull-down **Stat ⇒ Basic Statistics ⇒ Normality Tests** gives a dialog box with a choice of the Anderson-Darling test of normality, the Ryan-Joiner test of normality, or the Kolmogorov-Smirnov test of normality. The Kolmogorov-Smirnov test is shown in Figure 4-33.

The important number in Figure 4-33 is the *p*-value > 0.150. If this value is less than 0.05, then reject normality. Otherwise accept that the data came from a normally distributed population. The *p*-value given by the Anderson-Darling test is 0.369 and the *p*-value given by the Ryan-Joiner test is *p*-value > 0.10.

If the package STATISTIX is used, the following results are obtained:

The pull-down **Statistics ⇒ Randomness Normality tests ⇒ Shapiro-Wilk test** with the data from Figure 4-32 gives the following output:

```
Statistix 8.0
Shapiro-Wilk Normality Test

Variable        N        W          P
Pressure       30    0.9539     0.2144
```

The *p*-value here is 0.2144. It is reasonable that the data came from a normally distributed population.

EXAMPLE

Figure 4-35 gives the Kolmogorov-Smirnov results from the set of data given in Figure 4-34. The *p*-value is 0.014. This is less than 0.05. The chance of getting such a sample from a normal distribution is very small (0.014) and so, we reject that the data came from a normal distribution.

The *p*-value given by the Anderson-Darling test is less than 0.005 and the *p*-value given by the Ryan-Joiner test is *p*-value < 0.01. The STATISTIX analyses gave the following output:

```
Statistix 8.0
Shapiro-Wilk Normality Test

Variable        N        W          P
Pressure       30    0.8183     0.0001
```

15.31	15.15	14.64
14.66	15.66	15.18
14.59	15.62	15.18
14.71	14.92	14.44
15.80	15.45	14.89
14.95	15.64	14.44
14.47	14.57	15.71
14.26	14.30	14.76
15.32	15.11	14.22
14.87	14.54	15.04

Figure 4-32 Pressure readings (from normal population).

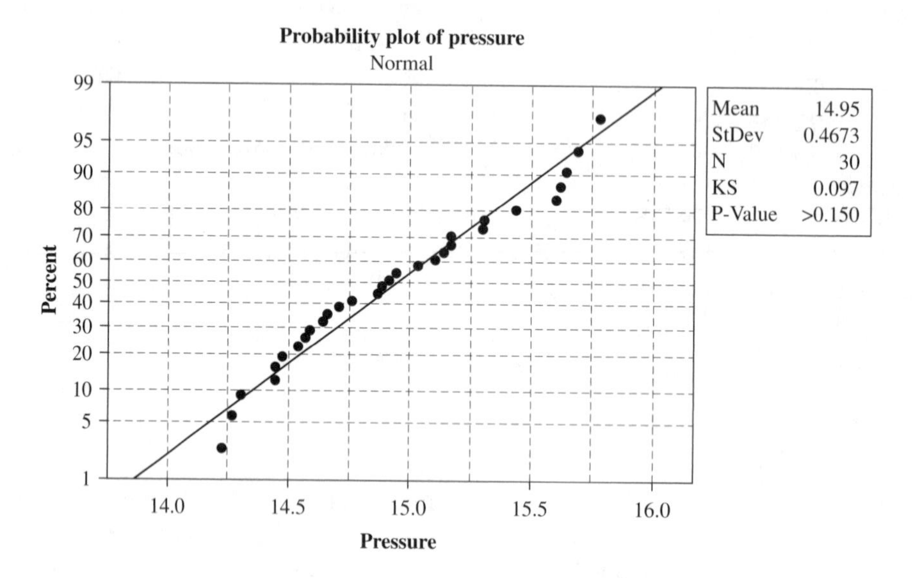

Figure 4-33 Kolmogorov-Smirnov test of normality for data in Figure 4-32.

4.56	46.26	48.16
12.55	14.51	24.29
5.89	22.33	37.22
21.04	6.75	4.21
1.04	4.11	20.43
4.22	4.19	17.84
20.79	6.34	5.96
20.20	8.22	1.17
13.79	59.01	0.25
4.01	14.94	5.04

Figure 4-34 Pressure readings (from non-normal population).

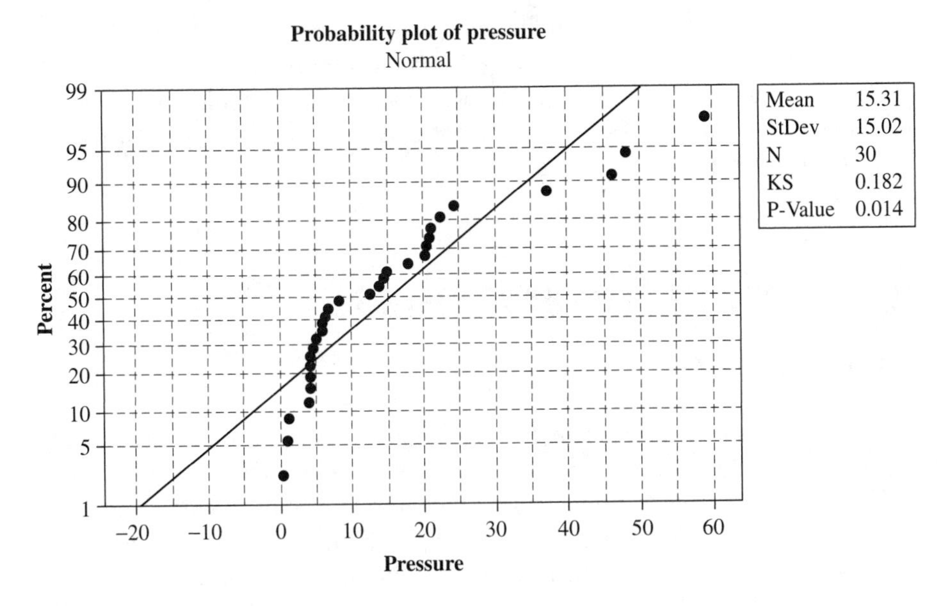

Figure 4-35 Kolmogorov-Smirnov test of normality for data in Figure 4-34.

Notice the p-value is 0.0001. The chance of getting this sample data from a normal distribution is very very small.

Figures 4-33 and 4-35 reveal the following:

If the data is from a normal population, the points tend to be close to the straight line. If the data is not from a normal population, the points do not fall close to the straight line. The p-values are reflective of how near the points are to the line.

Figure 4-36 gives dot plots for the data in Figures 4-32 and 4-34. The four tests of normality of sample data given in this section take the guess work out of testing for normality. They represent reasonable techniques that will lead to a conclusion.

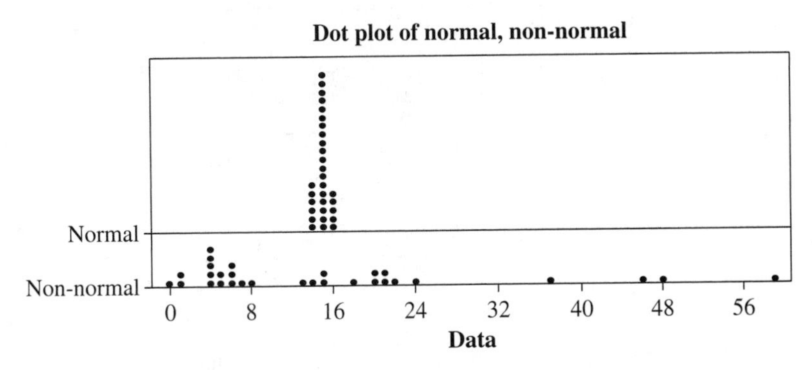

Figure 4-36 Dot plot of normal data and non-normal data.

Some practitioners recommend looking at a histogram, a box plot, or a dot plot. If the histogram, the box plot, or the dot plot reveals a normal shape, then conclude that you have normality. If the shape is not normal, then conclude you do not have normality. This, of course, leads to different conclusions from different practitioners for the same set of data.

4.11 Transforming Observations to Near Normality

What do you do if a statistical test, to be valid, requires the normality assumption, and you have concluded that your sample did not come from a normally distributed population? It is possible to transform your data by replacing it with logarithms of the data, or replacing it with square roots of the data. The transformed data may well pass the normality test, though the original data did not pass the normality test.

The data in Figure 4-34 is reproduced here. Recall that this data failed all the four normality tests as given in the previous section. First, transform this data using natural logs. This new data is shown in Figure 4-37.

The Anderson-Darling test gives a p-value of 0.102, The Ryan-Joiner gives a p-value of 0.049, the Kolmogorov-Smirnov gives a p-value of 0.065, and the p-value for the Shapiro-Wilk test is 0.0599. The transformed data basically passes the tests of normality.

Consider one more transformation. Suppose the fourth root transformation is used. That is, the data is replaced by taking the fourth root of each number and forming a new set of data consisting of fourth roots. This transformed data is shown in Figure 4-38.

1.52	3.83	3.87
2.53	2.67	3.19
1.77	3.11	3.62
3.05	1.91	1.44
0.04	1.41	3.02
1.44	1.43	2.88
3.03	1.85	1.79
3.01	2.11	0.16
2.62	4.08	−1.39
1.39	2.70	1.62

Figure 4-37 Data in Figure 4-34 transformed by replacing with natural logs.

1.46	2.61	2.63
1.88	1.95	2.22
1.56	2.17	2.47
2.14	1.61	1.43
1.01	1.42	2.13
1.43	1.43	2.06
2.14	1.59	1.56
2.12	1.69	1.04
1.93	2.77	0.71
1.42	1.97	1.50

Figure 4-38 Data in Figure 4-34 transformed by replacing with fourth roots.

The Anderson-Darling test gives a *p*-value of 0.321, the Ryan-Joiner gives a *p*-value > 0.1, the Kolmogorov-Smirnov gives a *p*-value > 0.15, and the *p*-value for the Shapiro-Wilk test is 0.538. The transformed data clearly passes the tests of normality. The second transformation is a better one than the first for this set of data.

Figure 4-39 shows how transforming the data changes it from a set of highly skewed data to more normal data. Transforming data is more of an art than a science. The more experience one has at transforming the data, the better one gets. Taking logarithms and roots tends to bring the data closer together. It takes the skew away and causes the data to move closer to the center as a general rule.

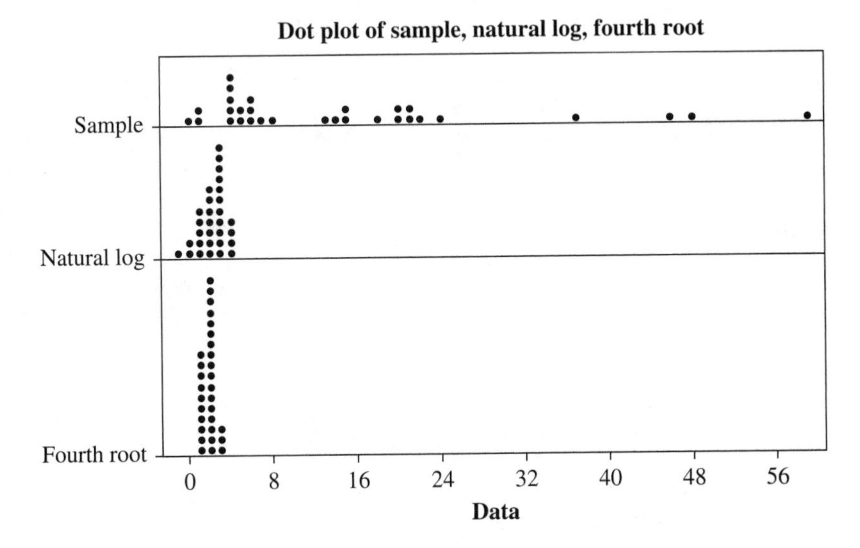

Figure 4-39 Dot plot of original data, natural log, and fourth root of original data.

4.12 Simulation

The need to form a simulated sample often arises in engineering applications. For one thing, it may be expensive to obtain a sample and for another, it may be very difficult to obtain a real sample.

EXAMPLE

Consider a random variable, X, having the following pdf:

$$f(x) = \begin{Bmatrix} 0.08x, & 0 < x < 5 \\ 0, & \text{elsewhere} \end{Bmatrix}$$

The variable has the following cumulative distribution function:

$$F(x) = \begin{Bmatrix} 0, & x < 0 \\ 0.04x^2, & 0 < x < 5 \\ 1, & x > 5 \end{Bmatrix}$$

Now, consider a uniform variable over the interval (0, 1). Most software packages have a uniform variable built into the package capable of giving uniform values over the unit interval. Let u be a uniformly generated value between 0 and 1. Set $F(x) = u$ and solve the resulting equation for x in terms of u. This gives the following:

$$0.04x^2 = u, \quad \text{or} \quad x = \sqrt{\frac{u}{0.04}}$$

EXAMPLE

Give a simulated random sample from the distribution in the above example of size 10. Using the following pull-down **Calc ⇒ Random Data ⇒ Uniform** in MINITAB gives the dialog box shown in Figure 4-40.

The following uniform numbers are created:

```
0.822  0.229  0.478  0.284  0.489  0.128  0.296  0.958  0.425  0.042
```

The following simulated sample is determined:

$$x = \sqrt{\frac{u}{0.04}}$$

This formula is applied to the uniform values. The following simulated random sample of x values is determined:

```
4.53  2.39  3.46  2.67  3.49  1.79  2.72  4.90  3.26  1.03
```

Figure 4-41 shows the density from which the above simulated random sample came.

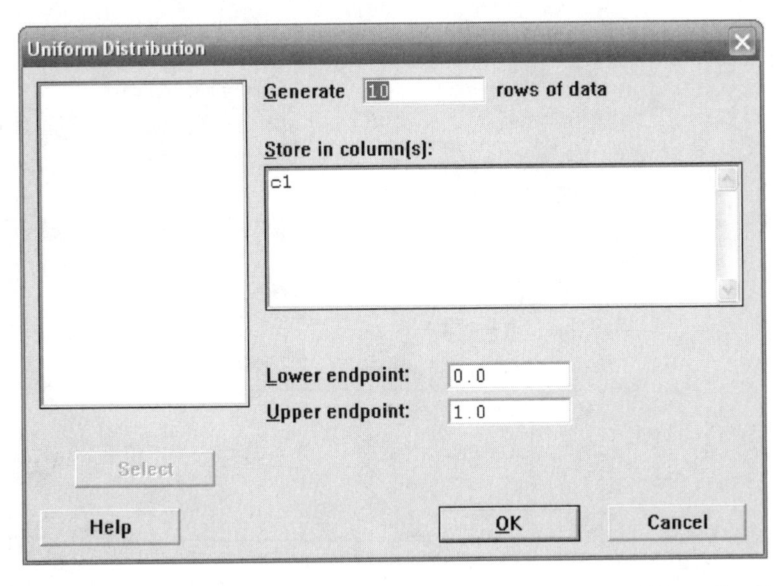

Figure 4-40 Generating a random sample of size 10 from the uniform distribution.

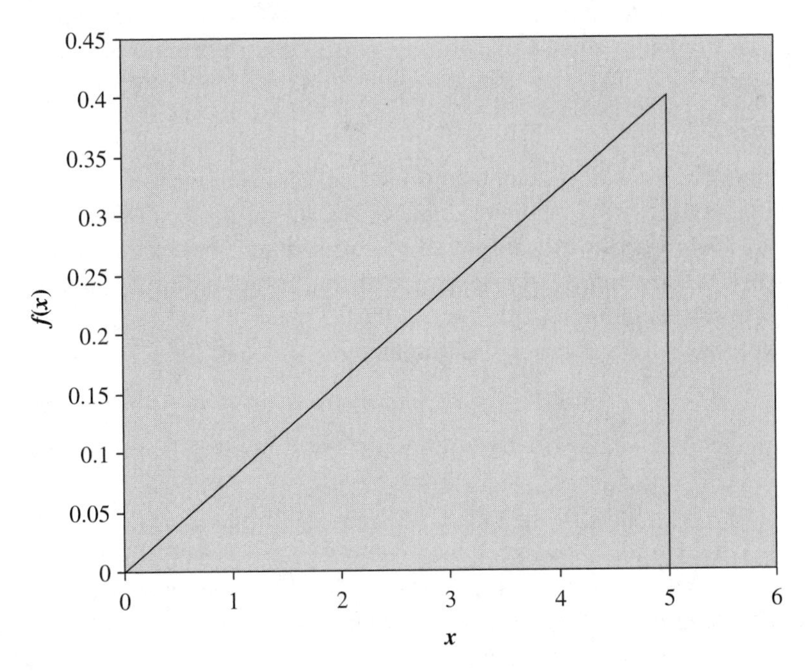

Figure 4-41 Graph of the density from which the simulated sample was obtained.

Binomial Distribution

Generate 15 rows of data

Store in column(s):

c1

Number of trials: 10

Probability of success: 0.5

Select

Help OK Cancel

Figure 4-42 Simulated binomial random variable with $n = 10$ and $p = 0.5$ ($n = 15$).

If this simulation was repeated, a different but typical sample would result.

In addition to the technique described above, MINITAB has the ability to simulate samples from some of the distributions that we have studied previously in the past two chapters.

EXAMPLE

Simulate the outcomes of tossing a coin 10 times and observe the number of heads to occur in the 10 tosses. Obtain a sample of size 15 of this random variable.

The pull-down **Calc ⇒ Random Data ⇒ Binomial** gives the dialog box shown in Figure 4-42. The request is for a sample of size 15 to be placed into column C1. The variable is the binomial based on $n = 10$ and $p = 0.5$.

The following random sample is simulated:

```
5   4   6   7   6   6   6   7   3   4   7   6   4   7   7
```

If the command is given again, the following new simulated sample is obtained:

```
3   6   4   3   5   4   4   6   5   4   8   5   2   5   4
```

EXAMPLE

A software engineer is interested in how long it takes to learn to use a new set of software. The engineer believes the time required to be proficient with the software is normally distributed with a mean equal to 20 hours and a standard deviation equal to 3.5 hours. Create a simulated sample for the times required for 25 employees of Ace manufacturing to become proficient with the software.

Software comes in the form of a compact disk.

The pull-down **Calc ⇒ Random Data ⇒ Normal** gives the dialog box which is filled out as shown in Figure 4-43. The following simulated sample was given by **MINITAB**

```
18.6   15.2   19.8   19.7   19.6   24.6   23.8   16.2   23.1   16.0
19.2   26.0   14.4   16.7   27.7   21.5   24.5   15.1   20.5   19.6
18.0   24.7   14.4   26.6   21.9
```

The data represents a typical sample of size 25.

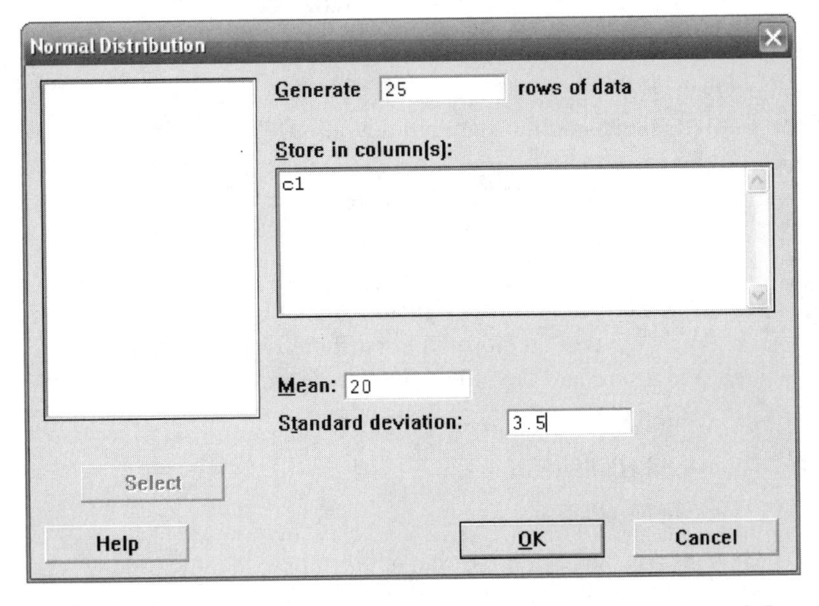

Figure 4-43 Simulated normal distribution with $\mu = 20$ and $\sigma = 3.5$.

Exercises

1. The voltage in an engineering experiment, X, varies according to the following probability density function:

$$f(x) = 4x^3 e^{-x^4}, x > 0$$

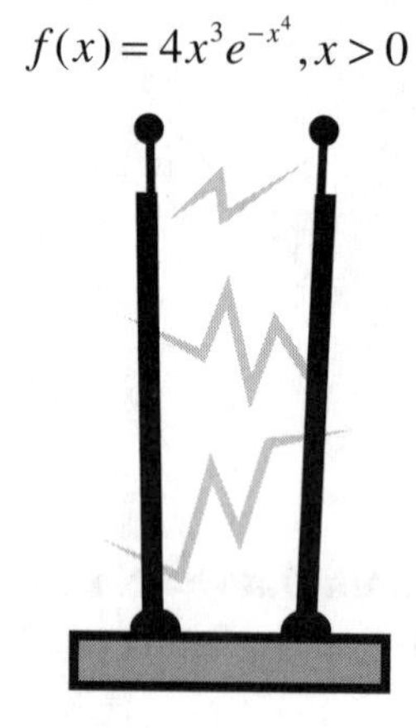

Voltage is a random variable. It is not constant.

   (a) Graph the function $f(x)$ over the interval from 0 to 2 using EXCEL.
   (b) Verify that $f(x)$ is a pdf by exact integration.
   (c) Use MAPLE to verify that $f(x)$ is a pdf.
   (d) Use EXCEL and the definition of integral as a limiting sum to verify that $f(x)$ is a pdf.
   (e) Find the mean and the standard deviation.
   (f) Find $P(X < 1)$.
   (g) Find $P(X > 2)$.
   (h) Find $P(0.5 < X < 1.5)$.

2. It is determined by acoustical engineers that MP3 players are played at decibel levels (X) that are normally distributed with a mean equal to 100 decibels and a standard deviation equal to 5 decibels. Find $P(90 < X < 105)$
   (a) By using EXCEL,
   (b) By using MINITAB,
   (c) By using MAPLE,
   (d) By using rectangular areas that approximate the integral.

The decibel level of an MP3 player is a random variable.

3. Engineering students at Midwestern University are asked the following question:

   "Do you think online degrees in engineering are realistic?"

   Suppose 15% believe this is true presently. A survey of 500 Midwestern engineering students is conducted.

   (a) What is the probability 100 or more answer yes?

   (b) What is the probability between 50 and 80, both inclusive, answer yes?

   (c) What is the probability less than 60 answer yes?

   Use the normal approximation to the binomial to answer.

4. The proportion of defective ammunition produced by the company Bullets Inc. has a defective proportion per day that follows a beta distribution with $\alpha = 1$ and $\beta = 19$.

The daily proportion defective follows a beta distribution.

(a) Give the density and graph the pdf of X, the proportion of defectives produced per day.

(b) Use MAPLE to verify that it is a pdf.

(c) Find the mean and standard deviation of X.

(d) Find the probability that the percent defective on a given day is 10% or less.

(e) Simulate a sample of 10 days for a beta distribution with $\alpha = 1$ and $\beta = 19$.

5. Three random variables have a joint distribution equal to $f(x, y, z) = cxy^2e^{-z}$ for $0 < X < 1$, $0 < Y < 2$, and $Z > 0$ and 0 elsewhere.

(a) Use your knowledge of calculus to prove that $c = 0.75$.

(b) Find the marginal densities of X, Y, and Z and determine if the three variables are independent.

(c) Find the probability that all three variables are between 0 and 1.

6. The following data is the total carbon in micrograms per cubic meter of air in an 8-hour workday at a salt mine in which diesel exhaust is present for 15 selected days:

14	410	3	25	129
180	17	58	7	9
26	147	37	187	55

(a) Determine which, if any, of the four normality tests this data passes. The four tests are

i. Anderson-Darling test

ii. Ryan-Joiner test

iii. Kolmogorov-Smirnov test

iv. Shapiro-Wilk test

(b) Perform the fourth root transformation and test the transformed data for normality. Give the transformed data and the four test results.

7. The errors in an industrial process have a distribution given by the following density:

$$f(x) = \begin{cases} -0.75(x^2 - 1), & -1 < x < 1 \\ 0, & \text{elsewhere} \end{cases}$$

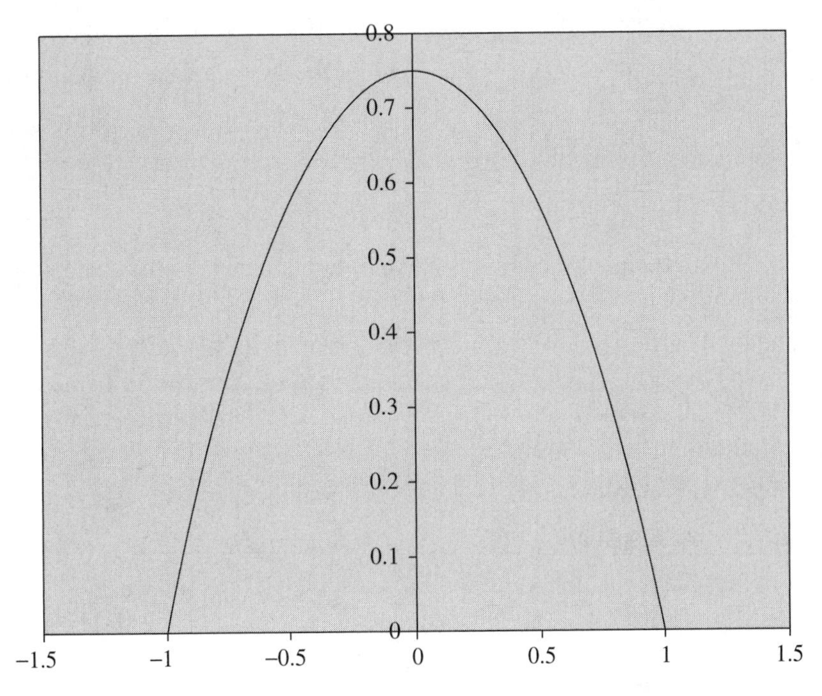

Figure 4-44 Distribution of errors in an industrial process.

An EXCEL generated graph of the density is shown in Figure 4-44.

(a) Find the mean.

(b) Find the standard deviation.

(c) Find $P(-.5 < x < .5)$.

8. Random variable X has the density

$$f(x) = \begin{cases} 0, & x < 0 \\ 2xe^{-x^2}, & x \geq 0 \end{cases}$$

(a) Plot the graph of the density.

(b) Select a simulated sample from this density of size 20.

9. The chi-square distribution is a gamma distribution with $\alpha = \nu/2$ and $\beta = 2$. The parameter ν is called the number of **degrees of freedom** for the chi-square distribution.

(a) Give the chi-square pdf by substituting these values for α and β into the gamma pdf.

(b) Take the expression for the mean of a gamma and substitute the above values for α and β into the mean of the gamma distribution to find the mean of the chi-square.

(c) Take the expression for the variance of a gamma and substitute the above values for α and β into the variance of the gamma distribution to find the variance of the chi-square.

(d) Draw graphs of the chi-square in separate panels of the same graph for 2, 5, 10, and 15 degrees of freedom.

10. The total carbon in micrograms per cubic meter of air in an 8-hour workday at a salt mine in which diesel exhaust is present, X, is the variable of interest. It is found that $\ln(X)$ has a normal distribution with mean equal to 3.5 and standard deviation equal to 1.5. This may also be stated as follows:

    X has a log normal distribution with parameters $\alpha = 3.5$ and $\beta = 1.5$.

    (a) Plot the distribution of X.

    (b) Find the mean and the standard deviation of the total carbon in an 8-hour shift.

    (c) Find the probability for $X < 150$.

Summary

1. A **probability density function (pdf)** shows how the characteristic is distributed. It has two properties:

   (a) The curve is always positive, $f(x) \geq 0$.

   (b) The total area under the curve is 1.

2. **MAPLE** is a software package that performs many different mathematical functions, including integration.
3. Probabilities are interpreted as areas under the pdf curve.
4. Another property of continuous random variables is that the probability of a single point is zero, that is $P(X = a) = 0$.
5. The **mean** is given by

$$\mu = \int_{-\infty}^{\infty} xf(x)\,dx$$

6. The **variance**, using the defining formula, is

$$\sigma^2 = \int_{-\infty}^{\infty} (x-\mu)^2 f(x)\,dx$$

7. An equivalent formula for the variance is

$$\sigma^2 = \int_{-\infty}^{\infty} x^2 f(x)\,dx - \mu^2$$

8. The **cumulative distribution function** is

$$F(x) = \int_{-\infty}^{x} f(t)\,dt$$

9. The **normal distribution** has the following probability density function:

$$f(x) = \frac{1}{\sqrt{2\pi}\sigma} e^{-\frac{(x-\mu)^2}{2\sigma^2}}, -\infty < x < \infty$$

10. The normal distribution may be used to approximate the binomial distribution when $np > 5$ and $nq > 5$.
11. Random variable X is **uniformly distributed** over the interval from a to b if it has the following pdf:

$$f(x) = \begin{cases} 0, & x < a \\ \dfrac{1}{b-a}, & a \le x \le b \\ 0, & x > b \end{cases}$$

The mean is $\mu = \frac{a+b}{2}$ and the variance is $\sigma^2 = \frac{(b-a)^2}{12}$.

12. Random variable X has a **log-normal distribution** if it has the following pdf:

$$f(x) = \begin{cases} \dfrac{1}{\sqrt{2\pi}\beta} x^{-1} e^{-\frac{(\ln x-\alpha)^2}{2\beta^2}}, & x > 0, \beta > 0 \\ 0, & \text{otherwise} \end{cases}$$

It has a mean given by $\mu = e^{\alpha+\beta^2/2}$ and the variance given by

$$\sigma^2 = e^{2\alpha+\beta^2}\left(e^{\beta^2} - 1\right)$$

13. The **gamma function** is defined as follows:

$$\Gamma(\alpha) = \int_0^\infty x^{\alpha-1}e^{-x}dx, \alpha > 0$$

14. The gamma function satisfies the following:

$$\Gamma(\alpha) = (\alpha - 1)\Gamma(\alpha - 1)$$

15. We find for any positive integer n, $\Gamma(n) = (n - 1)!$.
16. The **gamma distribution** with parameters α and β is defined as

$$f(x) = \frac{1}{\beta^\alpha \Gamma(\alpha)} x^{\alpha-1} e^{-x/\beta}, x > 0, \alpha > 0, \beta > 0$$

The mean of the gamma distribution is $\mu = \alpha\beta$ and the variance is $\sigma^2 = \alpha\beta^2$.

17. The exponential distribution has probability density function

$$f(x) = \frac{1}{\beta} e^{-\frac{x}{\beta}}, x > 0$$

and $\mu = \beta$ and $\sigma = \beta$. Sometimes the pdf for the exponential is expressed as $f(x) = \lambda e^{-\lambda x}$. The mean and standard deviation is $1/\lambda$.

18. The **beta distribution** has the following probability density function:

$$f(x) = \begin{cases} \dfrac{\Gamma(\alpha+\beta)}{\Gamma(\alpha)\Gamma(\beta)} x^{\alpha-1}(1-x)^{\beta-1}, & 0 < x < 1, \alpha > 0, \beta > 0 \\ 0, & \text{otherwise} \end{cases}$$

The mean of the beta distribution is $\mu = \frac{\alpha}{\alpha+\beta}$ and the variance is

$$\sigma^2 = \frac{\alpha\beta}{(\alpha+\beta)^2(\alpha+\beta+1)}$$

19. The **Weibull density function** has several forms:

First form $f(x) = \alpha\beta x^{(\beta-1)} e^{-\alpha x^{\beta}}, x > 0, \alpha > 0, \beta > 0$

The mean is $\mu = \alpha^{-\frac{1}{\beta}} \Gamma\left(1 + \frac{1}{\beta}\right)$ and the variance is

$$\sigma^2 = \alpha^{-\frac{2}{\beta}} \left\{ \Gamma\left(1 + \frac{2}{\beta}\right) - \left[\Gamma\left(1 + \frac{1}{\beta}\right)\right]^2 \right\}$$

Second form $f(x) = \frac{\alpha}{\beta} x^{\alpha-1} e^{-\frac{x^{\alpha}}{\beta}}, x > 0, \alpha > 0, \beta > 0$

For this functional form for the density, it can be shown that the mean and the variance are as follows:

$$\mu = \beta^{\frac{1}{\alpha}} \Gamma\left(1 + \frac{1}{\alpha}\right) \quad \text{and} \quad \sigma^2 = \beta^{\frac{2}{\alpha}} \left\{ \Gamma\left(1 + \frac{2}{\alpha}\right) - \left[\Gamma\left(1 + \frac{1}{\alpha}\right)\right]^2 \right\}$$

Third form $f(x) = \frac{\alpha}{\beta^{\alpha}} x^{\alpha-1} e^{-\left(\frac{x}{\beta}\right)^{\alpha}}, x > 0, \alpha > 0, \beta > 0$

For this functional form for the density, it can be shown that the mean and the variance are as follows:

$$\mu = \beta \Gamma\left(1 + \frac{1}{\alpha}\right) \quad \text{and} \quad \sigma^2 = \beta^2 \left\{ \Gamma\left(1 + \frac{2}{\alpha}\right) - \left[\Gamma\left(1 + \frac{1}{\alpha}\right)\right]^2 \right\}$$

20. Two continuous random variables have a **bivariate joint probability density function** if they have the following properties:

    a. $f(x, y) \geq 0$ for all x and y.

    b. $\int_{-\infty}^{\infty} \int_{-\infty}^{\infty} f(x,y)\, dx\, dy = 1$.

21. When $f(x,y) = f_1(x) f_2(y)$, we say that X and Y are **independent random variables.**

    The **conditional density function** for X given Y is

$$f_1(x|y) = \frac{f(x,y)}{f_2(y)}$$

and the conditional density function for Y given X is

$$f_2(y\,|\,x) = \frac{f(x,y)}{f_1(x)}$$

22. Several of the software packages have **normality tests**. These are tests that help answer the question: "Is it reasonable to assume that this data was taken from a population that is normally distributed?"
23. What do you do if a statistical test, to be valid, requires the normality assumption, and you have concluded that your sample did not come from a normally distributed population? It is possible to transform your data by replacing it with logarithms of the data, or replacing it with square roots of the data, or with fourth roots of data, and so forth. The transformed data may well pass the normality test, though the original data did not pass the normality test.
24. The need to form a **simulated sample** often arises in engineering statistics. For one thing, it may be expensive to obtain a sample and for another, it may be very difficult to obtain a real sample.

CHAPTER 5

Sampling Distributions

5.1 Populations and Samples

A **population** is a set of data that is of interest to the engineer or the scientist. The population may be described by a probability distribution function or a probability density function. There are **parameters** of the populations in which we are interested. The parameters are population measurements such as the mean, the standard deviation, the population proportion, and so forth. The parameters are usually unknown. For example, the normal population has two parameters that are usually unknown. Until the mean μ and standard deviation σ are known, the normal population's form is known but the exact density is unknown ($f(x)=\frac{1}{\sqrt{2\pi}\sigma}e^{-\frac{(x-\mu)^2}{2\sigma^2}}, -\infty < x < \infty$). We must estimate the mean and standard deviation before any probability statement concerning the population can be made.

The population is sampled in order to find out something about it. The most common sample is a random sample. A set of n observations constitutes a **simple random sample** if it is chosen in such a manner so that each subset of n elements has the same probability of being chosen. One way of selecting a simple random sample is to write the name of the population items on a piece of paper and pull

your sample from a box containing the names. A random number generator in a software package is the modern day technique of obtaining a random sample from a finite population. In a large population, the sample is chosen in as haphazard a way as possible since it may not be possible to number the elements.

The parameters such as the mean μ and standard deviation σ are estimated by the sample mean $\bar{x}$ and the sample standard deviation

$$s = \sqrt{\frac{\sum(x - \bar{x})^2}{(n-1)}}$$

which are calculated from the sample observations. The sample mean and the sample standard deviation are referred to as statistics. A **statistic** is a measurement made on the sample. A statistic is actually a random variable. Its value changes from sample to sample. The parameter is a constant and the statistic is a variable.

The **sampling distribution** of a sample statistic, calculated from a sample of n measurements, is the probability distribution of the statistic. We will find in the next section that under certain conditions $\bar{X}$ has a normal distribution. We also find that $\frac{(n-1)S^2}{\sigma^2}$ has a chi-square distribution with $(n - 1)$ degrees of freedom.

5.2 The Sampling Distribution of the Mean (σ Known)

The results in this section are best illustrated by an example. We illustrate the central limit theorem by an example and then we state the central limit theorem more generally.

EXAMPLE

We are interested in taking various sample sizes from the students at Midwestern University. The random variable of interest, x, is the credit hours the student is taking. The probability distribution of x is shown in Figure 5-1.

x	6	9	12	15
$p(x)$	0.25	0.25	0.25	0.25

Figure 5-1 Distribution of credit hours per student at Midwestern University.

The distribution is uniform with the possible values 6, 9, 12, and 15. The mean of the distribution is

$$\mu = 6(0.25) + 9(0.25) + 12(0.25) + 15(0.25) = 10.5$$

and the variance is

$$\sigma^2 = \sum x_i^2 p(x_i) - \mu^2 = 36(0.25) + 81(0.25) + 144(0.25) + 225(0.25) - 10.5^2 = 11.25$$

The population mean and variance are therefore known to be $\mu = 10.5$ and $\sigma^2 = 11.25$.

Now consider the distribution of the sample mean constructed by taking all samples of size $n = 2$, then $n = 3$, and so forth. A pattern will soon emerge. Start by considering all samples of size 2. In Figure 5-2, the sample mean will be represented by *x*bar. The symbols *x*bar and $\bar{x}$ will be used interchangeably throughout the discussion.

Figure 5-2 shows all possible samples from Midwestern University of size $n = 2$ in A1:B16. For example, 6 and 6 represents a sample in which the first selected student was taking 6 credit hours and so was the second. The mean of each sample is shown in C1:C16. Each of the sample means in column C is equally likely (since the samples are taken from a uniform distribution) and the distribution of *x*bar is

A	B	C	D	E	F	G	H
6	6	6		Distribution of sample mean for $n = 2$			
6	9	7.5					
6	12	9		*x*bar	*P*(*x*bar)	*x*bar**p*(*x*bar)	*x*bar^2**p*(*x*bar)
6	15	10.5		6	0.0625	0.375	2.25
9	6	7.5		7.5	0.125	0.9375	7.03125
9	9	9		9	0.1875	1.6875	15.1875
9	12	10.5		10.5	0.25	2.625	27.5625
9	15	12		12	0.1875	2.25	27
12	6	9		13.5	0.125	1.6875	22.78125
12	9	10.5		15	0.0625	0.9375	14.0625
12	12	12			sum = 1	sum = 10.5	sum = 115.88
12	15	13.5					
15	6	10.5					
15	9	12					
15	12	13.5					
15	15	15					

Figure 5-2 Constructing the distribution of *x*bar ($\bar{X}$), for $n = 2$.

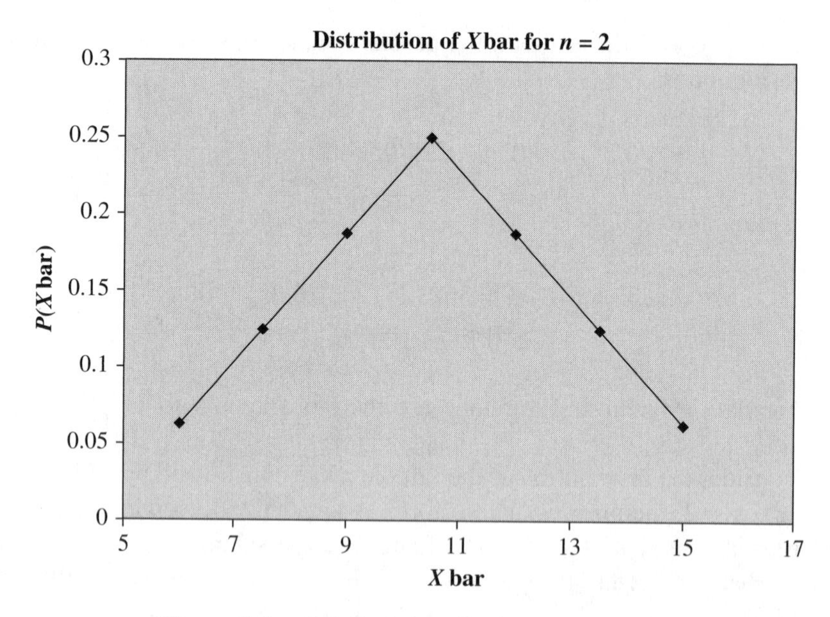

Figure 5-3 Graph of *x*bar distribution for $n = 2$.

given in columns E and F. The sample mean, 10.5 for example, occurs 4 times in 16 and has probability 4/16 or 0.25. A plot of the distribution of *x*bar is given in Figure 5-3. Column G of Figure 5-2 illustrates that

$$\mu_{\bar{x}} = \sum \bar{x}p(\bar{x}) = 10.5 = \mu$$

That is, the mean of the sample means equals the population mean for $n = 2$. A sample statistic whose expected value equals the parameter it is estimating is said to be an **unbiased estimator** of the parameter. ($E(\bar{x}) = \mu_{\bar{x}} = \mu$.) $\bar{x}$ is said to be an unbiased estimator of μ. The variance of the sample mean is

$$\sigma_{\bar{x}}^2 = \sum \bar{x}^2 P(\bar{x}) - \mu_{\bar{x}}^2 = 115.875 - 10.5^2 = 5.625$$

The computations are shown in Figure 5-2, for $n = 2$, $\mu_{\bar{x}} = \mu$, and $\sigma_{\bar{x}}^2 = \sigma^2 / 2$.

Now, suppose we list all samples of size 3 and see what we may discover. Figure 5-4 lists all possible samples of size 3. The 64 samples of size 3 are all listed as well as

A	B	C	D	E	F	G	H	I
First	Second	Third	Mean		First	Second	Third	Mean
6	6	6	6		12	6	6	8
6	6	9	7		12	6	9	9
6	6	12	8		12	6	12	10
6	6	15	9		12	6	15	11
6	9	6	7		12	9	6	9
6	9	9	8		12	9	9	10
6	9	12	9		12	9	12	11
6	9	15	10		12	9	15	12
6	12	6	8		12	12	6	10
6	12	9	9		12	12	9	11
6	12	12	10		12	12	12	12
6	12	15	11		12	12	15	13
6	15	6	9		12	15	6	11
6	15	9	10		12	15	9	12
6	15	12	11		12	15	12	13
6	15	15	12		12	15	15	14
9	6	6	7		15	6	6	9
9	6	9	8		15	6	9	10
9	6	12	9		15	6	12	11
9	6	15	10		15	6	15	12
9	9	6	8		15	9	6	10
9	9	9	9		15	9	9	11
9	9	12	10		15	9	12	12
9	9	15	11		15	9	15	13
9	12	6	9		15	12	6	11
9	12	9	10		15	12	9	12
9	12	12	11		15	12	12	13
9	12	15	12		15	12	15	14
9	15	6	10		15	15	6	12
9	15	9	11		15	15	9	13
9	15	12	12		15	15	12	14
9	15	15	13		15	15	15	15

Figure 5-4 Constructing the distribution of *x*bar for $n = 3$.

their means. From Figure 5-4, the distribution of *x*bar, for samples of size $n = 3$, may be built. The distribution of *x*bar is shown in Figure 5-5. We also see from Figure 5-5 that *x*bar is an unbiased estimator of μ for samples of size 3.

$$\mu_{\bar{x}} = \sum \bar{x}p(\bar{x}) = 10.5 = \mu$$

xbar	*p*(*x*bar)	*x*bar**p*(*x*bar)	*x*bar^2**p*(*x*bar)
6	0.015625	0.09375	0.5625
7	0.046875	0.328125	2.296875
8	0.09375	0.75	6
9	0.15625	1.40625	12.65625
10	0.1875	1.875	18.75
11	0.1875	2.0625	22.6875
12	0.15625	1.875	22.5
13	0.09375	1.21875	15.84375
14	0.046875	0.65625	9.1875
15	0.015625	0.234375	3.515625
	sum =1	sum = 10.5	sum = 114

Figure 5-5 Distribution of *x*bar for $n = 3$.

Referring again to Figure 5-5, we find that

$$\sigma_{\bar{x}}^2 = \sum \bar{x}^2 P(\bar{x}) - \mu_{\bar{x}}^2 = 114 - 10.5^2 = 3.75$$

A plot of the distribution of *x*bar for samples of size $n = 3$ is given in Figure 5-6.

Or the relationship between the population variance and the variance of the sample means is $\sigma_{\bar{x}}^2 = \sigma^2/3$.

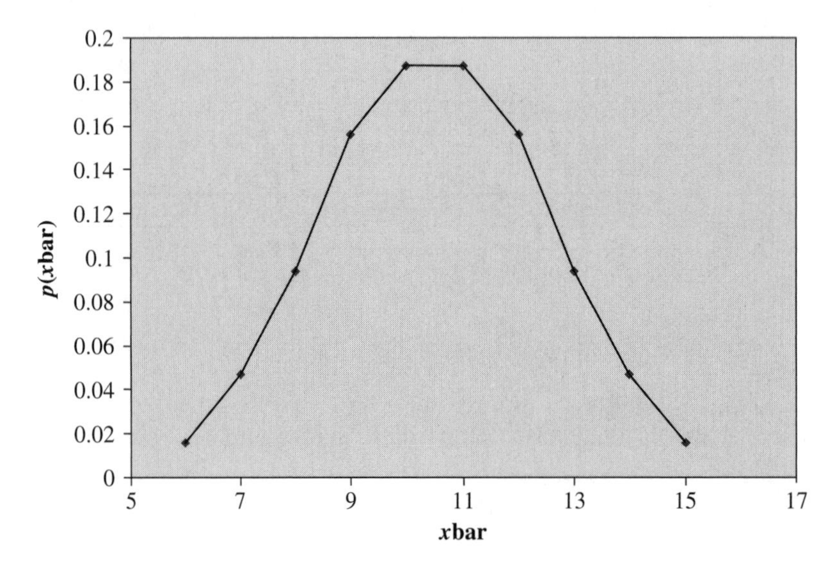

Figure 5-6 Graph of *x*bar distribution for $n = 3$.

C1	C2	C3	C4	C5	C6	C7	C8
x	*p*(*x*)	*x*1bar	*p*(*x*1bar)	*x*2bar	*p*(*x*2bar)	*x*3bar	*p*(*x*3bar)
6	0.25	6.0	0.0625	6	0.015625	6.00	0.003906
9	0.25	7.5	0.1250	7	0.046875	6.75	0.015625
12	0.25	9.0	0.1875	8	0.093750	7.50	0.039063
15	0.25	10.5	0.2500	9	0.156250	8.25	0.078125
		12.0	0.1875	10	0.187500	9.00	0.121094
		13.5	0.1250	11	0.187500	9.75	0.156250
		15.0	0.0625	12	0.156250	10.50	0.171875
				13	0.093750	11.25	0.156250
				14	0.046875	12.00	0.121094
				15	0.015625	12.75	0.078125
						13.50	0.039063
						14.25	0.015625
						15.00	0.003906

Figure 5-7 Original population distribution and sample mean distribution for $n = 2$, 3, and 4.

Figure 5-7 summarizes the discussion so far. The original population distribution for the credit hours taken by the students at Midwestern University is given in the first two columns and is labeled x and $p(x)$. The distribution for the sample mean based on samples of size $n = 2$ is labeled *x*1bar and *p*(*x*1bar). The distribution for the sample mean based on samples of size $n = 3$ is labeled *x*2bar and *p*(*x*2bar). The distribution for the sample mean based on samples of size $n = 4$ is labeled *x*3bar and *p*(*x*3bar). To make certain the student comprehends the discussion, it is suggested that he or she uses EXCEL and build the distribution given in C7 and C8.

The sampling distributions given in Figure 5-7 are graphed in Figure 5-8. A MINITAB plot of C2 versus C1, C4 versus C3, C6 versus C5, and C8 versus C7 are shown in separate panels of the same graph. Figure 5-8 shows the original population, and the distribution of *x*bar for $n = 2$, $n = 3$, and $n = 4$.

When this process is continued, the listing of all possible samples becomes very difficult. For $n = 2$, the number of samples that are possible is $4^2 = 16$, and for $n = 3$, the number of possible samples is $4^3 = 64$. For samples of size 10, the worksheet to list all samples would require $4^{10} = 1{,}048{,}576$ rows. But the same pattern would occur for samples of size 10. However, if we did continue we would find that the mean of the sample means is the mean of the population from which the samples are selected, that is, $\mu_{\bar{x}} = \mu$. We would find that

$$\sigma_{\bar{x}}^2 = \frac{\sigma^2}{10} = \frac{11.25}{10} = 1.125$$

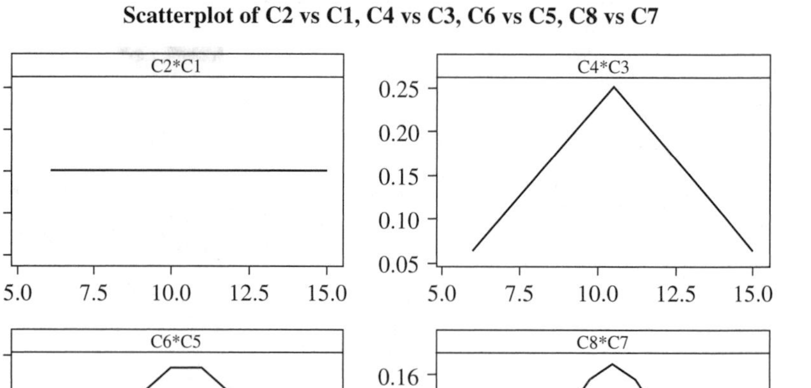

Figure 5-8 Original population distribution and sample mean distribution for $n = 2, 3,$ and 4.

We would also find that the graph of the probability distribution would move closer to a normal curve for $n = 10$ than for $n = 2$ or $n = 3$, and so on.

These results are referred to as the **central limit theorem**. The central limit theorem is one of the fundamental theorems of statistics. *It is extremely important that the student understand this theorem.*

Central limit theorem: Let $X_1, X_2, \ldots, X_n$ be a simple random sample from a population with mean μ and variance σ^2. Let $\bar{X} = \frac{\sum X_i}{n}$ be the sample mean. Then if n is sufficiently large, the sample mean has an approximately normal distribution with mean μ and standard deviation $\sigma_{\bar{x}} = \frac{\sigma}{\sqrt{n}}$. The standard deviation $\sigma_{\bar{x}} = \frac{\sigma}{\sqrt{n}}$ is called the **standard error of the mean** to keep it distinct from the population standard deviation, σ. The standard error of the mean measures the variability of the sample mean. The standard error of the mean depends on the variability of the original population and the sample size. How quickly the distribution of the sample mean approaches normality depends on the original population. In our opening example in this section, the original population was uniform and the normality appeared for fairly small n. *For most populations, if the sample size is 30 or greater, the central limit approximation is good.* Let us consider some additional examples of the implications of the central limit theorem.

EXAMPLE

The number of flaws on DVDs is a random variable.

The number of flaws on DVDs produced by a company has the following distribution: (Note that this distribution is not uniform.)

x	0	1	2	3
$p(x)$	0.60	0.25	0.10	0.05

Thirty-six of the DVDs are sampled from this population. What is the probability that the average number of flaws per DVD in this sample is less than 0.5? We are interested in $P(\bar{X} < 0.5)$. We may treat $\bar{X}$ as a normal random variable having mean μ and standard error = $\sigma/6$, where μ and σ are the population mean and standard deviation.

$$\mu = 0(0.60) + 1(0.25) + 2(0.10) + 3(0.05) = 0.6$$

and

$$\sigma^2 = 0 + 0.25 + 0.40 + 0.45 - 0.36 = 0.74$$

The standard error of the mean is

$$\sigma_{\bar{x}} = \frac{\sigma}{\sqrt{n}} = \frac{\sqrt{0.74}}{\sqrt{36}} = 0.143$$

The grey area under the normal curve shown in Figure 5-9 is given by the EXCEL command =NORMDIST(0.5,0.6,0.143,1) which is 0.242182. The probability is 0.242 that the sample mean will be 0.5 or less. Recall that the parameters in the NORMDIST function are: the first is the value of x, the second is the mean, the third is the standard error of the distribution, and the one in the fourth position tells EXCEL to accumulate the area from 0.5 to the left of 0.5.

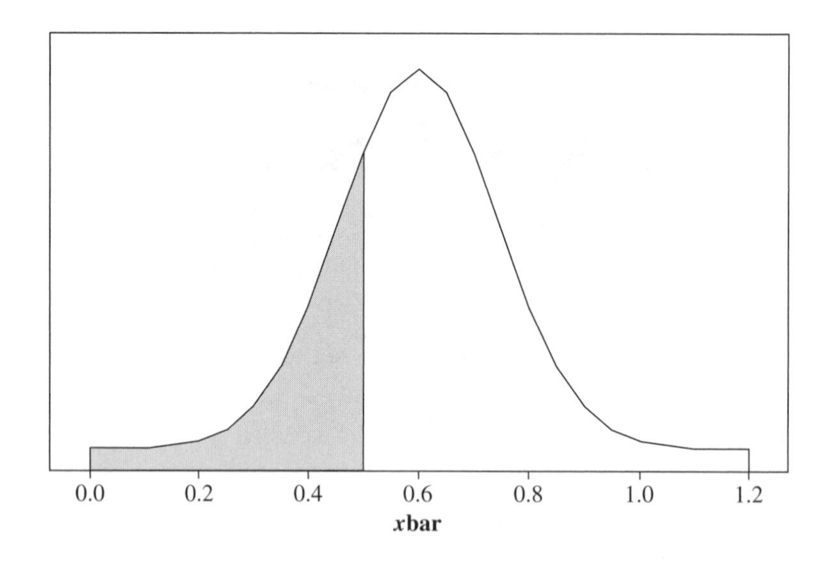

Figure 5-9 $P(x \text{ bar} < 0.5)$ is represented by the grey area under the curve.

EXAMPLE

Wood stoves emit a fair amount of fine particle pollution.

An engineer claimed that wood stoves emit an average of 40 million micrograms of fine particle pollution per hour. The standard deviation is 15 million micrograms per hour. To test the engineer's claim, thirty wood stoves were tested for fine particle pollution and it was found that the sample average for the thirty stoves was 50 million micrograms of fine particle pollution per hour. If the engineer's statement is correct, what is the probability of obtaining a sample of 30 with a mean of 50 or larger?

*X*bar has a normal distribution with mean equal to 40 and standard error equal to

$$\sigma_{\bar{x}} = \frac{\sigma}{\sqrt{n}} = \frac{15}{\sqrt{30}} = 2.74$$

assuming the engineer's statement is correct. The probability the sample mean exceeds 50, assuming the engineer is correct, is given by the EXCEL command =1-NORMDIST(50,40,2.74,1). This represents the area in the right hand tail past 50 of a normal curve centered at 40 and having a standard error equal to 2.74. This gives *P*(*x*bar > 50) = 0.000131. Since this probability involving *x*bar is extremely small, *x*bar > 50 would cast doubt on the engineer's claim.

5.3 The Sampling Distribution of the Mean (σ Unknown)

In the previous section, it was assumed that σ was known. When σ is unknown, *S* is calculated from the sample and is used in place of σ. The proof of the following theorem is beyond the scope of this book:

Theorem: When the sample is taken from a normal distribution, having population mean μ, the variable $t_v = (\bar{x} - \mu)/\frac{s}{\sqrt{n}}$ has a *t* distribution with ν **degrees of freedom** and $\nu = n - 1$. The Greek letter ν is used to represent the degrees of freedom. *S* is the standard deviation of the sample. If the sample does not come from a normal distribution, it is not known what distribution the variable t_v has. One of the four normality tests may be used to determine if it may be assumed that the sample was taken from a normal distribution. Note that different sample sizes determine different *t* distributions.

SOME PROPERTIES OF THE *t* DISTRIBUTION

1. The *t* distribution is a mound-shaped symmetrical distribution much like the normal distribution. However, its shape is dependent on the sample size. As the sample size is increased, it approaches the standard normal distribution. A **standard normal distribution** is a normal distribution with $\mu = 0$ and $\sigma = 1$. When the degrees of freedom is 30 or more, there is very little difference between a *t* distribution and a standard normal distribution. Figure 5-10 shows a standard normal curve as a solid line and a *t* distribution with $\nu = 5$ degrees of freedom as a dashed line. The *t* distribution is thicker in the tails than the normal curve. Both curves have a mean of 0. The standard deviation of the *t* distribution is greater than 1.
2. The total area under any *t* distribution is 1, with half the area to the right of 0 and half to the left of 0.
3. The *t* distribution was developed by William Gosset in 1908. Gosset worked for the Guinness Brewing Company in Dublin, Ireland. He was a chemist who was good at mathematics. The company forbade their employees to publish results of research, so Gosset published the results concerning the *t* distribution using the pen name **Student**. Hence the distribution is often referred to as the student *t* distribution.
4. Most statistics texts contain tables of the student *t* distribution with various degrees of freedom. However, most statistical software packages have the distribution as part of the software and the tables in books are not really necessary. Consider the following examples:

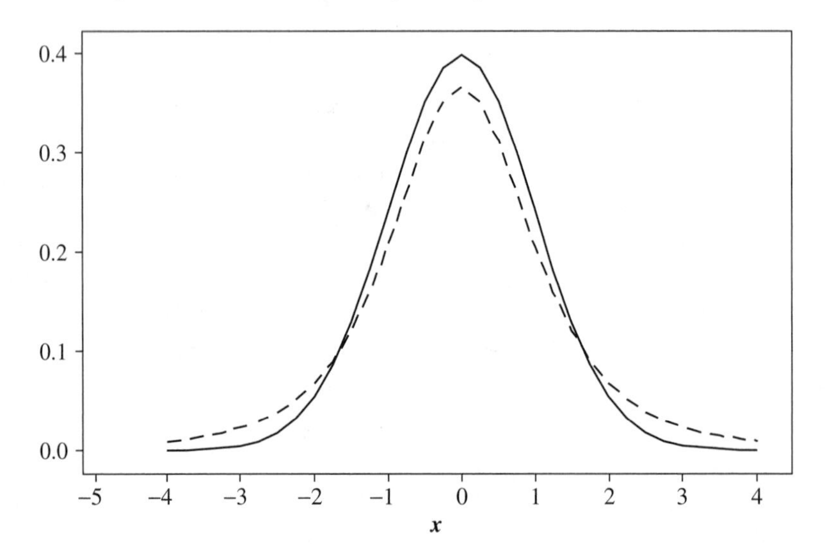

Figure 5-10 Standard normal as a solid curve and *t* with $\nu = 5$ as a dashed curve.

EXAMPLE

A sample of size 6 is selected from a normal distribution. Find the following probabilities using EXCEL:

a. $P(t_5 > 2)$
b. $P(t_5 < -2.5)$
c. $P(-1.5 < t_5 < 1.5)$

a. The paste function =TDIST of EXCEL gives the dialog box shown in Figure 5-11. When the dialog box is filled in as shown, it gives the area to the right of 2 under the student t distribution with 5 degrees of freedom. Figure 5-12 shows the area described above. The actual area is 0.05097.
b. The paste function =TDIST(2.5,5,1) gives the area to the right of 2.5, which equals 0.027245. But because of symmetry, this is the same as the area to the left of −2.5. Therefore, $P(t_5 < -2.5) = 0.027245$.
c. The paste function =TDIST(1.5,5,2) gives the area beyond −1.5 and 1.5, which equals 0.193904. This must be subtracted from 1 to give the correct answer.

$$P(-1.5 < t_5 < 1.5) = 1 - 0.193904, \text{ or } 0.806096$$

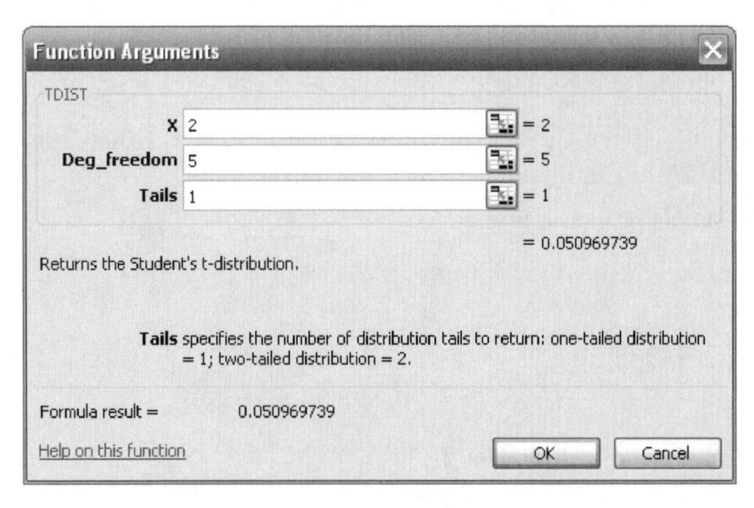

Figure 5-11 Dialog box for finding $P(t_5 > 2)$.

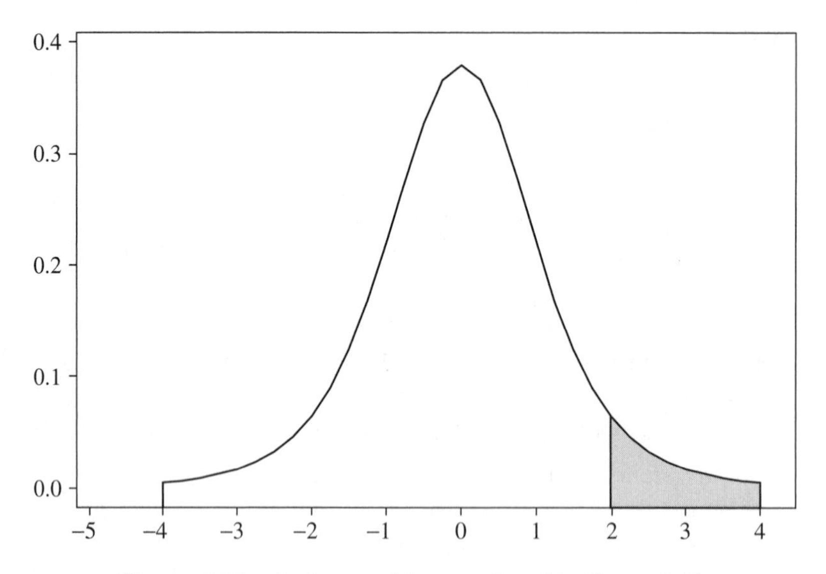

Figure 5-12 A picture of the area found in Figure 5-11.

EXAMPLE

Find the following probabilities using MINITAB:

a. $P(t_5 > 2)$
b. $P(t_5 < -2.5)$
c. $P(-1.5 < t_5 < 1.5)$

a. The pull-down **Calc ⇒ Probability Distributions ⇒ *t* Distribution** gives the dialog box in Figure 5-13. It is filled in as shown.

The output produced is as follows:

```
Student's t distribution with 5 DF

x    P(X <= x)
2       0.949030
```

$$P(t_5 > 2) = 1 - 0.949030 = 0.05097$$

b. The solution here is given directly.

```
Student's t distribution with 5 DF

   x    P(X <= x)
-2.5     0.0272450
```

$$P(t_5 < -2.5) = 0.027245$$

c. The solution here is obtained by subtraction.

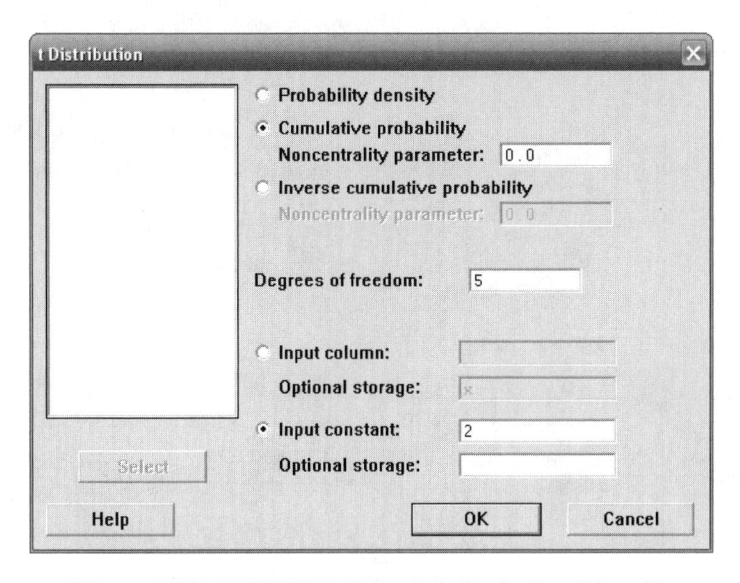

Figure 5-13 MINITAB dialog box for finding $P(t_5 < 2)$.

Cumulative Distribution Function

```
Student's t distribution with 5 DF

  x    P(X <= x)
1.5      0.903048
```

Cumulative Distribution Function

```
Student's t distribution with 5 DF

   x    P(X <= x)
-1.5     0.0969518
```

Subtracting, we have

$$P(-1.5 < t_5 < 1.5) = 0.903048 - 0.0969518 = 0.806096$$

EXAMPLE

Find the following probabilities using STATISTIX:

a. $P(t_5 > 2)$

b. $P(t_5 < -2.5)$

c. $P(-1.5 < t_5 < 1.5)$

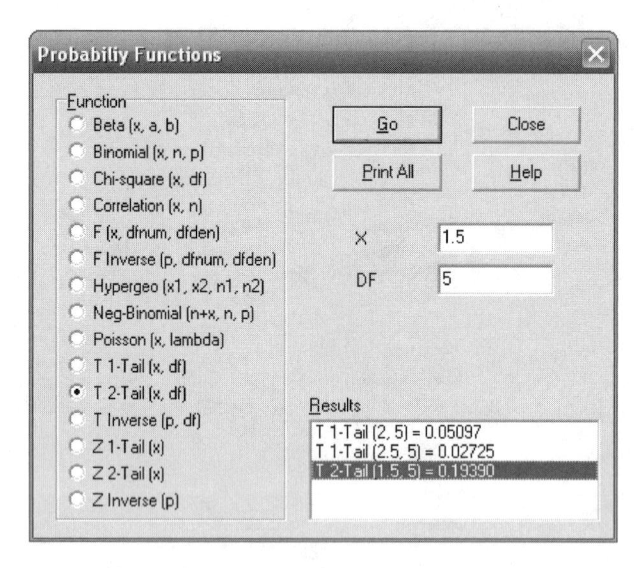

Figure 5-14 STATISTIX dialog box for finding Student's *t* probabilities.

The pull-down **Statistics ⇒ Probability Functions** produces the dialog box in Figure 5-14. The results of three different commands are shown in the lower portion of the dialog box.

Figure 5-14 shows that

a. $P(t_5 > 2) = 0.05079$.

b. $P(t_5 < -2.5) = 0.02725$ (using the symmetry of the distribution).

c. $P(-1.5 < t_5 < 1.5) = 1 - 0.19390 = 0.8061$.

PROPERTIES OF THE *t* DISTRIBUTION (CONTINUED)

5. The probability density function for a student *t* distribution, with ν degrees of freedom, is

$$f(x) = \frac{\Gamma\left(\frac{\nu+1}{2}\right)}{\sqrt{\pi\nu}\,\Gamma\left(\frac{\nu}{2}\right)}\left(1+\frac{t^2}{\nu}\right)^{-\frac{\nu+1}{2}}, -\infty < x < \infty$$

The mean is 0 and for $\nu > 2$, the variance of the student t distribution is

$$\sigma^2 = \frac{\nu}{\nu - 2}$$

EXAMPLE

High intensity lights on airplanes are critical and have lifetimes that are random variables.

An engineer wishes to test the claim that a new high intensity light on aircraft has a mean lifetime of 1250 hours. She tests 10 of the lights and finds the following lifetimes in hours:

962 1127 1089 1132 1282 955 1071 1319 1121 1141

First, she uses STATISTIX to perform the Shapiro-Wilk test of normality to see if the normality assumption is reasonable. She obtains the following results:

```
Shapiro-Wilk Normality Test

Variable        N         W          P
Lifetime       10    0.9232     0.3848
```

Because of the large p-value, she feels the normality assumption is reasonable.

Performing a descriptive statistics using STATISTIX gives the following results:

```
Descriptive Statistics

Variable          Mean          SD     SE Mean
Lifetime        1109.9      104.67      33.101
```

Assuming the claim is true, (i.e., $\mu = 1250$ hours) she calculates

$$t_v = \frac{\bar{x} - \mu}{\frac{s}{\sqrt{n}}} = \frac{1109.9 - 1250}{33.101} = -4.23$$

If the mean lifetime claim is true (that is $\mu = 1250$ hours), she has obtained a highly unlikely value for t_9. In fact the chance of obtaining such a low value for t_9 or one that is smaller is only 0.0011. This calculation is shown in the dialog box of Figure 5-15.

If the claim that $\mu = 1250$ hours is true, then there are only 11 chances out of 10,000 of obtaining such a low sample mean. She would likely reject the claim as being too high.

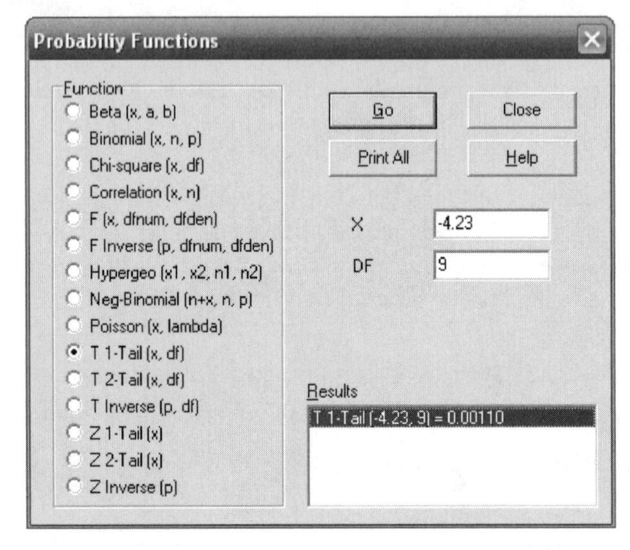

Figure 5-15 Calculation of $P(t_9 < -4.23)$.

5.4 The Sampling Distribution of the Proportion

Consider the population shown in Figure 5-16. It contains 15 defectives (D) and 35 nondefectives (N). The items are contained in 50 numbered locations. For example, location 1 contains a defective item and location 31 contains a nondefective item, and so forth. Samples are selected at random with replacement.

The population has 15 out of 50 or proportion $p = 0.30$ that are defective. Ten samples of size 5 are taken from the population using a random number generator. The 5 random numbers are shown in the columns of the following table:

9	14	16	14	31	2	49	4	48	47
32	1	26	43	14	1	28	5	21	50
13	28	22	46	41	10	25	16	33	25
18	1	13	39	14	42	3	38	5	7
45	41	11	33	27	39	15	19	42	46

These random numbers are used to select the defectives and the nondefectives from the population in Figure 5-16 as follows:

N	D	N	D	N	N	D	N	D	D
N	D	N	N	D	D	D	N	N	N
N	D	N	N	D	D	N	N	N	N
N	D	N	N	D	D	N	N	N	N
D	D	N	N	N	N	N	D	D	N

The 10 sample proportion defectives are: 0.20, 1, 0, 0.2, 0.6, 0.6, 0.4, 0.2, 0.4, and 0.2. The sample proportion defective is represented by $\hat{p}$ and the population proportion by p. Generally, p is constant but $\hat{p}$ is variable. The average value of $\hat{p}$

1D	11N	21N	31N	41D
2N	12D	22N	32N	42D
3N	13N	23D	33N	43N
4N	14D	24N	34N	44N
5N	15N	25N	35N	45D
6N	16N	26N	36D	46N
7N	17N	27N	37N	47D
8N	18N	28D	38N	48D
9N	19D	29N	39N	49D
10D	20D	30N	40N	50N

Figure 5-16 Population with each cell numbered and status of item.

for the 10 samples is 0.38 and the value of p is 0.30. Now, suppose this situation is magnified and the population is much larger and all possible samples are taken. What is the connection between $\hat{p}$ and p? In other words, what is the sampling distribution of $\hat{p}$?

If n is large enough to ensure that $\hat{p} \pm 3\sigma_{\hat{p}}$ does not include 0 or 1, then the distribution of $\hat{p}$ may be assumed to be normal. $\hat{P}$ is an unbiased estimator of p. The standard error of $\hat{p}$ is $\sigma_{\hat{p}} = \sqrt{\frac{pq}{n}}$ where $q = 1 - p$. It is customary to replace p and q by $\hat{p}$ and $\hat{q}$ when calculating $\sigma_{\hat{p}} = \sqrt{\frac{pq}{n}}$. Thus, when n is large enough, $\hat{p}$ has a sampling distribution that is normal.

EXAMPLE

A sample of 100 DVDs is taken and it is found that 10 have problems and will not go forward until some action is taken. It is of interest to know something about p, the population proportion that has this problem. The sample proportion $\hat{p}$ is 10/100 = 0.1 for this sample. Calculate the probability that $\hat{p}$ is as small as 0.10 or smaller if the population proportion is 0.15.

First let's ask if $\hat{p}$ may be treated as if it is normally distributed. Suppose we calculate $\hat{p} \pm 3\sigma_{\hat{p}}$ and see if it does not include 0 or 1. To calculate $\sigma_{\hat{p}} = \sqrt{\frac{pq}{n}}$, assume $p = \hat{p} = 0.1$ and $q = 0.9$, since these are our only estimates of p and q. The standard deviation of $\hat{p}$ is $\sigma_{\hat{p}} = \sqrt{\frac{.1(.9)}{100}} = 0.03$. $\hat{p} \pm 3\sigma_{\hat{p}}$ is $0.1 \pm 3(0.03)$, or (0.01, 0.19). This interval does not include 0 or 1 and the distribution of $\hat{p}$ may be assumed to be normal.

Assume $p = 0.15$ and calculate the chances that $\hat{p} \leq 0.1$. That is find $P(\hat{p} \leq .1)$ assuming $p = 0.15$. $\hat{p}$ may be treated as normally distributed with mean 0.15 and standard deviation

$$\sigma_{\hat{p}} = \sqrt{\frac{pq}{n}} = \sqrt{\frac{0.15(0.85)}{100}} = 0.0357$$

Using EXCEL, $P(\hat{p} \leq .1)$ is given by =NORMDIST(0.1,0.15,0.0357,1), which equals 0.0807.

Summarizing, if the sample size is large enough, then $\hat{p}$ may be assumed to have a normal distribution. If the population has proportion p_o that has the characteristic of interest, then under repeated sampling, $\hat{p}$ has mean equal to p_o and standard deviation

$$\sigma_{\hat{p}} = \sqrt{\frac{p_o q_o}{n}}$$

EXAMPLE

Ten percent of the cell phones manufactured by Cell phones Inc. have minor flaws. If you selected a sample of 50 of their phones, you would expect to find 5 of them with minor flaws. To check if $\hat{p}$ has an approximate normal distribution, check to see if $p_o \pm 3\sqrt{\frac{p_o q_o}{n}}$ does not contain 0 or 1 where $p_o = 0.1$. The interval is $0.1 \pm 3\sqrt{\frac{.1(.9)}{50}}$, or 0.1 ± 0.13, which is the interval (–0.03, 0.23). Since this interval contains 0, either increase the sample size and then assume normality or use small sample techniques, which utilize the binomial distribution.

5.5 The Sampling Distribution of the Variance

Consider a tire manufacturer that produces a tire that has lifetimes that are normally distributed with a mean lifetime equal to 60,000 miles and a standard deviation equal to 5000 miles. Samples of size 4 are taken each day for 30 days. The MINITAB simulated samples and their variances are shown in Figure 5-17. The first sample is 63, 68, 59, and 65. The sample variance for that sample is 14.3. This is repeated 30 times.

It can be proved that $(n - 1)S^2/\sigma^2$ has a chi-square sampling distribution with (4 – 1) degrees of freedom. The sample size n is 4, the population variance is $5^2 = 25$, and the values of S^2 are given in Figure 5-17. The variable $(n - 1)S^2/\sigma^2$ is equal to $3S^2/25$. If each value of S^2 is multiplied by 3 and divided by 25 and then plotted, a histogram resembling a chi-square distribution with 3 degrees of freedom will result. Figure 5-18 shows a plot of the 30 values of $3S^2/25$.

1	2	3	4	5	6	7	8	9	10	11	12	13	14	15
63	67	62	53	64	65	69	66	62	62	61	62	52	59	62
68	63	60	62	53	58	61	55	66	65	57	45	59	60	69
59	65	55	64	56	63	57	59	58	55	65	58	65	59	57
65	56	59	58	63	53	60	64	54	48	65	55	56	68	57
14.3	22.9	8.67	23.6	28.7	28.9	26.3	24.7	26.7	57.7	14.7	52.7	30	19	32.3
16	17	18	19	20	21	22	23	24	25	26	27	28	29	30
59	59	67	60	59	48	60	64	63	57	53	62	57	54	60
56	60	57	58	56	55	61	66	48	57	59	58	60	62	64
54	63	64	60	59	64	54	59	55	58	59	64	65	57	52
64	63	60	55	64	58	63	59	49	66	61	69	62	62	63
18.9	4.25	19.3	5.58	11	44.3	15	12.7	47.6	19	12	20.9	11.3	16	29.6

Figure 5-17 Thirty samples of size 4 and variances of tire mileages in thousands.

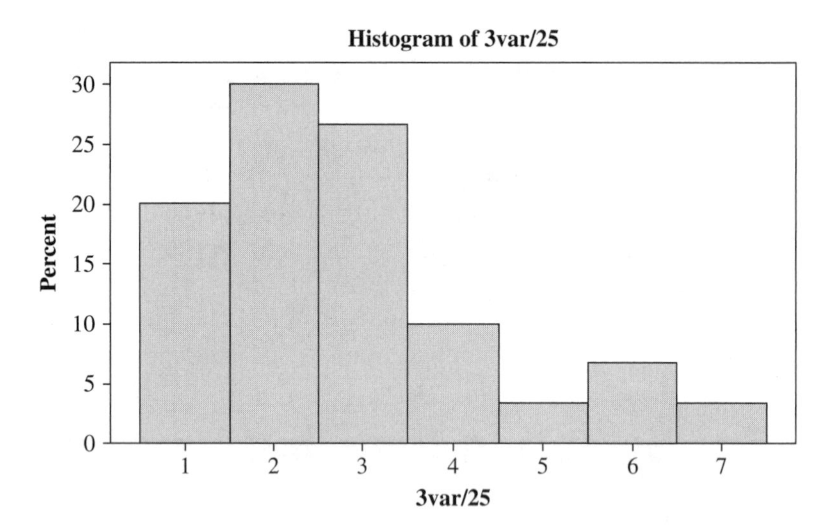

Figure 5-18 Histogram of the 30 values of $3S^2/25$.

If the number of samples is increased from 30 to a very large number, the shape of the histogram in Figure 5-18 approaches the shape of the probability density in Figure 5-19.

The above discussion illustrates the sampling distribution of S^2. In summary, this is what we know. Suppose a sample of size n is taken from a normal distribution having variance σ^2. The variable $(n - 1)S^2/\sigma^2$ has a chi-square distribution with $(n - 1)$ degrees of freedom. The normality assumption is necessary. That is, the

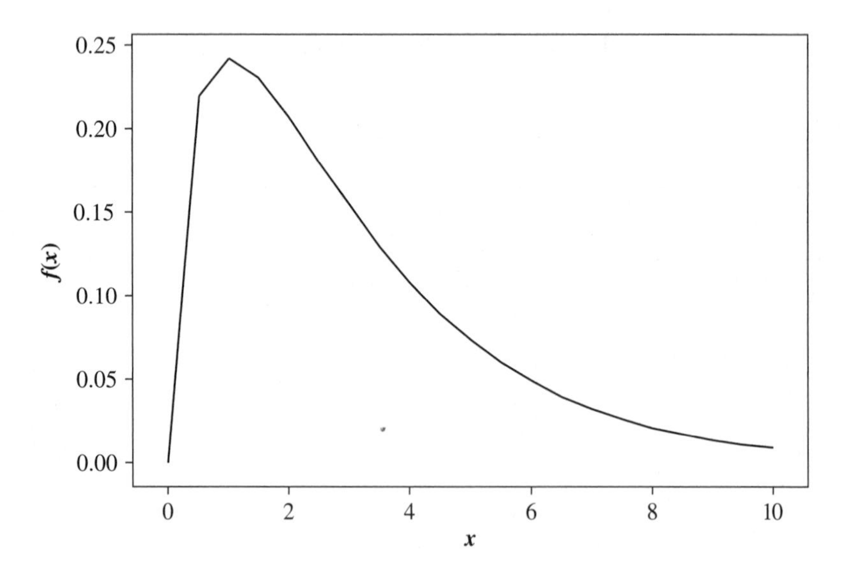

Figure 5-19 The chi-square distribution with 3 degrees of freedom.

sample has to come from a normal distribution for the sampling distribution to be chi-square. We have previously discussed how to check the normality assumption.

EXAMPLE

The standard deviation in years of experience at a large engineering consulting firm is stated to be 8 years. A sample of size $n = 10$ is taken and the following data (in years) is obtained: 20, 15, 27, 16, 10, 13, 17, 20, 15, and 20. If the 8-year claim is correct, what is the probability of getting a set of data as contradictory to the 8-year claim as this one or more or so?

First of all check out normality using the Shapiro-Wilk test of STATISTIX.

```
Shapiro-Wilk Normality Test

Variable        N        W         P
Years          10    0.9492    0.6591
```

The normality of the data is accepted because the p-value is greater than 0.05. The variance using STATISTIX is found to be

```
Statistix 8.0
Descriptive Statistics
Variable      Variance
Years           22.233
```

The quantity $(n - 1)S^2/\sigma^2$ is equal to 3.13. The probability of this value or one smaller from Figure 5-20 is $1 - 0.95892 = 0.04108$. There is only a 0.041 chance of

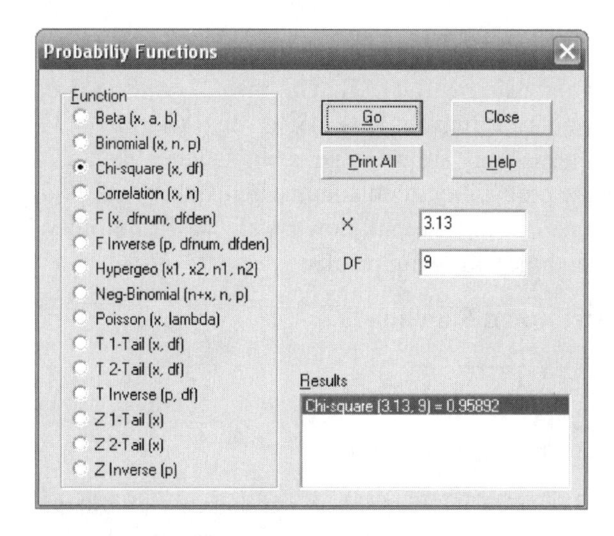

Figure 5-20 The chi-square dialog box.

getting this value or one that is smaller if the standard deviation of all years of experience is 8 years.

NOTE *Even though we take only one sample, if it is random, we know what values are likely for* $(n-1)S^2/\sigma^2$*, and what values are unlikely.*

EXAMPLE

An industrial engineer is concerned about the variability of fill in 340-gram containers of coffee. The process is deemed to be in control if σ is less than or equal 1 gram. Corrective action (the process is stopped and the variability reduced) is taken only if the standard deviation is judged to be greater than 1 gram. A sample of 20 of the 340-gram containers is selected at random form production and the following weights are found: (Use MINITAB to assist in your decision making.)

338 339 337 342 342 343 342 339 338 341 340 342 340 344 341
342 337 340 341 343

First, a Kolmogorov-Smirnov test is conducted for normality. When the Kolmogorov-Smirnov is performed, the result is a p-value > 0.15. This result tells us that we may assume normality. That is, we may assume that our sample came from a normally distributed population.

Descriptive Statistics: Weight

```
Variable     Mean    Variance
Weight     340.56        3.91
```

A descriptive statistics analysis where only the mean and the variance are requested is performed using MINITAB. The mean is a little over 340 grams, $S^2 = 3.91$, and the chi-square variable is $(n-1)S^2/\sigma^2$ or $19(3.91)/1 = 74.3$. Now the question is: "If $\sigma = 1$, how likely are we to get a chi-square equal to 74.3 or larger?" We wish to compute the probability of obtaining a chi-square with 19 degrees of freedom that is 74.3 or larger? Using the pull-down **Calc ⇒ Probability Distributions ⇒ Chi-Square** we find the following results:

Cumulative Distribution Function

```
Chi-Square with 19 DF

   x    P(X <= x)
74.3      1.00000
```

And $P(X \geq 74.3) = 0.00000$. If $\sigma \leq 1$, the probability of obtaining a chi-square as large as 74.3 is practically 0. Therefore, we conclude $\sigma > 1$. The process would be stopped and corrective action would be taken.

Exercises

1. The number of TVs per household has the distribution shown in the following table:

x	1	2	3	4	5
$p(x)$	0.2	0.2	0.2	0.2	0.2

How many TVs does your household have?

(a) For this population, find the mean number of TVs per household and find the standard deviation of the number of TVs per household.

(b) What shape describes this population distribution?

(c) Give the distribution of the sample mean for $n = 2$, and find $\mu_{\bar{x}}$ and $\sigma_{\bar{x}}$.

(d) Give the distribution of the sample mean for $n = 3$, and find $\mu_{\bar{x}}$ and $\sigma_{\bar{x}}$.

(e) Give the distribution of the sample mean for $n = 4$, and find $\mu_{\bar{x}}$ and $\sigma_{\bar{x}}$.

2. Plot the distributions in problem 1 parts: c, d, and e. Also plot the distributions overlaid on the same graph. What do these plots show?

   Exercises 1 and 2 were concerned with helping establish the theory in the central limit theorem. Exercise 3 will be interested in applying the theory of the central limit theorem. In practice, we do not take every possible sample. We only take one sample. The central limit theorem tells us what the most probable outcomes are and what the least probable outcomes are with respect to that sample. We make statistical decisions based on high or low probabilities. In the real world we cannot know the complete population. We can only infer about the population based on what we see in the sample.

3. A city claims that the mean fluoride is 2.5 milligrams per liter in its drinking water. The standard deviation is known to be 1.0 milligrams per liter. In order to check the claim, the EPA collects a sample of size 35 and finds the mean amount of fluoride to be 3.1 milligrams per liter. The EPA does not expect the sample mean to be the same as the claimed population mean. The claim that $\mu = 2.5$ is in doubt, since the sample mean was 3.1 based on a sampling of 35 locations randomly chosen around the city. The EPA assumes that $\mu = 2.5$ and will calculate the probability of obtaining a sample whose mean is 3.1 or larger based on 35 observations. The probability $P(\bar{x} > 3.1$ assuming that $\mu = 2.5)$ will be calculated and if that probability is less than 0.05, the EPA will reject the city engineers' claim that $\mu = 2.5$. Calculate this probability and state your conclusion.

4. Suppose in problem 3 that a small sample is used instead of a large sample. A sample of size $n = 10$ is obtained from water supplies around the city. The sample values are as follows:

   3.7 2.0 1.8 2.6 3.8 1.6 2.8 4.5 3.3 1.3

   Test to see if 2.5 is a reasonable value for the population mean. If there is less than a 5% chance (that a sample with a mean equal to the one obtained from this sample), then reject the claim that $\mu = 2.5$. Calculate the value of t_9 and base your decision on this calculated value. Check out the one assumption that is critical when working with the t distribution.

5. In addition to the rule given in section 5.4 to test if $\hat{p}$ may be assumed to have an approximate normal distribution, there is another rule that is often used. The first rule given in Section 5.4 states that if the interval $\hat{p} \pm 3\sqrt{\frac{\hat{p}\hat{q}}{n}}$ does not contain 0 or 1, then $\hat{p}$ may be approximated by a normal distribution. The second rule states that if $n\hat{p} > 5$ and $n\hat{q} > 5$, where $\hat{q} = 1 - \hat{p}$, then the distribution of $\hat{p}$ may be approximated by a normal distribution.

Using both the rules, determine if the normal approximation is appropriate in the following cases:

(a) $n = 150,\ \hat{p} = 0.10$

(b) $n = 500,\ \hat{p} = 0.05$

(c) $n = 50,\ \hat{p} = 0.15$

(d) $n = 1000,\ \hat{p} = 0.001$

6. A quality engineer is interested in estimating the proportion of personal digital assistants (PDAs) that have minor surface flaws. She selects 100 at random from the production process and finds that 15 have minor surface flaws. The estimate of the proportion from the process with a minor flaw is 0.15. Since $n\hat{p} > 5$ and $n\hat{q} > 5$, the distribution may be approximated by the normal distribution. Because of the property of the normal distribution that 95% of the distribution is within 2 standard deviations of the mean of the distribution, there is a probability close to 0.95 that $\hat{p}$ will be within 2 standard errors of the true unknown value of p.

Note *To attain 0.95 probability, 1.96 standard errors would be needed. We will address this in the next chapter.*

The standard error of $\hat{p}$ is $\sigma_{\hat{p}} = \sqrt{\frac{pq}{n}}$. This standard error is estimated by

$$\sigma_{\hat{p}} = \sqrt{\frac{\hat{p}\hat{q}}{n}} = \sqrt{\frac{0.15(0.85)}{100}} = 0.036$$

and 2 standard errors is 0.072. At this point, we estimate that 15% of the PDAs have minor flaws. We are approximately 95% sure that our **error of estimate** is off by not more than 7.2%. This means the proportion defective may be as low as 7.8% or as high as 22.2%. Suppose, the management directs the engineer to reduce the 95% error of estimate from 7.2% to 5%. What sample size would need to be taken to have a 5% error of estimate?

7. A population consists of 6 items: A, B, C, and D are defective and E and F are nondefective. The parameter p is the proportion defective in the population and equals $4/6 = 0.67$. Sample size is $n = 4$ and the statistic PHAT is the sample proportion defective. Give the sampling distribution of PHAT and find the mean of PHAT.

8. The number of TVs per household has the distribution shown in the following table:

x	1	2	3	4	5
$p(x)$	0.2	0.2	0.2	0.2	0.2

   (a) For this population, find the variance and the standard deviation of the number of TVs per household.

   (b) Give the sampling distribution of the sample variance and the sampling distribution of the sample standard deviation for $n = 2$, and find the expected value of both.

   (c) Give the sampling distribution of the sample variance and the sampling distribution of the sample standard deviation for $n = 3$, and find the expected value of both.

   (d) Give the sampling distribution of the sample variance and the sampling distribution of the sample standard deviation for $n = 4$, and find the expected value of both.

9. Plot the sampling distributions of S^2 for $n = 2$, 3, and 4 overlaid on the same graph. Plot the sampling distributions of S for $n = 2$, 3, and 4 overlaid on the same graph.

10. Too much variability in shaft lengths results in some being too short and some being too long. It is found that if σ is not more than 3.5 millimeters, the shaft lengths are acceptable. A sample of shafts is taken and their lengths are measured. The results in millimeters are as follows:

    103.704 98.311 98.895 97.373 98.506 96.565 97.506 104.004
    96.411 96.923

    Compute the probability that if $\sigma \leq 3.5$, that there would be as much variability as there is in the lengths.

Summary

1. A **population** is a set of data that is of interest to the engineer or the scientist.

2. The **parameters** are population measurements such as the mean, the standard deviation, the population proportion, and so forth.

3. A set of n observations constitutes a **simple random sample** if it is chosen in such a manner that each subset of n elements has the same probability of being chosen.
4. A **statistic** is a measurement made on the sample.
5. The **sampling distribution** of a sample statistic, calculated from a sample of n measurements, is the probability distribution of the statistic.
6. A sample statistic whose expected value equals the parameter it is estimating is said to be an **unbiased estimator**.
7. **Central limit theorem**: Let $X_1, X_2, \ldots, X_n$ be a simple random sample from a population with mean μ and variance σ^2. Let $\bar{X} = \frac{\sum X_i}{n}$ be the sample mean. Then if n is sufficiently large, the sample mean has an approximately normal distribution with mean μ and standard deviation $\sigma_{\bar{x}} = \frac{\sigma}{\sqrt{n}}$. The standard deviation $\sigma_{\bar{x}} = \frac{\sigma}{\sqrt{n}}$ is called the **standard error of the mean** to keep it distinct from the population standard deviation, σ.
8. When the sample is taken from a normal distribution, having population mean μ, the variable $t_v = (\bar{x} - \mu)/\frac{s}{\sqrt{n}}$ has a t distribution with v **degrees of freedom** and $v = n - 1$.
9. A **standard normal distribution** is a normal distribution with $\mu = 0$ and $\sigma = 1$.
10. If n is large enough to insure that $\hat{p} \pm 3\sigma_{\hat{p}}$ does not include 0 or 1 then the **sampling distribution of** $\hat{P}$ may be assumed to be normal. $\hat{p}$ is an unbiased estimator of p. The standard error of $\hat{p}$ is $\sigma_{\hat{p}} = \sqrt{\frac{pq}{n}}$ where $q = 1 - p$. It is customary to replace p and q by $\hat{p}$ and $\hat{q}$ when calculating $\sigma_{\hat{p}} = \sqrt{\frac{pq}{n}}$. Thus, when n is large enough, $\hat{p}$ has a sampling distribution that is normal.
11. Suppose a sample of size n is taken from a normal distribution having variance σ^2. The variable $(n - 1)S^2/\sigma^2$ has a sampling distribution that is a chi-square distribution with $(n - 1)$ degrees of freedom. The normality assumption is necessary. That is, the sample has to come from a normal distribution for the sampling distribution to be chi-square.

CHAPTER 6

Inferences Concerning Means

6.1 Point Estimation

Some definitions and properties of estimators were discussed in the previous chapter. It is time to formalize these discussions. First, let us develop some properties of expectation and variance. Assume random variable X is continuous. Let $Y = a + bX$ and consider the expectation (or mean) of Y and the variance of Y.

$$E(Y) = E(a + bX) = \int_{-\infty}^{\infty} (a + bx) f(x)\,dx = a\int_{-\infty}^{\infty} f(x)\,dx + b\int_{-\infty}^{\infty} xf(x)\,dx = a + bE(x)$$

$$\begin{aligned} \text{Var}\,(a + bX) &= \int_{-\infty}^{\infty} (a + bx - (a + bE(\text{x})))^2 f(x)\,dx \\ &= \int_{-\infty}^{\infty} (bx - bE(x))^2 f(x)\,dx = b^2 \text{Var}(X) \end{aligned}$$

Note *If X is discrete rather than continuous, sums replace integrals and the proofs are similar. Notation*: μ or $E(X)$ is used to represent the mean of X. σ^2 or $\mathrm{Var}(X)$ is used to represent the variance of X.

A **point estimator** is a single number used to estimate a parameter. $\bar{X}$ is a point estimator of the parameter μ. Before the sample is taken, $\bar{X}$ is regarded as a linear combination of random variables. After the sample is taken and the numbers averaged, $\bar{x}$ is a single number having no distribution. If the relationship $E(a+bX)=a+bE(X)$ is extended, it becomes $E(a_0+a_1X_1+\cdots+a_nX_n)=a_0+a_1E(X_1)+\cdots+a_nE(X_n)$. This is true because the integral of the sum is the sum of the integrals. Applying this to $\bar{X}$, we have

$$E(\bar{X})=E\left(\frac{X_1+\cdots+X_n}{n}\right)=\frac{1}{n}(E(X_1)+\cdots+E(X_n))=\mu$$

since each expected value of X_i is μ and there are n μ's. When the expected value of a statistic is equal to the parameter the statistic is intended to estimate, the statistic is said to be an **unbiased estimator** of the parameter. Therefore $\bar{X}$ is an unbiased estimator of the population mean, μ.

It can be shown that S^2 is an unbiased estimator of σ^2. $\frac{(n-1)S^2}{\sigma^2}$ has a chi-square distribution with $(n-1)$ degrees of freedom as stated in the last chapter. Now, $E\left(\frac{(n-1)S^2}{\sigma^2}\right)=n-1$ since the mean of a chi-square is its degrees of freedom.

$$E\left(\frac{(n-1)S^2}{\sigma^2}\right)=\frac{(n-1)}{\sigma^2}E(S^2)=n-1$$

Solving for $E(S^2)$, we obtain $E(S^2)=\frac{n-1}{n-1}\sigma^2=\sigma^2$ and hence S^2 is an unbiased estimator of σ^2.

Similarly, it can be shown that $E(\hat{P})=P$, that is, the sample proportion is an unbiased estimator of the population proportion. It can be shown that S is a biased estimator of σ. Reiterating a point made earlier, whether using $\bar{X}$ as an estimator of μ, $\hat{P}$ as an estimator of P or S^2 as an estimator of σ^2, one can view the estimator as a random variable or the estimate as a single number. Before taking the sample, we know certain properties about the estimator. After the sample has been taken we have confidence in the point estimate because of the properties the estimator has.

Recalling the central limit theorem, we know that for large n ($n \geq 30$), $\bar{X}$ has a normal distribution with mean μ and standard error, $\sigma_{\bar{x}}=\frac{\sigma}{\sqrt{n}}$. The random variable $(\bar{X}-\mu)/\frac{\sigma}{\sqrt{n}}$ has a standard normal distribution that is represented by the letter z.

The notation $z_{\alpha/2}$ represents the standard normal value, with area $\alpha/2$ under the standard normal curve to its right. There will be another value $-z_{\alpha/2}$ with $\alpha/2$ area to its left. There will be $1 - \alpha$ area under the standard normal curve between $-z_{\alpha/2}$ and $z_{\alpha/2}$. An example will help clarify this notation.

EXAMPLE

Suppose $1 - \alpha = 0.90$, then $\alpha = 0.10$ and $\alpha/2 = 0.05$. Using EXCEL, =NORMSINV(0.05) gives −1.645 and =NORMSINV(0.95) gives 1.645. These are two values (−1.645 and 1.645) with $1 - \alpha = 0.90$ between them.

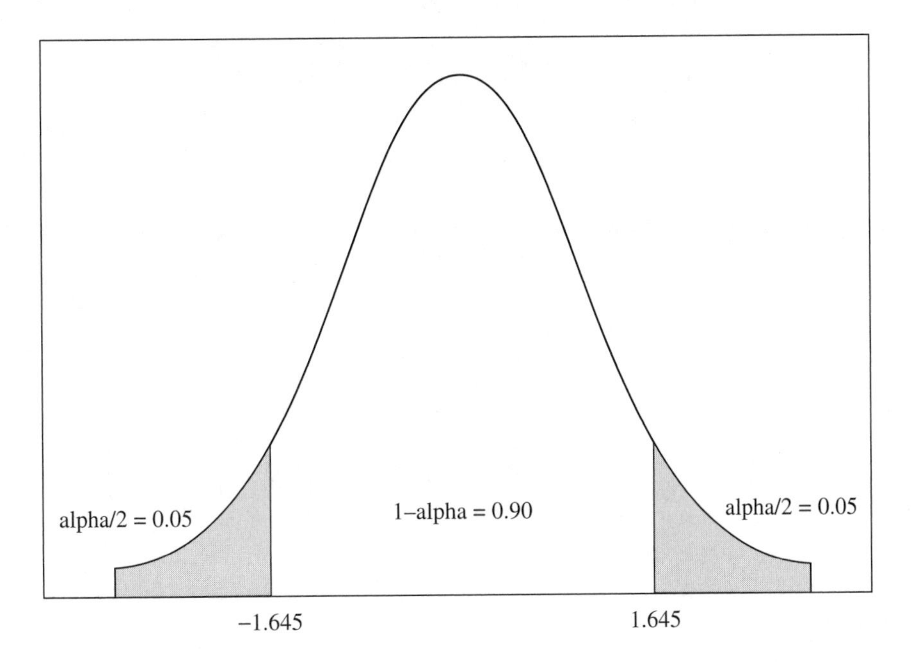

Note that $-z_{0.05} = -1.645$ and that $z_{0.05} = 1.645$.
We have that

$$-z_{\alpha/2} < \frac{\overline{X} - \mu}{\frac{\sigma}{\sqrt{n}}} < z_{\alpha/2} \text{ with probability } 1 - \alpha$$

The inequality is equivalent to $|\overline{X} - \mu| < z_{\alpha/2}\frac{\sigma}{\sqrt{n}}$. The quantity $|\overline{X} - \mu|$ is called the **maximum error of estimate**. It is the absolute difference between the population mean and the estimate of it. The maximum error of estimate E is $z_{\alpha/2}\frac{\sigma}{\sqrt{n}}$ or we have $E = z_{\alpha/2}\frac{\sigma}{\sqrt{n}}$. If the value of $z_{\alpha/2}$ is determined for $1 - \alpha = 0.95$, it is found that $z_{0.025} = 1.96$ and if the value of $z_{\alpha/2}$ is determined for $1 - \alpha = 0.99$, it is found that $z_{0.005} = 2.575$. (The student should confirm these values using EXCEL.)

EXAMPLE

The Internet is a source of information for engineers.

It is of interest to find the average time spent on the Internet per week by engineers. Fifty engineers are randomly selected. On the basis of experience and other studies it is felt that $\sigma = 5$ hours. What can be asserted with probability 0.95 about the maximum error of estimate of the survey?

$$E = 1.96 \frac{5}{\sqrt{50}} = 1.39$$

The survey will have, with probability 0.95, an error of at most 1.39 hours.

What is the difference between confidence and probability?

Consider the past example further. To speak of taking samples and that the probability is 0.95 that the maximum error will be 1.39 is talking about the probabilistic properties of this technique. Now if the sample is actually taken and $\bar{X} = 20.5$ hours is found, then $-z_{\alpha/2} < \bar{X} - \mu / \frac{\sigma}{\sqrt{n}} < z_{\alpha/2}$ is equivalent to $19.11 < \mu < 21.89$ and this statement does not have 95% probability of being true. The statement is either false or true and has probability 0 or 1. Rather than saying the probability is 95% that $19.11 < \mu < 21.89$, we say that we are 95% confident about the statement $19.11 < \mu < 21.89$. In general probability statements are made about future values of the random variable. Confidence statements are made once the data has been collected.

Suppose we wish to specify the maximum error to be a specific value and ask the question: "What sample size will give this error?" The answer is found in the solution of the equation $E = z_{\alpha/2} \frac{\sigma}{\sqrt{n}}$ for n. If both sides of the equation is squared and solved for n, the solution is

$$n = \left[\frac{z_{\alpha/2}\sigma}{E} \right]^2$$

EXAMPLE

It is of interest to find the average time spent on the Internet per week by engineers. Use 5 as the estimated value of σ. Find the sample size needed to be 95% confident that the sample mean will be within 1 hour of the population mean.

$$n = \left[\frac{z_{\alpha/2}\sigma}{E}\right]^2 = \left[\frac{1.96(5)}{1}\right]^2 = 96.04$$

Use a sample of size 97. Always round up to the next integer.

Now consider that only small samples are available and that sampling from a normal population is a reasonable assumption. Also assume the sample standard deviation is used in place of the population standard deviation. The random variable $t_v = (\bar{x} - \mu)/\frac{s}{\sqrt{n}}$ may be shown to have a Student's t distribution with $v = n - 1$ degrees of freedom. The probability will be $(1 - \alpha)$ that $|\bar{X} - \mu| < t_{\alpha/2}\frac{S}{\sqrt{n}}$ or that the maximum error of estimate is $E = t_{\alpha/2}\frac{S}{\sqrt{n}}$. Now, when the sample is actually taken, and the numbers are computed, it can be asserted with $(1 - \alpha) \times 100\%$ confidence that the maximum error of estimate is $E = t_{\alpha/2}\frac{s}{\sqrt{n}}$.

EXAMPLE

A computer engineer studies the mean time between failures for a disk drive.

A computer engineer wished to estimate the mean time between failures of a disk drive. The engineer recorded the time between failures for a sample of 10 disk drive failures. Give the estimated maximum error of estimate with 99% confidence. The following times between failures were recorded in hours:

1870 2140 1711 2039 1894 1677 1640 1700 1969 1750

First, the data is tested to see if it is reasonable to assume that the data came from a normally distributed population. The Kolmogorov-Smirnov test gives a p-value > 0.15. Thus, it is reasonable to assume normality.

A descriptive statistics routine gives the sample mean equal to 1839.1 hours, a standard deviation equal to 170.3 hours, and a standard error equal to 53.9 hours. The 0.01 Student's t value with 9 degrees of freedom is given by =TINV(0.01,9) = 3.25. The TINV function takes two parameters. The first is the area in the two tails. The second is the degrees of freedom.

The 99% confidence maximum error of estimate is 3.25(53.9) = 175.2 hours.

6.2 Interval Estimation

Rather than giving a point estimate, it is often preferable to give an interval. First, consider the case when n is large and σ is known. If the inequality

$$-z_{\alpha/2} < \frac{\overline{X} - \mu}{\frac{\sigma}{\sqrt{n}}} < z_{\alpha/2}$$

is solved for μ using simple algebra, we get

$$\overline{x} - z_{\alpha/2}\frac{\sigma}{\sqrt{n}} < \mu < \overline{x} + z_{\alpha/2}\frac{\sigma}{\sqrt{n}}$$

This is generally referred to as the **large sample confidence interval** or the **z-interval**. If σ is unknown, it is replaced by the sample standard deviation and the interval takes on the following form:

$$\overline{x} - z_{\alpha/2}\frac{s}{\sqrt{n}} < \mu < \overline{x} + z_{\alpha/2}\frac{s}{\sqrt{n}}$$

The interval is referred to as a **confidence interval**. The $(1 - \alpha)$ is called the **degree of confidence**. The endpoints of the interval are called **confidence limits**.

EXAMPLE

Drill bits have average lifetimes.

The lifetime of drill bits is of interest. The lifetime of the drill bit is defined as the number of holes drilled before failure. The data in Figure 6-1 gives the lifetimes of 40 drill bits.

174	149	169	150
101	145	134	104
137	131	113	161
124	123	135	124
147	139	137	144
106	128	131	137
143	118	135	109
153	122	111	137
132	140	149	138
141	150	140	118

Figure 6-1 Lifetimes of 40 drill bits.

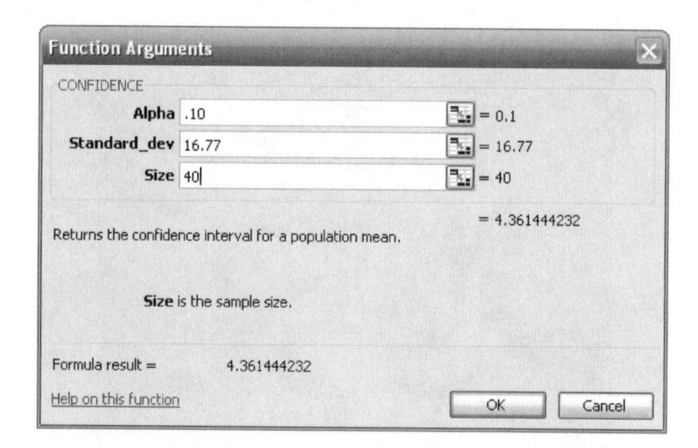

Figure 6-2 EXCEL dialog box for computing *E*.

Find a 90% confidence interval for the mean lifetime using EXCEL, MINITAB, SAS, SPSS, and STATISTIX.

EXCEL

Figure 6-2 is a dialog box to be filled out for finding a $(1 - \alpha)$ confidence interval for the population mean. If you are interested in a 90% confidence interval, α is equal to 0.10. The standard deviation is found to be 16.77. The sample size is 40. The EXCEL function CONFIDENCE finds

$$E = z_{\alpha/2}\frac{\sigma}{\sqrt{n}} = 1.645(16.77)/\sqrt{40} = 4.361$$

To find the 90% interval, add and subtract E from the sample mean which equals 134.42. The 90% confidence interval thus extends from 134.42 – 4.36 = 130.06 to 134.42 + 4.36 = 138.78.

MINITAB

Give the pull-down **Stat ⇒ Basic Statistics ⇒ 1-Sample** z. This gives Figure 6-3, which is filled out as shown.

Under options, choose confidence level equal to 90%. This gives the following output:

One-Sample *z*: Lifetime

```
The assumed standard deviation = 16.77

Variable   N     Mean      SD    SE Mean          90% CI
Lifetime  40  134.420  16.766     2.652  (130.058, 138.781)
```

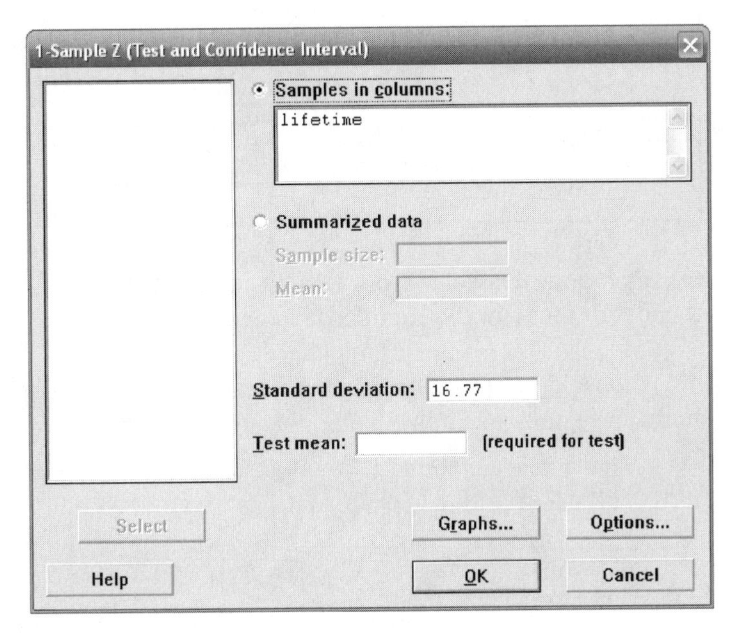

Figure 6-3 MINITAB dialog box for computing confidence interval for μ.

SAS

```
                    One-Sample z-Test for a Mean
Sample Statistics for Lifetime
     N          Mean              SD          Std. Error
-------------------------------------------------
    40        134.48          16.75              2.65
With a specified known standard deviation of 16.77

90% Confidence Interval for the Mean
                   Lower Limit    Upper Limit
                   ----------- -----------
                      130.11          138.84
```

The 90% confidence interval is shown above.

SPSS

The pull-down **Analyze ⇒ Compare Means ⇒1-Sample *t*-Test** gives the 1-sample *t*-test dialog box. In that box choose options and as an option choose a 90% confidence interval. This gives the following output:

One-Sample Statistics

	N	Mean	Std. Deviation	Std. Error Mean
lifetime	40	134.48	16.750	2.648

One-Sample Test

	Test Value = 0					
					90% Confidence Interval of the Difference	
	t	df	Sig. (2-tailed)	Mean Difference	Lower	Upper
lifetime	50.776	39	.000	134.475	130.01	138.94

The reason the interval differs some from EXCEL and MINITAB is that t rather than z is used in the formula for the confidence interval.

STATISTIX

In STATISTIX, choose descriptive statistics. That gives the descriptive statistics dialog box shown in Figure 6-4. In the dialog box, choose confidence interval and 90% coverage.

The following output is given:

```
Statistix 8

Descriptive Statistics

Variable       Lo 90% CI    Up 90% CI
Lifetime          130.01       138.94
```

The package STATISTIX is, like SPSS, using the t distribution, and so the answer is a little different from MINITAB and EXCEL.

When working with a small sample from a normally distributed population and using S as an estimate of σ, we have the following as a confidence interval for μ:

$$\bar{x} - t_{\alpha/2}\frac{S}{\sqrt{n}} < \mu < \bar{x} + t_{\alpha/2}\frac{S}{\sqrt{n}}$$

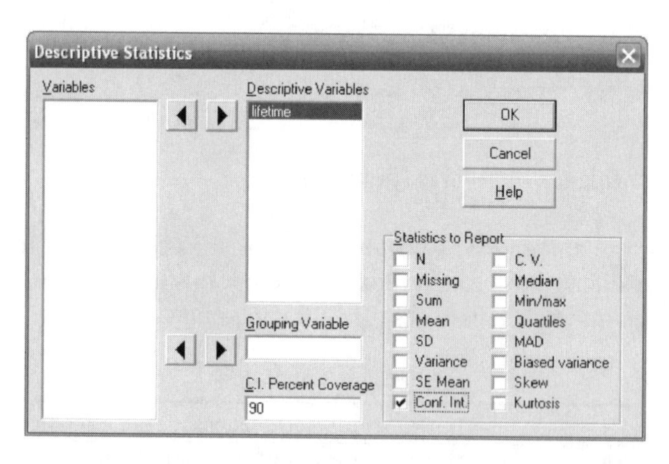

Figure 6-4 STATISTIX dialog box for 90% confidence interval.

174	149	169	150
101	145	134	104
137	131	113	161

Figure 6-5 Lifetimes of 12 drill bits.

This is generally referred to as the **small sample confidence interval** or the ***t*-interval**. When the sample size is small, the *t*-interval must be used. If a *z*-value rather than a *t*-value is used in the *t*-interval, the interval will be too narrow.

EXAMPLE

The lifetime of drill bits is of interest. The lifetime of the drill bit is defined as the number of holes drilled before failure. The data in Figure 6-5 is the lifetimes of 12 drill bits. Find a 95% confidence interval for μ using EXCEL, MINITAB, SAS, SPSS, and STATISTIX.

First, test for normality. Remember, the sample is small and the theory underlying the confidence interval requires the normality assumption.

```
Statistix 8.0
Shapiro-Wilk Normality Test

Variable         N        W          P
Lifetime        12     0.9524     0.6725
```

Using the Shapiro-Wilk test of STATISTIX, we do not reject normality because of the large *p*-value.

Figure 6-6 shows the EXCEL computations for the 95% confidence interval for the population mean. The drill bit lifetimes are shown in A2:A13, the average is

	A	B	C	D
1	lifetime			
2	174		139	= AVERAGE(A2:A13)
3	101		23.93932	= STDEV(A2:A13)
4	137		6.910686	= C3/SQRT(12)
5	149		2.200985	= TINV(0.05,11)
6	145			
7	131		123.7897	= C2-C4*C5
8	169		154.2103	= C2+C4*C5
9	134			
10	113			
11	150			
12	104			
13	161			

Figure 6-6 EXCEL computations for 95% confidence interval for μ.

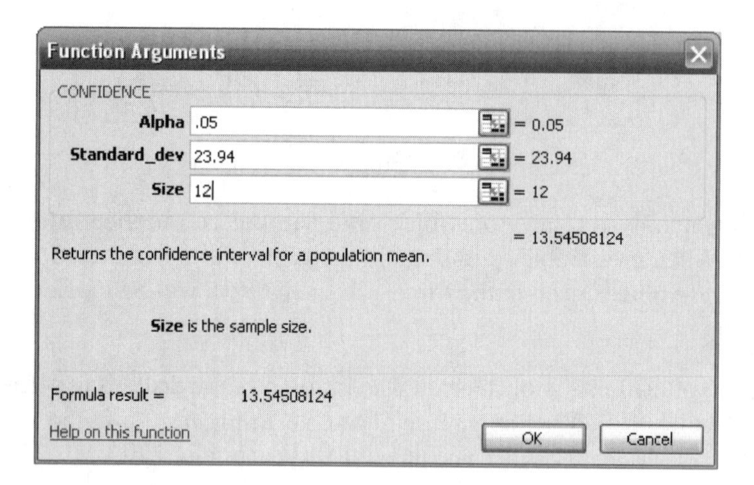

Figure 6-7 An incorrect use of an EXCEL function.

computed in C2, the standard deviation in C3, the standard error in C4, the 95% t-value in C5, the lower 95% confidence limit in C7, and the upper 95% confidence limit in C8.

Figure 6-7 illustrates an incorrect use of EXCEL. The dialog box shown in this figure is intended only for large samples, i.e., for samples > 30. The quantity E is computed and given as 13.545. Here,

$$E = z_{\alpha/2}\frac{\sigma}{\sqrt{n}} = 1.96(6.911) = 13.545$$

The small sample confidence interval requires

$$E = t_{\alpha/2}\frac{S}{\sqrt{n}} = 2.200985(6.911) = 15.211$$

The interval using the value of E in Figure 6-7 would be too narrow. It would give (125.455, 152.545) rather than the correct interval (123.7897, 154.2301). *Remember, it is the responsibility of the user of the software to make sure they are using it correctly.*

If MINITAB is used to calculate the 95% confidence interval, enter the 12 data values in column C1 and use the pull-down **Stat ⇒ Basic Statistics ⇒ 1-Sample *t*.** The output is

One-Sample *t*: Lifetime

```
Variable   N     Mean      SD       SE Mean         95% CI
Lifetime   12  139.000   23.939     6.911   (123.790, 154.210)
```

You have to remember that small sample is the reason you perform the 1-sample *t*. The STATISTIX confidence interval is the same as the one you computed using EXCEL. The STATISTIX output is

```
Statistix 8.0
Descriptive Statistics

Variable      Lo 95% CI    Up 95% CI
Lifetime         123.79       154.21
```

The SAS solution is

```
                        One-Sample t-test for a Mean
Sample Statistics for Lifetime
       N            Mean            SD            Std. Error
    ------------------------------------------------------------
      12          139.00          23.94             6.91
95% Confidence Interval for the Mean
      Lower limit:           123.79
      Upper limit:           154.21
```

The SAS solution is the same as EXCEL, MINITAB, and STATISTIX.
The SPSS solution is

One-Sample Statistics

	N	Mean	Std. Deviation	Std. Error Mean
lifetime	12	139.00	23.939	6.911

One-Sample Statistics

	Test Value = 0					
					95% Confidence Interval of the Difference	
	t	df	Sig. (2-tailed)	Mean Difference	Lower	Upper
lifetime	20.114	11	.000	139.00	123.79	154.21

The SPSS solution is reached by using the pull-down **Analyze ⇒ Compare Means ⇒1-Sample *t*-Test**.

EXAMPLE

Have MINITAB select 20 simulated samples of size 10 from a normal population with mean $\mu = 139$ and standard deviation equal to 24. Form 80% confidence intervals for μ from each of the 20 samples. Determine the percentage of the 20 confidence intervals that actually contain the mean (139) of the population.

Samples									
1	2	3	4	5	6	7	8	9	10
107	139	150	150	119	126	60	131	130	126
173	149	112	171	104	105	151	163	154	140
106	123	102	148	143	154	137	155	162	129
116	181	123	133	149	177	165	124	121	97
97	126	152	105	171	117	149	182	170	201
134	118	136	111	188	153	134	105	111	160
156	161	153	159	153	114	149	107	134	155
134	136	162	131	164	191	135	92	158	131
152	86	158	139	150	140	130	131	135	83
137	129	166	106	111	104	139	121	135	182
Samples									
11	12	13	14	15	16	17	18	19	20
138	141	135	128	157	107	114	173	161	108
118	175	141	176	127	108	110	128	135	153
187	134	197	142	142	146	169	172	121	151
142	200	150	168	94	111	115	111	157	158
99	133	130	130	110	127	158	151	105	137
161	147	125	141	152	162	134	143	156	126
153	132	138	161	135	104	140	171	147	129
152	121	122	156	106	109	99	138	114	97
152	198	100	147	130	124	155	136	122	137
109	109	166	167	124	117	147	144	153	128

Figure 6-8 Twenty samples ($n = 10$) from a normal population with $\mu = 139$ and $\sigma = 24$ are simulated by MINITAB.

The 20 samples of size 10 each are shown in Figure 6-8. For each of the samples, Student's t interval is calculated.

$$\left(\bar{x} - t_{\alpha/2}\frac{S}{\sqrt{n}}, \bar{x} + t_{\alpha/2}\frac{S}{\sqrt{n}}\right)$$

The 20 intervals are shown in Figure 6-9. The intervals that do not contain the value of the mean (139) are shown with an asterisk beside the sample number in Figure 6-9.

We find that 15% of the 80% confidence intervals do not contain the population mean. This means that 85% contain the mean. We say that the confidence interval captures the population mean if the population mean is found inside the interval. In practice, you will not know whether a given interval captures the mean or not. Knowing that theory predicts that 80% of the intervals will capture the mean is what gives the statistician the 80% confidence that he or she has. If this simulation study were carried out over thousands of times, we would see the percentage

Sample	Confidence interval	Sample	Confidence interval
1	(120.422, 141.978)	11	(129.594, 152.606)
2	(123.549, 146.051)	12	(135.247, 162.753)
3	(131.718, 151.082)	13	(128.827, 151.973)
4	(125.393, 145.207)	14*	(144.372, 158.828)
5	(133.440, 156.960)	15*	(118.897, 136.503)
6	(124.882, 151.318)	16*	(113.173, 129.827)
7	(122.514, 147.286)	17	(123.782, 144.418)
8	(118.796, 143.404)	18	(137.742, 155.658)
9	(132.648, 149.352)	19	(128.195, 146.005)
10	(124.696, 156.104)	20	(123.914, 140.886)

Figure 6-9 Seventeen of the 20 intervals contain the mean (139).

approach, the 80% figure. The student is invited to verify the confidence intervals given in Figure 6-9.

6.3 Basic Concepts of Testing Hypotheses

We shall introduce the concept of testing hypotheses with a very simple example involving a coin. Then in later sections we will move into engineering examples to test means, proportions, and variances.

EXAMPLE

Suppose we have a nickel and we think that it is loaded or biased so that there is not a 50/50 chance of heads/tails. Furthermore, we suspect that it is biased so that a head is more likely to occur than a tail. We will use the following decision-making test procedure: Let p be the probability of a head turning up when the coin is tossed once. We set up two hypotheses. One is called the **null hypothesis** and the other is called the **research** or **alternative hypothesis**. Our research hypothesis denoted by H_a: is that $p > 0.5$ or that the coin is biased in favor of a head occurring. This research hypothesis is called a **one-sided** or **one-tailed test**. The null hypothesis is that the coin is not biased or $H_o: p = 0.5$. The hypothesis system is $H_o: p = 0.5$ versus $H_a: p > 0.5$. In order to test this hypothesis we agree to flip the coin 16 times and if 12 or more heads occur, then we will reject the null hypothesis that the coin is fair ($p = 0.5$). Obtaining 12 or more heads will support our belief that the coin is biased. If the coin is unbiased, we would expect somewhere near 8 heads to occur. If X represents the number of heads to occur when the coin is tossed 16 times, then X

Decision	Null true, $p = 0.5$	Null False, $p > 0.5$
Reject null	Type I error	No error made
Do not reject null	No error made	Type II error

Figure 6-10 Type I and type II error defined.

has a binomial distribution. The **rejection region** is defined to be the outcomes for which we shall reject the null hypothesis that the coin is fair. The **nonrejection region** is the outcomes for which we shall not reject the null hypothesis. There are two other research hypotheses that we shall be concerned with. The hypothesis H_a: $p < 0.5$ would mean that we were interested in the research hypothesis that the coin was biased so that a head was less likely to occur than a tail. This is also called a one-tail test. Finally, there is a **two-tail test**. It is expressed as H_a: $p \neq 0.5$. This says we suspect that the coin is biased but we are not sure which way.

We now have rejection region: $X \geq 12$ and nonrejection region: $X < 12$. The variable X is called our **test statistic**.

There are two types of errors that are possible using this decision-making procedure. They are illustrated in Figure 6-10.

A **type I error** occurs when the null is true but it is rejected. A **type II** error occurs when the null is false but it is not rejected. The probability of making a type I error is represented by the Greek letter α and the probability of making a type II error is represented by the Greek letter β. For this example, $\alpha = P(X \geq 12$ when $p = 0.5)$. Using EXCEL to evaluate this, we find α =1-BINOMDIST(11,16,0.5,1) = 0.038. To calculate β, we choose $p > 0.5$. The value of β depends on which value is chosen for p. The type II error is $P(X \leq 11$ for $p > 0.5)$. Figure 6-11 gives different values of β calculated for different values of p calculated using EXCEL. To build the results shown in Figure 6-11, place the values in the column under p from A1 to A9. Enter =BINOMDIST(11,16,A1,1) in B1 and perform a click-and-drag from B1 to B9.

The quantities in the column labeled $1 - \beta$ are called **power values**. The power of a test is the probability of rejecting the null for various values of p in the alternative. For example, the row where $p = 0.75$, $\beta = 0.369814$, and $1 - \beta = 0.630186$

P	β	$1-\beta$
0.55	0.914691	0.085309
0.6	0.833433	0.166567
0.65	0.710793	0.289207
0.7	0.550096	0.449904
0.75	0.369814	0.630186
0.8	0.201755	0.798245
0.85	0.079051	0.920949
0.9	0.017004	0.982996
0.95	0.000857	0.999143

Figure 6-11 Calculations of type II error and power values for H_a: $p > 0.5$.

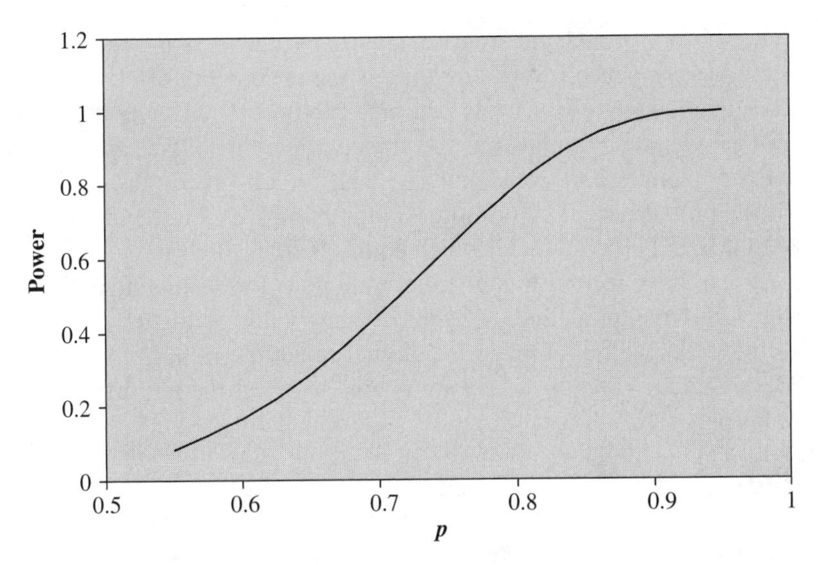

Figure 6-12 Power curve for the hypothesis test H_o: $p = 0.5$ versus H_a: $p > 0.5$.

tells us if the coin is biased so that the probability of a head is 0.75, the probability of not rejecting the null hypothesis is 0.369, and the probability of rejecting the null is 0.630. The power curve is plotted in Figure 6-12.

It is not possible to make perfect decisions about the coin by flipping it 16 times. The coin could be balanced but you might obtain an unusually large number of heads leading you to make a type I error. Or the coin might be biased, but you obtain somewhere near 8 heads, leading you to make a type II error. What the discipline of statistics helps you do is make the right decision often and to know what the probabilities of correct decisions are.

When conducting a test of hypothesis, the following are the recommended steps to follow:

1. State the null and research hypothesis: H_o: $p = 0.5$ versus H_a: $p > 0.5$
2. Give the rejection and nonrejection regions:

   Rejection region: $X = 12$, 13, 14, 15, and 16.

   Nonrejection region: $X = 0$, 1, 2, 3, 4, 5, 6, 7, 8, 9, 10, and 11.
3. Calculate α and give the power curve shown in Figure 6-11.

   α =1-BINOMDIST(11,16,0.5,1) = 0.038

   (Step 4 is performed only after the other steps have been completed.)
4. Perform the experiment and give the value of the test statistic, and your conclusion.

After step 4 is performed, the observed significance level or ***p*-value** is computed. The observed significance level or ***p*-value** is the probability of observing a value of the test statistic that is at least as contradictory to the null hypothesis and supportive of the research hypothesis as the one computed from the sample data. Suppose the experiment is conducted and 11 heads are observed out of 16 flips. The *p*-value is the probability of 11 or more heads. Using EXCEL, the *p*-value is given by =1-BINOMDIST(10,16,0.5,1), which equals 0.105.

When researchers follow the *p*-value technique for testing hypotheses, they follow the following procedure: Compare the *p*-value with the α-value. If the *p*-value < α, reject the null. Otherwise, do not reject the null. In the present case, $\alpha = 0.038$ and the *p*-value > α. You are unable to reject the null hypothesis using the *p*-value approach.

EXAMPLE

Suppose the research hypothesis is that the coin is biased so that a head is less likely to occur than a tail. In this case the research hypothesis is $H_a: p < 0.5$. In this case, the left-hand tail becomes the rejection region. This time the rejection region is chosen to be $X \leq 5$. The probability of a type I error is $P(X \leq 5 \text{ when } p = 0.5)$. Using EXCEL, this probability is given by =BINOMDIST(5,16,0.5,1) which equals 0.105.

The value of β depends on which value is chosen for *p*. The type II error is $P(X \geq 6 \text{ for } p < 0.5)$. Figure 6-13 gives different values of β calculated for different values of *p* using EXCEL. To build the results shown in Figure 6-13, place the values in the column under *p* from A1 to A9. Enter =1-BINOMDIST(5,16,A1,1) in B1 and perform a click-and-drag from B1 to B9.

Summarizing:

1. State the null and research hypothesis: $H_o: p = 0.5$ versus $H_a: p < 0.5$.
2. Give the rejection and nonrejection regions:

   Rejection region: $X = 0, 1, 2, 3, 4$, or 5.

   Nonrejection region: $X = 6, 7, 8, 9, 10, 11, 12, 13, 14, 15$, or 16.

p	β	$1-\beta$
0.05	8.08995E-05	0.9999191
0.1	0.003296751	0.996703249
0.15	0.023544381	0.976455619
0.2	0.081687888	0.918312112
0.25	0.189654573	0.810345427
0.3	0.340217674	0.659782326
0.35	0.510036428	0.489963572
0.4	0.671159587	0.328840413
0.45	0.80240244	0.19759756

Figure 6-13 Calculations of type II error and power values for $H_a: p < 0.5$.

3. Calculate α and give the power curve shown in Figure 6-13.

   α=BINOMDIST(5,16,0.5,1), which equals 0.105.

The hypothesis test may now be carried to completion.

EXAMPLE

Suppose the research hypothesis is that the coin is biased but we have no idea which way it is biased. In this case the research hypothesis is H_a: $p \neq 0.5$. The rejection region is in both tails of the distribution. This time the rejection region is chosen to be $X \leq 3$ or $X \geq 13$. The probability of a type I error is $P(X \leq 3$ or $X \geq 13$ when $p = 0.5)$. Using EXCEL, this probability is given by BINOMDIST(3,16,0.5,1)+(1-BINOMDIST(12,16,0.5,1)), or 0.021.

The value of β depends on which value is chosen for p. The type II error is $P(4 \leq X \leq 12$ for $p \neq 0.5)$. Figure 6-14 gives different values of β calculated for different values of p using EXCEL. To build the results shown in Figure 6-14, place the values in the column under p from A1 to A19. Enter =BINOMDIST(12,16,A1,1)-BINOMDIST(3,16,A1,1) in B1 and perform a click-and-drag from B1 to B19.

The power curve for this test is shown in Figure 6-15.

P	β	$1-\beta$
0.05	0.007003908	0.992996092
0.1	0.068406174	0.931593826
0.15	0.21010929	0.78989071
0.2	0.401865427	0.598134573
0.25	0.595009107	0.404990893
0.3	0.754110535	0.245889465
0.35	0.865935913	0.134064087
0.4	0.933914808	0.066085192
0.45	0.968418793	0.031581207
0.5	0.978729248	0.021270752
0.55	0.968418793	0.031581207
0.6	0.933914808	0.066085192
0.65	0.865935913	0.134064087
0.7	0.754110535	0.245889465
0.75	0.595009107	0.404990893
0.8	0.401865427	0.598134573
0.85	0.21010929	0.78989071
0.9	0.068406174	0.931593826
0.95	0.007003908	0.992996092

Figure 6-14 Calculations of type II error and power values for H_a: $p \neq 0.5$.

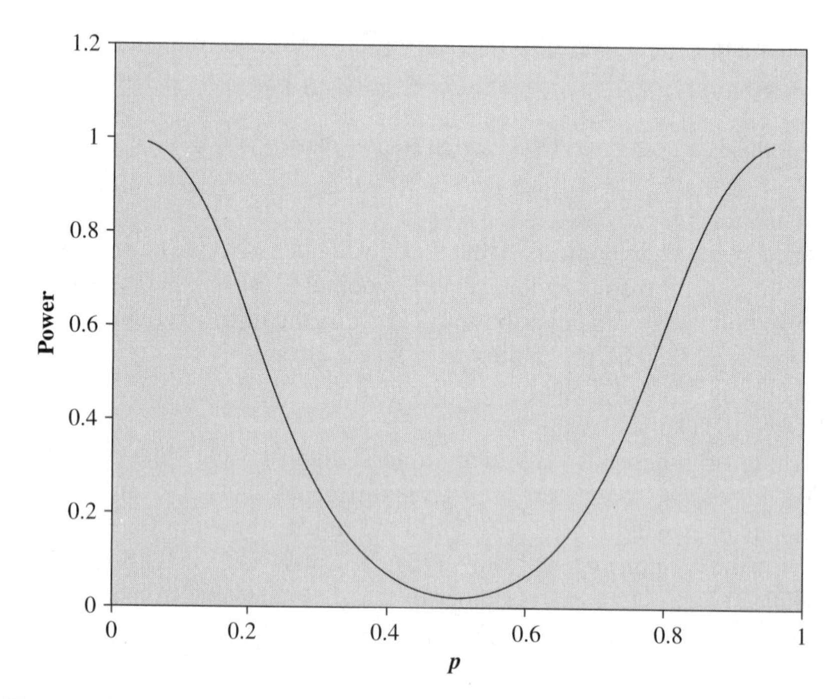

Figure 6-15 Power curve for the hypothesis test H_o: $p = 0.5$ versus H_a: $p \neq 0.5$.

Summarizing:

1. State the null and research hypothesis: H_o: $p = 0.5$ versus H_a: $p \neq 0.5$
2. Give the rejection and nonrejection regions:

   Rejection region: X = 0, 1, 2, 3, 13, 14, 15, or 16.

   Nonrejection region: X = 4, 5, 6, 7, 8, 9, 10, 11, 12.
3. Calculate α and give the power curve shown in Figure 6-15.

   α =BINOMDIST(3,16,0.5,1)+(1-BINOMDIST(12,16,0.5,1)), or 0.021.

Figure 6-16 gives the three cases we have considered. The upper left graph gives the complete binomial distribution for $n = 16$ and $p = 0.5$. The upper right shows the rejection region that was chosen for the research hypothesis that $p > 0.5$. The lower left shows the rejection region that was chosen for the research hypothesis that $p < 0.5$. The lower right shows the rejection that was chosen for the research hypothesis that $p \neq 0.5$. This graph shows the reasons for the terms: one-tail test and two-tail test. In the one-tail test either the upper tail or the lower tail of the binomial distribution was used for the rejection region. For this reason, one sometimes sees the terms **upper-tail test** and **lower-tail test**. When both tails of the distribution

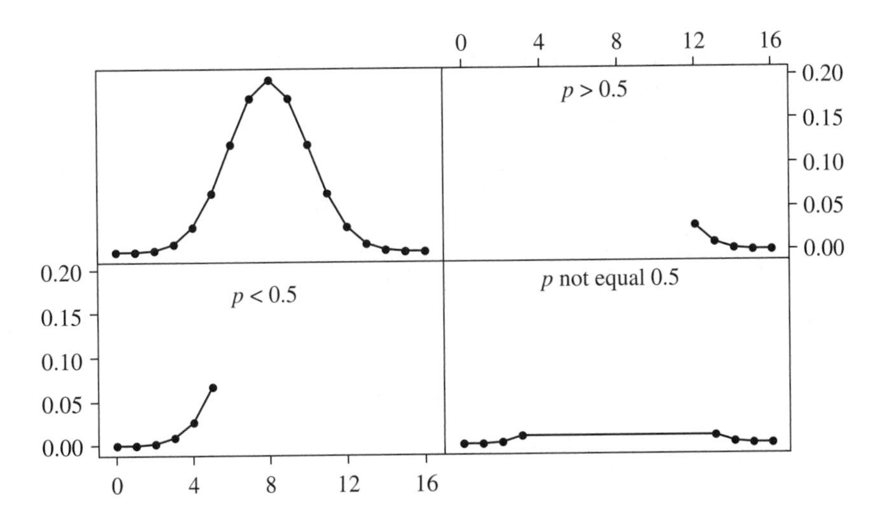

Figure 6-16 The complete binomial distribution and the one-sided and two-sided rejection regions.

are used, this is called the **two-tail test** because both tails of the binomial distribution are being utilized.

The fact that the experiment is performed and outcomes in the tails occur, are the evidences that the coins might be biased. However, these outcomes could occur even with a fair coin. This is what the curves in Figure 6-16 show. They show the probabilities of getting outcomes in the tails with a fair coin.

The student is encouraged to read and understand the logic involved in this section. The logic in this section is the same as that involved with tests involving means, proportions, and variances in later sections. There are quite a number of terms and definitions in this section that are used throughout the remainder of the book. Remember, it is not the mathematical underpinnings that are hard in statistics but the concepts and ideas that are difficult.

6.4 Testing Hypothesis About a Single Mean

Testing Hypotheses about a mean is divided into two cases: the large sample case and the small sample case. We shall discuss the large sample case first and the small sample case second. When samples are larger than 30, the test of hypothesis is based on the central limit theorem. Reviewing, the central limit theorem states that if the sample size, n, is larger than 30, $\overline{X}$, regarded as a variable, has a normal distribution with mean μ and standard error $\sigma_{\bar{x}} = \frac{\sigma}{\sqrt{n}}$, where μ is the mean of the population from which the sample is taken, and σ is the standard deviation of the population

from which the sample is taken. Furthermore, the variable Z where $Z = \frac{\bar{X} - \mu}{\sigma_{\bar{x}}}$ has a standard normal distribution, that is, a normal distribution with mean 0 and standard deviation 1. (In the large sample case, if σ is not known it is estimated by S.)

In testing hypotheses about μ, if the null hypotheses is that $\mu = \mu_o$, then the test statistic is $Z = \frac{\bar{X} - \mu_o}{\sigma_{\bar{x}}}$. If the test statistic takes on a highly unusual value, then the null hypothesis is rejected. There are three possible research hypotheses: $\mu < \mu_o$, $\mu > \mu_o$, or $\mu \neq \mu_o$. If the p-value $< \alpha$, the null is rejected. If not, the null is not rejected.

EXAMPLE

It is hypothesized that a machine is filling cereal boxes with 453 grams of cereal (on the average). The research hypothesis is H_a: $\mu \neq 453$. The test is to be performed with $\alpha = 0.05$. A sample of 40 boxes gave the results shown in Figure 6-17. The MINITAB analysis is shown below.

```
Test of μ = 453 vs μ ≠ 453
The assumed standard deviation = 2.4

Variable   N     Mean      SD     SE Mean          95% CI              Z       P
C1        40   450.440   2.379     0.379    (449.696, 451.184)    -6.75   0.000
```

The standard deviation of the weights are calculated and it is found that $S = 2.4$. The routine, called the z-test, is requested by the pull-down **Stat ⇒ Basic Statistics ⇒ 1-Sample** z. The calculated test statistic is $z = -6.75$. The p-value is 0.000. It would be safe to conclude that the fill is below 453 on the average.

The EXCEL solution is shown in Figure 6-18. The data is shown in A1:D10. The computation of the test statistic is shown in F1:F4. The computation of the p-value is shown in F6.

452.2	450.7	453.8	450.3
451.0	447.5	451.1	447.9
449.6	452.9	448.5	449.6
450.7	450.0	449.4	449.0
451.4	453.8	451.8	451.9
451.3	454.2	452.1	448.3
453.4	448.5	448.4	451.1
448.2	449.3	450.7	455.0
442.3	451.8	449.9	447.9
450.3	447.3	451.3	453.2

Figure 6-17 Weights of 40 cereal boxes selected from production.

A	B	C	D	E	F	G
452.2	450.7	453.8	450.3		450.44	AVERAGE(A1:D10)
451	447.5	451.1	447.9		2.379161671	STDEV(A1:D10)
449.6	452.9	448.5	449.6		0.37617849	F2/SQRT(40)
450.7	450	449.4	449		−6.8052801	(F1−453)/F3
451.4	453.8	451.8	451.9			
451.3	454.2	452.1	448.3		1.01049E−11	2*NORMSDIST(−6.805)
453.4	448.5	448.4	451.1			
448.2	449.3	450.7	455			
442.3	451.8	449.9	447.9			
450.3	447.3	451.3	453.2			

Figure 6-18 EXCEL computation of test statistic and *p*-value.

EXAMPLE

It is hypothesized that the mean time per week that engineers spend on the Internet looking up job related information is 15 hours. The research hypothesis is that the mean is greater than 15 hours. Survey results are shown in Figure 6-19.

```
One-Sample t-Test

Null hypothesis: μ = 15
Alternative hypothesis: μ > 15
                                            95% Conf. Interval
Variable          Mean         SE         Lower       Upper        t     DF        p
Hours           14.857     0.4224        13.999      15.715    -0.34     34   0.6314
```

The STATISTIX output shown above gives a *t*-value rather than a *z*. When *n* is larger than 30, the *t* distribution and the *z* distribution may be used interchangeably. Because of the large *p*-value (0.6314), there does not exist any evidence to reject the null. Note that we have not proved that $\mu = 15$. However, the data does not tend to make us doubt that $\mu = 15$.

When samples are small, the distribution of the test statistic changes and an assumption is made that must be checked. For samples of size 30 or smaller, it is assumed that the sample is taken from a normal distribution. The test statistic $t_v = (\bar{x} - \mu)/\frac{s}{\sqrt{n}}$ has a Student's *t* distribution with $v = (n - 1)$ degrees of freedom. There are three possible research hypotheses: $\mu < \mu_o$, $\mu > \mu_o$, or $\mu \neq \mu_o$. If the *p*-value $< \alpha$, the null is rejected. If not, the null is not rejected.

13	12	19	17	14	14	14
14	15	15	16	17	20	17
14	12	19	14	13	15	12
17	11	14	11	19	18	15
16	12	15	17	16	11	12

Figure 6-19 Hours per week spent on the Internet.

EXAMPLE

Operating an automobile and talking on a cell phone can be dangerous.

Highway engineers were studying the causes of rising highway death rates. One study looked at the time spent on cell phones per week while operating an automobile. The time spent on cell phones per week in hours for 10 randomly chosen individuals who drive and use cell phones were recorded in a study. The data was

5.2 5.5 4.9 2.9 4.1 4.4 2.7 6.8 4.3 3.5

The research hypothesis of interest was that $\mu < 7$. α was selected to be 0.01. Conduct the test of hypothesis using STATISTIX. First, the Shapiro-Wilk test of normality was conducted. The p-value was 0.9011 indicating a quite high probability that the sample could have come from a normal distribution. Normality of the sample was not rejected.

```
Shapiro-Wilk Normality Test

Variable        N         W          P
Time           10    0.9711     0.9011
```

The STATISTIX pull-down menu **Statistics ⇒ 1-, 2-, Multi-Sample Tests ⇒ 1-Sample *t*-Test** gave the 1-sample *t*-test dialog box. Time was moved into

variable slot, 7 was put into the hypothesis slot, and "Less than" into the alternative hypothesis slot. The results were:

```
One-Sample t-Test

Null Hypothesis: μ = 7
Alternative Hypothesis: μ < 7
                                95% Conf. Interval
Variable      Mean       SE        Lower      Upper       t     DF        p
Time        4.4300    0.3930    3.5410     5.3190    -6.54     9    0.0001
```

The computed test statistic was $t = -6.54$ with 9 degrees of freedom. There is 1 chance out of 10,000 that a sample with a t-value of –6.54 or smaller could be obtained if the mean of the population equals 7 hours. The decision is to reject the null hypothesis in favor of the research hypothesis.

Suppose the engineer is faced with this problem but has only EXCEL available. Figure 6-20 shows the variable name, **time**, in cell A1 of an EXCEL worksheet. The data is in A2:A11. The computation of the test statistic is shown in B2:B5 and the computation of the p-value is shown in B7.

Note that the EXCEL analysis results are the same as the STATISTIX analysis.

The SAS t-test for this example is as follows.

```
Sample Statistics for Time
        N          Mean            SD         Std. Error
     -------------------------------------------------
       10          4.43           1.24            0.39
   Hypothesis Test
      Null hypothesis:    Mean of time => 7
      Alternative:        Mean of time < 7
             t Statistic       DF       Prob > t
             ---------------------------------
               -6.539            9        <.0001
   99 % Confidence Interval for the Mean
                   (Upper Bound Only)
         Lower limit:           - infinity
         Upper limit:              5.54
```

Time	Computation	Function
5.2	4.43	AVERAGE(A2:A11)
5.5	1.242801	STDEV(A2:A11)
4.9	0.393008	C3/SQRT(10)
2.9	–6.5393	(C2-7)/C4
4.1		
4.4	5.32E-05	TDIST(6.5393,9,1)
2.7		
6.8		
4.3		
3.5		

Figure 6-20 EXCEL t-test.

The SPSS *t*-test gives the following output:

One-Sample Test

	Test Value = 7					
					95% Confidence interval of the difference	
	t	df	Sig. (2-tailed)	Mean difference	Lower	Upper
Time	−6.539	9	.000	−2.5700	−3.459	−1.681

Note that SPSS always performs a two-tailed test. The *p*-value is designated Sig. (2-tailed). If the test is 1-tailed, take half of the *p*-value in order to get a 1-tailed *p*-value.

The MINITAB output is

One-Sample *t*: Time

```
Test of μ = 7 vs μ < 7

                                                      99%
                                                     Upper
Variable   N      Mean       SD    SE Mean    Bound      t      p
Time      10   4.43000  1.24280   0.39301  5.53885  -6.54  0.000
```

6.5 The Relationship between Tests of Hypothesis and Confidence Intervals

Many statisticians/engineers use a technique that involves confidence intervals to do tests of hypotheses. This technique will be explained using examples from the previous section. We will introduce one-sided confidence intervals so that one-tailed tests may be illustrated also. Some of the software packages give one-sided as well as two-sided confidence intervals if requested.

EXAMPLE

It is hypothesized that a machine is filling cereal boxes with 453 grams of cereal (on the average). That is H_o: $\mu = \mu_0 = 453$. The research hypothesis is H_a: $\mu \neq 453$. The test is to be performed with $\alpha = 0.05$. Consider the MINITAB output for this example.

```
Test of μ = 453 vs μ≠453
The assumed standard deviation = 2.4

Variable   N     Mean     SD   SE Mean          95% CI            z      p
C1        40  450.440  2.379    0.379   (449.696, 451.184)   -6.75  0.000
```

The hypothesis is rejected because the p-value < $\alpha = 0.05$. Note also that μ_o = 453 is not contained in the 95% confidence interval for μ (449.696, 451.184). This will always be true. If a null hypothesis is being tested at α and a $1 - \alpha$ confidence interval for μ is formed, then the p-value < α, if and only if the null hypothesis value μ_o is not contained in the $1 - \alpha$ confidence interval for μ.

EXAMPLE

It is hypothesized that the mean time per week that engineers spend on the Internet looking up job related information is 15 hours. The research hypothesis is that the mean is greater than 15 hours. α is chosen to be 0.10.

One-Sample *z*: Hours

```
Test of μ = 15 vs μ > 15
The assumed standard deviation = 2.5
                                                         90%
                                                        Lower
Variable    N      Mean       SD      SE Mean   Bound        z      p
Hours      35   14.8571   2.4987   0.4226   14.3156   -0.34   0.632
```

In this example the p-value > $\alpha = 0.10$ and we do not reject. All 10% of the area under the standard normal curve is put into the lower tail for this lower bound confidence interval rather than being split half below and half above. The 90% lower bound confidence interval is (14.3156, ∞). Since the interval contains $\mu_o = 15$, the null would not be rejected if the confidence interval method were used.

EXAMPLE

Highway engineers were studying the causes of rising highway death rates. One study looked at the time spent on cell phones per week while operating an automobile. The time spent on cell phones per week in hours for 10 randomly chosen individuals who drive and use cell phones were recorded in a study. The data was

5.2 5.5 4.9 2.9 4.1 4.4 2.7 6.8 4.3 3.5

The research hypothesis is H_a: $\mu < 7$ and the null hypothesis is H_o: $\mu = 7$. Test at $\alpha = 0.01$.

One-sample *t*: Time

```
Test of μ = 7 vs μ < 7
                                                          99%
                                                         Upper
Variable    N      Mean        SD       SE Mean    Bound        t       p
Time       10   4.43000   1.24280   0.39301   5.53885   -6.54   0.000
```

Since the p-value $< \alpha = 0.01$, the null is rejected. The upper bound confidence interval is $(-\infty, 5.53885)$. Since 7 is not contained in this interval, the null is rejected. All 1% of the confidence is placed into the upper part of the t curve.

6.6 Inferences Concerning Two Means—Independent and Large Samples

Suppose we wish to compare two populations either by setting a confidence interval on the difference in their means or by testing hypotheses about the differences in their means. Population 1 has mean μ_1 and variance σ_1^2, and population 2 has mean μ_2 and variance σ_2^2. Samples of sizes n_1 and n_2, both larger than 30 and independent of each other, are taken and the difference in the sample means are considered. Theoretical studies have shown that the difference in sample means, $\bar{X}_1 - \bar{X}_2$, has a normal distribution with a mean equal to $\mu_1 - \mu_2$ and a variance equal to $\frac{\sigma_1^2}{n_1} + \frac{\sigma_2^2}{n_2}$. Furthermore, $\left\{\bar{X}_1 - \bar{X}_2 - (\mu_1 - \mu_2)\right\} \Big/ \sqrt{\frac{\sigma_1^2}{n_1} + \frac{\sigma_2^2}{n_2}}$ has a standard normal distribution. To test H_o: $(\mu_1 - \mu_2) = D_o$ versus any one of the three alternative hypothesis H_a: $(\mu_1 - \mu_2) < D_o$, H_a: $(\mu_1 - \mu_2) > D_o$, or H_a: $(\mu_1 - \mu_2) \neq D_o$, use the test statistic

$$Z = \frac{\bar{X}_1 - \bar{X}_2 - D_o}{\sqrt{\frac{\sigma_1^2}{n_1} + \frac{\sigma_2^2}{n_2}}}$$

The interval

$$(\bar{X}_1 - \bar{X}_2) \pm z_{\alpha/2}\sqrt{\frac{\sigma_1^2}{n_1} + \frac{\sigma_2^2}{n_2}}$$

is a $(1 - \alpha)$ confidence interval for $\mu_1 - \mu_2$. In most real world applications, the population variances are unknown and the sample variances replace them. The only assumptions needed are that the samples are independent and larger than 30.

EXAMPLE

A study was conducted to compare male and female engineers entering the work force. Their GPAs were recorded at graduation and the hypotheses that $\mu_1 - \mu_2 = 0$ versus $\mu_1 - \mu_2 \neq 0$ were tested at $\alpha = 0.05$. The GPAs for 40 males and 40 females are shown in Figure 6-21.

Male				Female			
3.18	3.18	3.02	3.15	3.29	3.73	2.98	3.32
3.40	2.71	2.76	3.38	3.12	3.72	3.64	3.16
3.19	2.88	2.58	3.40	3.01	3.10	3.56	3.54
3.13	2.86	3.18	3.22	3.64	3.82	3.26	2.99
3.24	2.57	2.92	3.37	3.24	3.39	2.94	3.18
3.30	3.06	3.29	3.42	3.11	2.74	3.33	3.40
2.67	2.69	2.74	3.30	3.44	3.41	3.22	3.80
2.75	3.09	3.05	3.21	3.57	3.30	3.62	3.05
2.74	2.95	2.89	3.30	3.64	3.14	2.88	3.08
2.94	2.77	3.18	3.42	2.80	3.43	3.39	3.49

Figure 6-21 GPA's for male and female engineers.

Compare the two groups using the independent samples test and using some of the standard software.

EXCEL

The pull-down **Tools ⇒ Data Analysis ⇒ *z*-Test: 2-Sample for Means** gives the dialog box shown Figure 6-22. The ranges tell where the data is identified. The variances are computed separately and entered in the dialog box. The hypothesized difference is 0, or H_o: $(\mu_1 - \mu_2) = 0$ and the research hypothesis is H_a: $(\mu_1 - \mu_2) \neq 0$. The α-value is 0.05.

This dialog box gives the output shown in Figure 6-23.

z-Test: Two Sample for Means
Input
Variable 1 Range: $A$1:$A$41
Variable 2 Range: $B$1:$B$41
Hypothesized Mean Difference: 0
Variable 1 Variance (known): 0.066
Variable 2 Variance (known): 0.079
Labels
Alpha: 0.05
Output options
Output Range: $D$5
New Worksheet Ply:
New Workbook
OK
Cancel
Help

Figure 6-22 EXCEL dialog box for two large independent samples *z*-test.

	Male	Female
Mean	3.052	3.312
Known variance	0.066	0.079
Observations	40	40
Hypothesized mean		
Difference	0	
z	–4.31421	
P(*Z*<=*z*) one-tail	8.01E-06	
z Critical one-tail	1.644854	
P(*Z*<=*z*) two-tail	1.6E-05	
z Critical two-tail	1.959964	

Figure 6-23 EXCEL output for the 2-sample *z*–test.

Recall the null hypothesis H_o: $(\mu_1 - \mu_2) = 0$ versus the research hypothesis H_o: $(\mu_1 - \mu_2) \neq 0$ may be tested in three different ways.

1. The two-tailed rejection region is $Z < -1.96$ or $Z > 1.96$. The computed test statistic is $z = -4.31$. Since this computed test statistic is less than -1.96 and is therefore in the rejection region, the null is rejected for $\alpha = 0.05$.
2. Using the *p*-value approach, compare the *p*-value $= 1.6E - 05 = 0.000016$ with the preset $\alpha = 0.05$. Since the *p*-value is less than α, reject the null.
3. The confidence interval approach calculates the 95% two-sided confidence interval and if it does not contain $D_o = 0$, reject the null. The 95% confidence interval is $(\bar{X}_1 - \bar{X}_2) \pm z_{\alpha/2}\sqrt{\frac{\sigma_1^2}{n_1} + \frac{\sigma_2^2}{n_2}}$, or $-0.26 \pm 1.96(0.0602)$, or -0.26 ± 0.118, or $(-0.378, -0.142)$. Since 0 is not in this interval, reject the null.

The same conclusion will always be reached no matter which of the three methods of testing hypothesis is used.

When testing two means the two-sample *t*-test to be discussed in the next section gives basically the same solution when both samples exceed 30 as the *z*-test of this section. The MINITAB 2-sample *t*-test using the data in Figure 6-21 and the pull-down **Stat ⇒ Basic Statistics ⇒ 2-Sample *t*** is as follows:

Two-Sample *t*-Test and CI: Male, Female

```
Two-Sample t for Male vs Female

         N    Mean     SD     SE Mean
Male    40   3.052   0.258     0.041
Female  40   3.312   0.281     0.044

Difference = mu (male) - mu (female)
Estimate for difference:   -0.259750
95% CI for difference:   (-0.379727, -0.139773)
t-test of difference = 0 (vs not =): t-value = -4.31  p-value = 0.000  DF = 78
Both use pooled SD = 0.2695
```

Comparing the output from Figure 6-23 and the MINITAB output, we see that EXCEL gives the test statistic as $Z = -4.31421$ and MINITAB gives the test statistic as t-value = –4.31. The two are practically the same. The 95% confidence interval computed in part 3 is (–0.378, –0.142). The MINITAB 95% confidence interval is (–0.379727, –0.139773). The p-value is 0 to three places in both cases.

The STATISTIX solution is as follows:

```
Statistix 8.0
Two-Sample t-Tests for GPA by Sex

Sex              Mean        N          SD          SE
F              3.3118       40      0.2810      0.0444
M              3.0520       40      0.2576      0.0407
Difference     0.2597

Null hypothesis: difference = 0
Alternative hypothesis: difference <> 0
                                                 95% CI for Difference
Assumption                 t        DF         p        Lower       Upper
Equal variances         4.31        78    0.0000       0.1398      0.3797
Unequal variances       4.31      77.4    0.0000       0.1398      0.3797

Test for equality          F        DF           p
      of variances      1.19     39,39      0.2948
```

The STATISTIX data file gives the GPA in one column and the sex of the individual in the other column. The pull-down **Statistics ⇒ 1-, 2-, and Multi-Sample Tests ⇒ 2-Sample *t*-Test** is used to perform the analysis. Under the t column the value 4.31 is seen as the same that EXCEL and MINITAB produced above. The 95% confidence interval is basically the same as given by EXCEL and MINITAB.

The SPSS and SAS analysis give basically the same output.

6.7 Inferences Concerning Two Means—Independent and Small Samples

EQUAL VARIANCES MODEL

When the sample sizes are small, it is necessary to make certain assumptions to compare the two population means, μ_1 and μ_2. The assumptions are:

1. The two samples are selected in an independent and random manner.
2. The two populations are normally distributed.
3. The population variances are equal, i.e., $\sigma_1^2 = \sigma_2^2 = \sigma^2$.

The common population variance is estimated by pooling the two sample variances together. That estimate is

$$S_p^2 = \frac{(n_1 - 1)S_1^2 + (n_2 - 1)S_2^2}{n_1 + n_2 - 2}$$

and it is called the **pooled estimate of variance**. It can be shown that the quantity $\left\{\bar{X}_1 - \bar{X}_2 - (\mu_1 - \mu_2)\right\} \Big/ \sqrt{S_p^2\left(\frac{1}{n_1} + \frac{1}{n_2}\right)}$ has a Student's t distribution with $(n_1 + n_2 - 2)$ degrees of freedom. To test H_o: $(\mu_1 - \mu_2) = D_o$ versus any one of the three alternative hypotheses:
H_a: $(\mu_1 - \mu_2) < D_o$, H_a: $(\mu_1 - \mu_2) > D_o$, or H_a: $(\mu_1 - \mu_2) \neq D_o$, use the test statistic

$$t = \frac{\bar{X}_1 - \bar{X}_2 - D_o}{\sqrt{S_p^2\left(\frac{1}{n_1} + \frac{1}{n_2}\right)}}$$

The interval $(\bar{X}_1 - \bar{X}_2) \pm t_{\alpha/2}\sqrt{S_p^2\left(\frac{1}{n_1} + \frac{1}{n_2}\right)}$ is a $(1 - \alpha)$ confidence interval for $\mu_1 - \mu_2$. When using this distribution theory to test hypothesis and set confidence intervals, the assumption of normality for both populations must be checked out using one of the techniques discussed earlier in this book. The assumption of equal variances must also be checked out. The next chapter discusses techniques for testing equal variances in more detail.

EXAMPLE

An engineer wishes to compare the mean lifetimes of two bulb types. She selects samples of size 10 from each and determines the lifetimes of the 20 bulbs. The lifetimes (in hours) are shown in Figure 6-24.

Bulb type 1	Bulb type 2
2164	1641
2040	1827
1926	1847
1911	1800
1973	1932
2064	1937
2018	2023
2097	1836
2105	1891
1972	1715

Figure 6-24 Bulb lifetimes, equal variability.

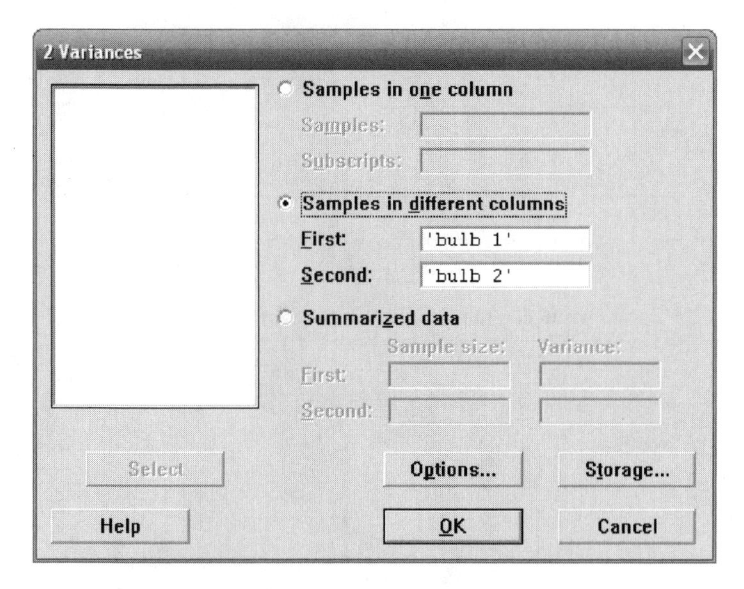

Figure 6-25 MINITAB dialog box for testing equal variances.

The Kolmogorov-Smirnov test for normality for bulb type 1 gives a p-value > 0.150 and for bulb type 2 gives a p-value > 0.15. The normality assumption is reasonable for both populations. The equal variances dialog box is shown in Figure 6-25. The columns, where the data for bulb type 1 and bulb type 2 may be found, are given in the dialog box. The output for the test of equal variances is shown in Figure 6-26. The F-test is valid if both sets of data come from normal populations. We have already tested and accepted that this is reasonable. The null hypothesis is that the variances are equal. This hypothesis is not rejected since the p-value is much greater than 0.05.

Test for Equal Variances: Bulb 1, Bulb 2

```
95% Bonferroni Confidence Intervals for Standard Deviations

          N     Lower        SD      Upper
Bulb 1   10   53.9591    82.491    166.108
Bulb 2   10   72.6845   111.117    223.752

F-test (normal distribution)
Test statistic = 0.55, p-value = 0.388

Levene's test (any continuous distribution)
Test statistic = 0.29, p-value = 0.598
```

The assumptions are reasonable for the test of equal means that assumes normality and equal variances. We proceed with the test $H_o: (\mu_1 - \mu_2) = 0$ versus $H_a: (\mu_1 - \mu_2) \neq 0$. The MINITAB pull-down menu to use is **Stat ⇒ Basic Statistics ⇒ 2-Sample *t***. The dialog box for the 2-sample t is shown in Figure 6-27.

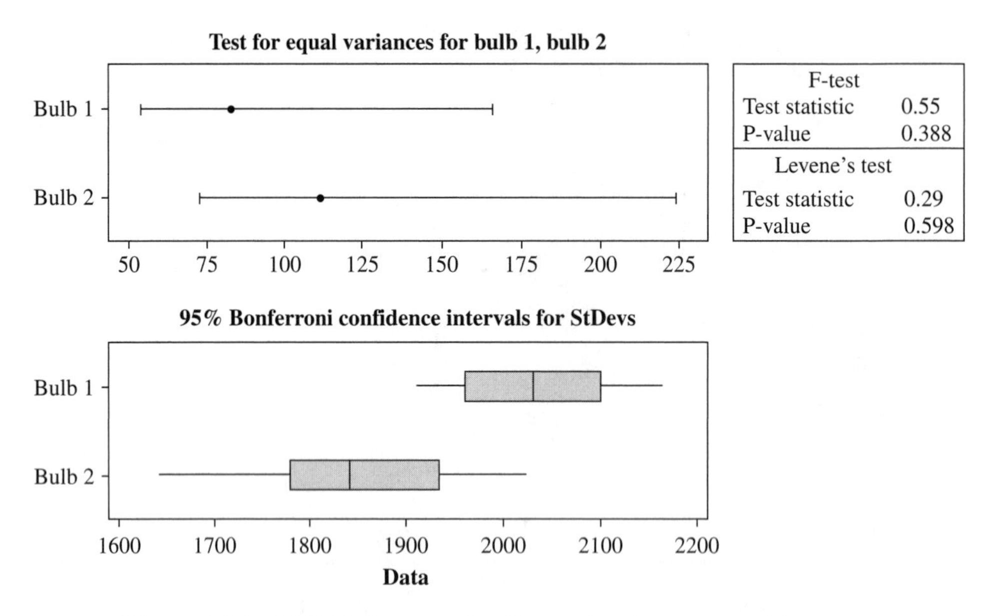

Figure 6-26 MINITAB output for the equal variances test with graphs.

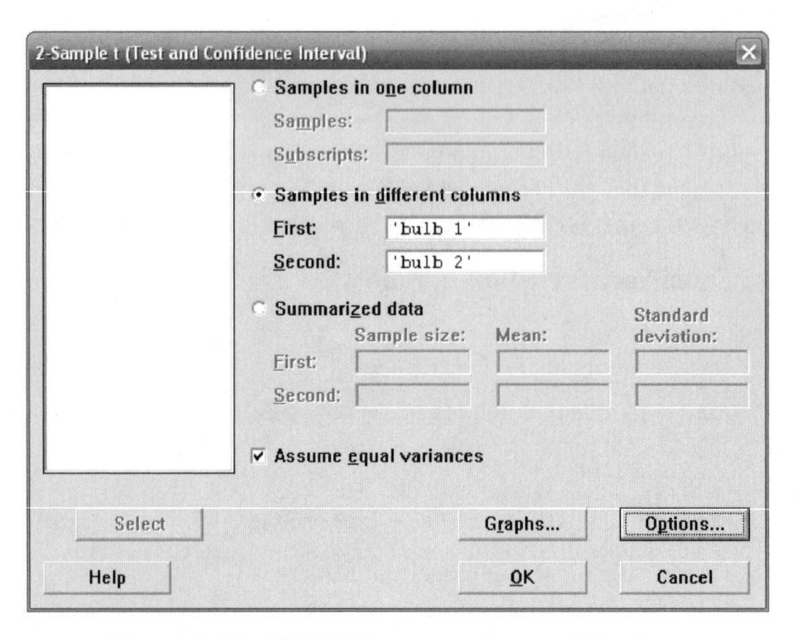

Figure 6-27 MINITAB two-sample t test dialog box.

Two-Sample *t*-Test and CI: Bulb 1, Bulb 2

```
Two-Sample t for Bulb 1 vs Bulb 2

         N    Mean      SD    SE Mean
Bulb 1   10   2026.9    82.5     26
Bulb 2   10   1845      111      35

Difference = mu (bulb 1) - mu (bulb 2)
Estimate for difference: 181.899
95% CI for difference: (89.957, 273.841)
t-test of difference = 0 (vs not =): t-value = 4.16   p-value = 0.001   DF = 18
Both use pooled SD = 97.8565
```

The point estimate for the difference in means is that bulb type 1 lasts about 182 hours on the average longer than bulb type 2. The hypothesis of no difference in mean lifetimes is rejected at $\alpha = 0.05$. Again, any one of the three methods of testing may be used. The three summaries are as follows:

1. The two-tailed rejection region is $t < -2.100922$ or $t > 2.100922$. The computed test statistic is $t = 4.16$. Since this computed test statistic is greater than 2.100922 and is therefore in the rejection region, the null is rejected for $\alpha = 0.05$. This *t*-value may be found using any of the software packages.
2. Using the *p*-value approach, compare the *p*-value = 0.001 with the preset $\alpha = 0.05$. Since the *p*-value is less than α, reject the null.
3. The confidence interval approach calculates the 95% two-sided confidence interval and if it does not contain $D_o = 0$, reject the null. The 95% confidence interval is (89.957, 273.841). Since 0 is not in this interval, reject the null.

NOTE *Any one of the three methods of testing may be used. You will always reach the same conclusion no matter which method you use.*

Now consider the output from other packages.

EXCEL

The pull-down **Tools ⇒ Data Analysis** leads to a dialog box where *t*-Test: 2-Sample Assuming Equal Variances may be chosen. This produces another dialog box which gives the analysis shown in Figure 6-28. This is the same analysis given by MINITAB.

t-Test: two-sample assuming equal variances		
	Variable 1	*Variable 2*
Mean	2027	1844.9
Variance	6774.444	12309.21
Observations	10	10
Pooled Variance	9541.828	
Hypothesized Mean		
Difference	0	
df	18	
t Stat	4.168494	
P(*T*<=*t*) one-tail	0.000289	
t Critical one-tail	1.734064	
P(*T*<=*t*) two-tail	0.000577	
t Critical two-tail	2.100922	

Figure 6-28 EXCEL analysis assuming equal variances.

STATISTIX

```
Statistix 8.0

Two-Sample t-Tests for Lifetime by Type

      Type        Mean       N          SD          SE
         1      2027.0      10      82.307      26.028
         2      1844.9      10      110.95      35.084
Difference      182.10

Null hypothesis: difference = 0
Alternative hypothesis: difference <> 0
                                                    95% CI for Difference
Assumption                  t        DF         p        Lower       Upper
Equal variances          4.17        18    0.0006       90.322      273.88
Unequal variances        4.17      16.6    0.0007       89.765      274.43

Test for equality           F        DF            p
      of variances       1.82       9,9       0.1935
```

The STATISTIX output provides a test for the equality of variances. The large *p*-value indicates that the equal variances model should be chosen. When the equal variances portion of the output is compared with the EXCEL and the MINITAB outputs it is seen to be the same.

SPSS

The SPSS output is:

Independent Samples Test

		Levene's Test for Equality of Variances		t-test for Equality of Means						
									95% Confidence Interval of the Difference	
		F	Sig.	*t*	df	Sig. (2-tailed)	Mean difference	Std. Error Difference	Lower	Upper
lifetime	Equal variances assumed	.293	.595	4.168	18	.001	182.10000	43.68484	90.32155	273.87845
	Equal variances not assumed			4.168	16.603	.001	182.10000	43.68484	89.76504	274.43496

Note that SPSS also gives both the equal and unequal variances model.

SAS

```
        Two-Sample t-Test for the Means of Lifetime within Type
Sample Statistics
     Group              N       Mean          SD       Std. Error
     ----------------------------------------------------------
     1                 10       2027        82.307       26.028
     2                 10       1844.9      110.95       35.084

Hypothesis Test
     Null hypothesis:     Mean 1 - Mean 2 = 0
     Alternative:         Mean 1 - Mean 2 ^= 0
     If Variances Are     t statistic       DF          Pr > t
     -----------------------------------------------------
     Equal                    4.168          18          0.0006
     Not equal                4.168       16.60          0.0007
```

SAS gives the choice of models just as SPSS. Once you decide on the equal or unequal variances model, select the output for that model.

UNEQUAL VARIANCES MODEL

The solution to this case is called the **Smith-Satterthwaite test**. I will give the equations, but the engineer will not have to be burdened with them. He can thank the software manufacturers.

To test H_o: $(\mu_1 - \mu_2) = D_o$ versus any one of the three alternative hypotheses H_a: $(\mu_1 - \mu_2) < D_o$, H_a: $(\mu_1 - \mu_2) > D_o$, or H_a: $(\mu_1 - \mu_2) \neq D_o$, use the test statistic

$$t = \frac{\bar{X}_1 - \bar{X}_2 - D_o}{\sqrt{\frac{S_1^2}{n_1} + \frac{S_2^2}{n_2}}}$$

The degrees of freedom are

$$v = \frac{\left(s_1^2 / n_1 + s_2^2 / n_2\right)^2}{\frac{\left(s_1^2 / n_1\right)^2}{n_1 - 1} + \frac{\left(s_2^2 / n_2\right)^2}{n_2 - 1}}$$

The interval $(\bar{X}_1 - \bar{X}_2) \pm t_{\alpha/2} \sqrt{\frac{S_1^2}{n_1} + \frac{S_2^2}{n_2}}$ is a $(1 - \alpha)$ confidence interval for $\mu_1 - \mu_2$ and

$$v = \frac{\left(s_1^2 / n_1 + s_2^2 / n_2\right)^2}{\frac{\left(s_1^2 / n_1\right)^2}{n_1 - 1} + \frac{\left(s_2^2 / n_2\right)^2}{n_2 - 1}}$$

EXAMPLE

An engineer wishes to compare the mean lifetimes of two bulb types. She selects samples of size 10 from each and determines the lifetimes of the 20 bulbs. The lifetimes (in hours) are shown in Figure 6-29.

The Kolmogorov-Smirnov test for normality for bulb type 1 gives a p-value > 0.150 and for bulb type 2 gives a p-value > 0.15. The normality assumption is reasonable for both populations.

Bulb type 1	Bulb type 2
1726	1787
1779	1973
1824	1803
1718	1903
1898	1971
1827	2378
1813	2226
1806	2040
1896	2200
1803	1675

Figure 6-29 Bulb lifetimes, unequal variability.

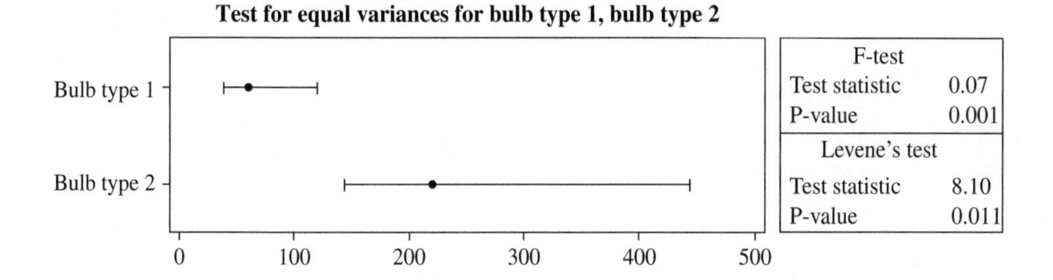

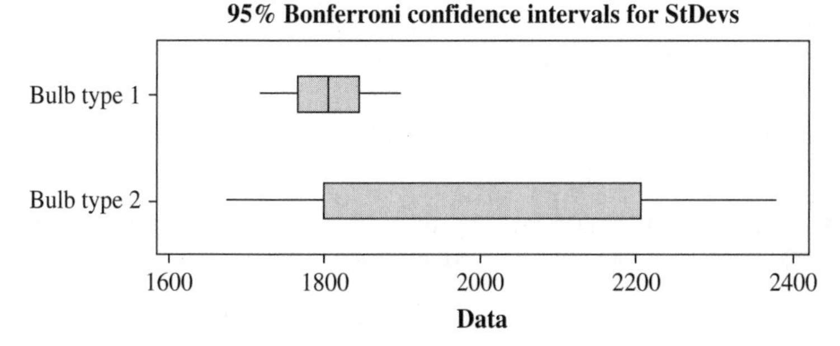

Figure 6-30 MINITAB output for the equal variances test.

The output for the test of equal variances is shown in Figure 6-30. The F-test is valid if both sets of data come from normal populations. We have already tested and accepted that both sets of data came from normal populations. The null hypothesis is that the variances are equal. This hypothesis is rejected since the p-value is much smaller than 0.05.

We have two small samples and we have concluded that the two populations do not have equal variances. We will test the equality of means first using MINITAB.

Two-Sample *t*-Test and CI: Bulb Type 1, Bulb Type 2

```
Two-Sample t for Bulb Type 1 vs Bulb Type 2

                                    SE
              N    Mean      SD   Mean
Bulb type 1  10  1809.0    59.8     19
Bulb type 2  10    1996     220     70

Difference = mu (bulb type 1) - mu (bulb type 2)
Estimate for difference:  -186.600
95% CI for difference:  (-347.328,-25.872)
t-test of difference = 0 (vs not =): t-value = -2.59  p-value = 0.027  DF = 10
```

The point estimate for the difference in means is that bulb type 2 lasts about 187 hours on the average longer than bulb type 1. The hypothesis of no difference

t-Test: two-sample assuming unequal variances		
	type 1	*type 2*
Mean	1809	1995.6
Variance	3574.444	48460.93
Observations	10	10
Hypothesized mean difference	0	
df	10	
t Stat	–2.5868	
P(*T*<=*t*) one-tail	0.013549	
t Critical one-tail	1.812461	
P(*T*<=*t*) two-tail	0.027098	
t Critical two-tail	2.228139	

Figure 6-31 EXCEL analysis assuming unequal variances.

in mean lifetimes is rejected at $\alpha = 0.05$. Again, any one of the three methods of testing may be used. The three summaries are as follows:

1. The two-tailed rejection region is $t < -2.228$ and $t > 2.228$. The computed test statistic is $t = -2.59$. Since this computed test statistic is therefore in the rejection region, the null is rejected for $\alpha = 0.05$. This *t*-value may be found using any of the software packages.
2. Using the *p*-value approach, compare the *p*-value = 0.027 with the preset $\alpha = 0.05$. Since the *p*-value is less than α, reject the null.
3. The confidence interval approach calculates the 95% two-sided confidence interval and if it does not contain $D_o = 0$, reject the null. The 95% confidence interval is (–347.328, –25.872). Since 0 is not in this interval, reject the null.

EXCEL

STATISTIX

The data file is structured for STATISTIX as shown in Figure 6-32. The lifetimes are referred to as the dependent variable and the bulb type as a categorical variable by the package STATISTIX. The dialog box is shown in Figure 6-33.

The STATISTIX output is as follows:

```
Statistix 8.0
Two-Sample t-Tests for Lifetime by Type

      Type        Mean      N         SD         SE
         1      1809.0     10     59.787     18.906
         2      1995.6     10     220.14     69.614
Difference     -186.60

Null hypothesis: difference = 0
Alternative hypothesis: difference <> 0
```

Lifetime	Type
1726	1
1779	1
1824	1
1718	1
1898	1
1827	1
1813	1
1806	1
1896	1
1803	1
1787	2
1973	2
1803	2
1903	2
1971	2
2378	2
2226	2
2040	2
2200	2
1675	2

Figure 6-32 Data file with lifetimes in one column and bulb type in the other.

```
                                                       95% CI for Difference
Assumption                t         DF          p          Lower        Upper
Equal variances       -2.59         18     0.0186        -338.15      -35.049
Unequal variances     -2.59       10.3     0.0265        -346.65      -26.546

Test for equality           F         DF            p
      of variances      13.56        9,9       0.0003
```

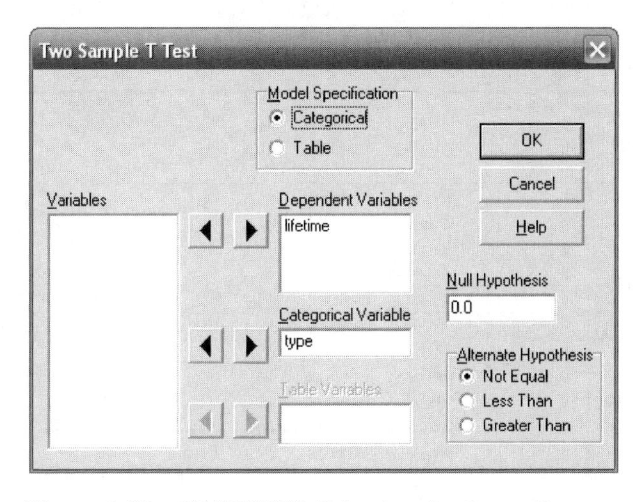

Figure 6-33 STATISTIX dialog box for 2-sample *t*-test.

The STATISTIX output gives the information needed to make a decision regarding equality of variance. Since the p-value equals 0.0003, I would reject the hypothesis of equal variances. I would then go to the assumption of unequal variances. The t-value is computed using the formula

$$t = \frac{\bar{X}_1 - \bar{X}_2 - D_o}{\sqrt{\frac{S_1^2}{n_1} + \frac{S_2^2}{n_2}}}$$

The degrees of freedom is computed using

$$v = \frac{\left(s_1^2 / n_1 + s_2^2 / n_2\right)^2}{\frac{\left(s_1^2 / n_1\right)^2}{n_1 - 1} + \frac{\left(s_2^2 / n_2\right)^2}{n_2 - 1}}$$

It is recommended that the degrees of freedom be rounded down when using this formula. The degrees of freedom would be 10. The information needed to make a decision regarding the null hypothesis of no difference in the means, using any one of the three methods, is contained in the output. Note that you can get the rejection region by using 10 degrees of freedom and Student's t distribution or you can compare the p-value = 0.0265 with $\alpha = 0.05$ or you can see that the confidence interval (−346.65, −26.546) does not contain $D_o = 0$.

SPSS

SPSS expects the data file as shown in Figure 6-32. The test for equality of variances uses Levene's test, which gives a different p-value than STATISTIX does. Otherwise, the output is the same as STATISTIX.

Independent Samples Test

		Levene's Test for Equality of Variances		*t*-Test for Equality of Means						
									95% Confidence Interval of the Difference	
		F	Sig.	*t*	df	Sig. (2-tailed)	Mean difference	Std. Error Difference	Lower	Upper
lifetime	Equal variances assumed	9.884	.006	−2.587	18	.019	−186.60000	72.13555	−338.151	−35.04883
	Equal variances not assumed			−2.587	10.320	.026	−186.60000	72.13555	−346.654	−26.54595

SAS

```
        Two-Sample t-Test for the Means of Lifetime within Type
   Sample Statistics
        Group              N       Mean          SD       Std. Error
        ---------------------------------------------------------
        1                 10       1809        59.787       18.906
        2                 10     1995.6        220.14       69.614

   Hypothesis Test
         Null hypothesis:     Mean 1 - Mean 2 = 0
        Alternative:         Mean 1 - Mean 2 ^= 0
        If Variances Are     t statistic       DF          Pr > t
        -------------------------------------------------------
        Equal                    -2.587         18          0.0186
        Not equal                -2.587      10.32          0.0265

  95% Confidence Interval for the Difference between Two Means
            Lower Limit    Upper Limit
            ------------   ------------
             -338.15         -35.05
```

SAS expects the data file to be in the form shown in Figure 6-32. The output is basically the same as SPSS and STATISTIX.

6.8 Inferences Concerning Two Means—Dependent Samples

In the previous two sections the two samples were purposely selected independently of one another. If a group of n objects are available for a study that was to have treatment 1 applied to n_1 units and treatment 2 applied to n_2 units and $n = n_1 + n_2$, we were interested in whether the mean for treatment 1 differs from the mean for treatment 2. The n objects are numbered 1 through n. For example, suppose 50 rats are available and they are numbered 1 through 50. Twenty-five numbers between 1 and 50 are randomly selected. Those rats with the numbers selected go into one group and receive treatment 1 or diet 1. The remaining rats go into group 2 and receive diet 2. The lifetime of each rat is recorded and at the end of the experiment, the mean age of each group is determined. It is of interest to know if the mean lifetimes of the two groups are equal or not. The analysis given in the previous section would be applied to the data.

Now, imagine a different scenario. Rats are paired up according to age, sex, and other factors that are important to longevity. For each pair, one of the rats is selected and a coin is flipped. If a head turns up, that rat is put on diet 1. If a tail turns up, that rat is placed on diet 2. The other member of the pair is placed on the other diet. In this way the treatments are randomly assigned to the pairs. The age at death is recorded for each member of each pair. It is of interest if the diets differ with respect to mean age reached by the rats. The two samples are dependent or paired. The analysis of such an experiment is what concerns us in this section.

EXAMPLE

Suppose 20 rats have been paired according to certain salient characteristics. The members of each pair have been randomly assigned to diet 1 or diet 2. There are 10 pairs. The experiment lasts until all 20 rats are dead. The age of each rat is recorded at death. The data is given in Figure 6-34.

The null hypothesis is that the mean difference is 0, or H_o: $\mu_d = 0$ versus one of three alternatives: H_a: $\mu_d < 0$, H_a: $\mu_d > 0$, or H_a: $\mu_d \neq 0$. The analysis is the same as for testing a single mean in Section 6.4. The small samples test is the *t*-test and the large samples test is the *z*-test. The differences are analyzed. The test statistic is $t = \frac{\bar{d} - 0}{s_d / \sqrt{n}}$ where *n* is the number of pairs. The Kolmogorov-Smirnov test confirms that normality is not rejected.

Using MINITAB, the results are:

One-Sample *t*: C3

```
Test of μ = 0 vs μ ≠ 0

Variable   N     Mean       SD       SE Mean          95% CI              t      p
C3        10   -54.6759  73.7114   23.3096   (-107.4059, -1.9460)   -2.35  0.044
```

The mean difference is $\bar{d} = -54.7$. The differences were formed by taking diet 1 lifetime minus diet 2 lifetime. The minus means diet 2 lifetime is larger on the average by 54.7 days than diet 1 lifetime. The difference is significant since the *p*-value < 0.05.

Diet 1	Diet 2	Difference
1271	1276	–5
1311	1359	–48
1221	1312	–91
1275	1247	28
1221	1279	–58
1289	1388	–99
1291	1296	–5
1393	1343	50
1224	1402	–178
1223	1365	–142

Figure 6-34 Ages at death (in days) of each rat for each diet.

EXAMPLE

Traffic engineers installed traffic control devices at 35 different intersections around the city. They recorded the average number of accidents per week before and after installing the devices. Do the following summary data supports the hypothesis that the devices reduced the number of accidents? $n = 35$, $\bar{d} = 10.5$, and $s_{\bar{d}} = 3.7$. The differences are formed by taking before and subtracting after.

The test statistic is

$$z = \frac{\bar{d} - 0}{s_d / \sqrt{n}} = \frac{10.5}{3.7} = 2.84$$

Positive differences tend to support the hypothesis that the devices reduced accidents.

MINITAB allows for the case where your data has been summarized. Figure 6-35 is the dialog box for testing a single mean for large samples. The dialog box asks for the sample size, the mean, the standard deviation, and the test mean. The sample size

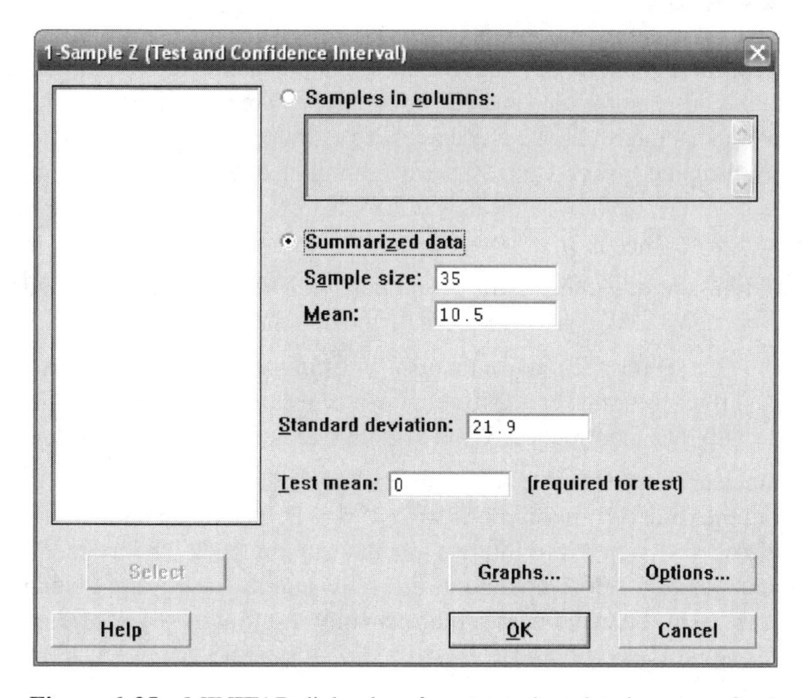

Figure 6-35 MINITAB dialog box for z-test when data is summarized.

is 35, the mean difference is 10.5 and the test mean is 0. Even though the standard deviation is not given, it can be figured. The formula for $s_{\bar{d}}$ is $s_{\bar{d}} = \frac{s_d}{\sqrt{n}}$. Solving for the standard deviation gives $s_d = \sqrt{n}(s_{\bar{d}})$. The output is

One-Sample *z*

```
Test of μ = 0 vs μ > 0
The assumed standard deviation = 21.9

                                95%
                               Lower
N      Mean     SE Mean      Bound      Z       P
35    10.5000    3.7018     4.4111    2.84   0.002
```

The null may be rejected because the p-value = $0.002 < \alpha = 0.05$. Or the one-sided confidence interval is (4.4111, ∞) and it does not contain the null value which is 0.

Exercises

1. Scientists are concerned about the declining population of brook trout. The brook trout requires cold and clean water. They are particularly sensitive to water temperatures above 68 degrees. To estimate the water temperature in streams in the eastern United States, a sampling of 75 locations was performed and it was found that the mean was 70.4 degrees and the standard deviation was 1.8.
   (a) Find the maximum error of estimate with 98% confidence when the sample mean is used to estimate the population mean.
   (b) After seeing the maximum error of estimate, it is decided to sample further to bring the maximum error of estimate down to 0.25 degrees with 99% confidence. What sample size is needed to do this?
2. A nuclear power plant is adjacent to a large reservoir and utilizes the water from the reservoir in the production of power. The mean temperature of the reservoir water is of interest to a nuclear engineer. Twenty random selections of water temperatures are made and the results are given in Figure 6-36. The data passes the normality test. Set a 95% confidence interval on μ using EXCEL, MINITAB, SAS, SPSS, and STATISTIX.
3. The balance of a die is in question. In particular, the probability of a 6 appearing on a given roll is believed not to be 1/6 = 0.167. The null hypothesis is H_o: $p = 0.167$ and the alternative is H_a: $p \neq 0.167$.

65.7	68.4
64.9	66.4
69.8	69.2
70.3	66.6
67.3	69.9
64.9	70.0
67.7	68.5
71.5	65.7
68.7	69.2
67.5	65.7

Figure 6-36 Twenty reservoir water temperatures.

The rejection region is chosen to be $X \leq 2$ or $X \geq 10$, where X = the number of times a 6 appears in 36 rolls of the die. Answer the following questions using EXCEL:

(a) Find α.

(b) Construct a graph of type 2 errors and a power curve.

(c) If the experiment is conducted and $X = 9$, find the p-value.

4. A large sample is used to test H_o: $\mu = 3$ milligrams per liter versus H_a: $\mu \neq$ 3 milligrams per liter, where μ is the mean fluoride in drinking water for a region. A sample of size 50 has a sample mean equal to 3.4 and a standard deviation equal to 1.5. α is set equal to 0.05. Use EXCEL and test using

   (a) The rejection region method

   (b) The confidence interval method

   (c) The p-value method

5. A small sample is used to test H_o: $\mu = 3$ milligrams per liter versus H_a: $\mu \neq$ 3 milligrams per liter, where μ is the mean fluoride in drinking water for a region. A sample of size 13 has a sample mean equal to 3.9 and a standard deviation equal to 1.5. α is set equal to 0.05. Use EXCEL and test the hypothesis using

   (a) The rejection region method

   (b) The confidence interval method

   (c) The p-value method

6. An industrial engineer wishes to test if two processes are producing products of the same weight on the average. She wishes to test H_o: $(\mu_1 - \mu_2) = 0$ versus the alternative hypothesis H_a: $(\mu_1 - \mu_2) \neq 0$, where μ_1 represents

the mean weight of products from process 1 and μ_2 represents the mean weight of products from process 2. She takes a sample of size 65 from process 1 and finds that the mean equals 16.10 ounces and the standard deviation equals 0.25 ounces. The sample of size 65 from process 2 has mean equal to 16.35 ounces and standard deviation equal to 0.55. Use MINITAB and test the hypothesis using

(a) The rejection region method

(b) The confidence interval method

(c) The p-value method ($\alpha = 0.05$.)

7. An industrial engineer wishes to test if two processes are producing products of the same weight on the average. She wishes to test $H_o: (\mu_1 - \mu_2) = 0$ versus the alternative hypothesis $H_a: (\mu_1 - \mu_2) \neq 0$ where μ_1 represents the mean weight of products from process 1 and μ_2 represents the mean weight of products from process 2. She takes a sample of size 10 from process 1 and finds that the mean equals 16.10 ounces and the standard deviation equals 0.25 ounces. The sample of size 15 from process 2 has mean equal to 16.35 ounces and standard deviation equal to 0.55. Use MINITAB and test the hypothesis using

   (a) The rejection region method

   (b) The confidence interval method

   (c) The p-value method ($\alpha = 0.05$)

   Assume normality of both populations and equal variances. Solve using MINITAB.

8. Nine automobiles were tested under stop-and-go as well as highway driving. The miles per gallon were measured and the results in Figure 6-37 were obtained.

Vehicle	Stop-and-go	Highway	Difference
1	25	29	4
2	30	33	3
3	27	35	8
4	25	30	5
5	23	26	3
6	27	33	6
7	20	28	8
8	31	33	2
9	29	34	5

Figure 6-37 Paired comparison of mpg for stop-and-go versus highway driving.

Process 1	Process 2
84.3	77.0
75.3	72.3
76.0	90.4
75.2	101.9
78.1	60.8
74.3	89.1
78.5	70.0
70.3	20.8
75.0	26.0
77.4	91.4

Figure 6-38 Pressures from two processes.

Test the null hypothesis H_o: $(\mu_1 - \mu_2) = 5$ versus H_a: $(\mu_1 - \mu_2) < 5$ at $\alpha = 0.05$ using EXCEL, MINITAB, and STATISTIX.

9. It is desired to compare the mean pressures in two processes. The engineer making the comparison is certain that the variability in pressure is not the same for the two processes. The sample data collected is shown in Figure 6-38.

   Calculate the degrees of freedom using EXCEL for the *t*-test that is used to compare means when equal variability is not assumed. Compare these computed degrees of freedom with the degrees of freedom given by MINITAB.

10. The mean lifetimes of the two bulbs are compared. Normality and equal variances are assumed. When the variances are equal, the two sample variances are pooled to form an estimate of the common population variance as follows:

$$S_p^2 = \frac{(n_1 - 1)S_1^2 + (n_2 - 1)S_2^2}{n_1 + n_2 - 2}$$

and the pooled standard deviation is the square root of this quantity. This pooled variance is part of the *t* variable as follows:

$$t = \frac{\overline{X}_1 - \overline{X}_2 - D_o}{\sqrt{S_p^2\left(\frac{1}{n_1} + \frac{1}{n_2}\right)}}$$

Use EXCEL to compute the pooled standard deviation for the following sample data and compare your results with the MINITAB output:

Bulb type 1	Bulb type 2
2164	1641
2040	1827
1926	1847
1911	1800
1973	1932
2064	1937
2018	2023
2097	1836
2105	1891
1972	1715

Summary

1. A **point estimator** is a single number used to estimate a parameter.
2. The notation $z_{\alpha/2}$ represents the **standard normal value**, with an area $\alpha/2$ under the standard normal curve to its right.
3. There will be $1 - \alpha$ area under the standard normal curve between $-z_{\alpha/2}$ and $z_{\alpha/2}$.
4. The quantity $|\bar{X} - \mu|$ is called the **maximum error of estimate**. It is the absolute difference between the population mean and our estimate of it.
5. Suppose we wish to specify the maximum error to be a specific value and ask the question: "What sample size will give this error?" The answer is found in the solution of the equation

$$E = z_{\alpha/2}\frac{\sigma}{\sqrt{n}} \quad \text{for } n$$

   If both sides of the equation are squared and solved for n, the solution is

$$n = \left[\frac{z_{\alpha/2}\sigma}{E}\right]^2$$

6. The interval

$$\bar{x} - z_{\alpha/2}\frac{\sigma}{\sqrt{n}} < \mu < \bar{x} + z_{\alpha/2}\frac{\sigma}{\sqrt{n}}$$

is generally referred to as the **large sample confidence interval** or the **z-interval**. If σ is unknown, it is replaced by the sample standard deviation and the interval takes on the following form:

$$\bar{x} - z_{\alpha/2}\frac{s}{\sqrt{n}} < \mu < \bar{x} + z_{\alpha/2}\frac{s}{\sqrt{n}}$$

The interval is referred to as a **confidence interval**. The $(1 - \alpha)$ is called the **degree of confidence**. The endpoints of the interval are called **confidence limits**.

7. When working with a small sample from a normally distributed population and using S as an estimate of σ, we have the following as a confidence interval for μ: The degrees of freedom to use with the $t_{\alpha/2}$ value is $(n - 1)$.

$$\bar{x} - t_{\alpha/2}\frac{S}{\sqrt{n}} < \mu < \bar{x} + t_{\alpha/2}\frac{S}{\sqrt{n}}$$

   This is generally referred to as the **small sample confidence interval** or the **t-interval**.

8. The **null hypothesis** is a hypothesis about one or more population parameters. The **alternative or research hypothesis** is a hypothesis that will be accepted if it is decided to reject the null hypothesis. The **test statistic** is computed from the sample data and we know the distribution of the test statistic. The **rejection region** is the values of the test statistic for which the null hypothesis will be rejected.
9. Rejecting the null hypothesis when it is true is a **type I error**. The probability of making a type I error is represented by the first letter of the Greek alphabet, α. Not rejecting the null hypothesis when it is false is a **type II error**. The probability of making a type II error is represented by the second letter of the Greek alphabet, β. The power of a statistical test, $(1 - \beta)$, is the probability of rejecting the null hypothesis when it is false.
10. When the entire rejection region is located in only one tail of the distribution, this is called a **one-tailed statistical test**. When the rejection region is located in both the upper and the lower tails of the distribution, this is called a **two-tailed statistical test**.
11. The **observed significance level** or **p-value** is the probability of observing a value of the test statistic that is at least as contradictory to the null hypothesis and supportive of the research hypothesis as the one computed from the sample data.

12. For statistical tests involving the binomial distribution, null hypothesis involving the balance of a coin, the balance of a die, and the like:

One-tailed test

H_o: $p = 0.5$ (the coin is balanced)

H_a: $p < 0.5$

$p > 0.5$

Two-tailed test

H_o: $p = 0.5$ (the coin is balanced)

H_a: $p \neq 0.5$

X = the number of heads to occur in n tosses of the coin.

Rejection region (one-tailed)

$0 \leq X \leq c$

$d \leq X \leq n$

$P(0 \leq X \leq c) = \alpha$ Left tail

$P(d \leq X \leq n) = \alpha$ Right tail

Rejection region (two-tailed)

$0 \leq X \leq e$ or $f \leq X \leq n$

$P(0 \leq X \leq e$ or $f \leq X \leq n) = \alpha$

Right and left tail

13. Large sample test about a population mean:

One-tailed test

H_o: $\mu = \mu_o$

H_a: $\mu < \mu_o$

$\mu > \mu_o$

Two-tailed test

H_o: $\mu = \mu_o$

H_a: $\mu \neq \mu_o$

$$\textbf{Test statistic: } Z = \frac{\bar{X} - \mu_o}{\sigma_{\bar{x}}}$$

Rejection region (one-tailed)

$Z < -z_\alpha$

$Z > z_\alpha$

Rejection region (two-tailed)

$|Z| > z_{\alpha/2}$

14. Small sample test about a population mean:

One-tailed test

H_o: $\mu = \mu_o$

H_a: $\mu < \mu_o$

$\mu > \mu_o$

Two-tailed test

H_o: $\mu = \mu_o$

H_a: $\mu \neq \mu_o$

$$\textbf{Test statistic: } t = \frac{\bar{X} - \mu_o}{\frac{s}{\sqrt{n}}}$$

Rejection region

$t < -t_{\alpha}$

$t > t_{\alpha}$

Rejection region

$|t| > t_{\alpha/2}$

Assumption: Normality assumption is satisfied, that is, the sample comes from a normal population.

15. Large sample test about $(\mu_1 - \mu_2)$: Independent samples

One-tailed test

$H_o: (\mu_1 - \mu_2) = D_o$

$H_a: (\mu_1 - \mu_2) < D_o$

$(\mu_1 - \mu_2) > D_o$

Two-tailed test

$H_o: (\mu_1 - \mu_2) = D_o$

$H_a: (\mu_1 - \mu_2) \neq D_o$

$$\textbf{Test statistic: } Z = \frac{\bar{X}_1 - \bar{X}_2 - D_o}{\sqrt{\frac{\sigma_1^2}{n_1} + \frac{\sigma_2^2}{n_2}}}$$

(Sample variances are substituted for unknown population variances when necessary.)

Rejection region

$Z < -z_{\alpha}$

$Z > z_{\alpha}$

Rejection region

$|Z| > z_{\alpha/2}$

16. Small sample test about $(\mu_1 - \mu_2)$: Independent samples, equal variances model

One-tailed test

$H_o: (\mu_1 - \mu_2) = D_o$

$H_a: (\mu_1 - \mu_2) < D_o$

$(\mu_1 - \mu_2) > D_o$

Two-tailed test

$H_o: (\mu_1 - \mu_2) = D_o$

$H_a: (\mu_1 - \mu_2) \neq D_o$

$$\textbf{Test statistic: } t = \frac{\bar{X}_1 - \bar{X}_2 - D_o}{\sqrt{S_p^2\left(\frac{1}{n_1} + \frac{1}{n_2}\right)}}$$

where

$$S_p^2 = \frac{(n_1 - 1)S_1^2 + (n_2 - 1)S_2^2}{n_1 + n_2 - 2}$$

Rejection region

$t < -t_\alpha$

$t > t_\alpha$

and $\nu = n_1 + n_2 - 2$

Rejection region

$|t| > t_{\alpha/2}$

Assumptions:

(a) Both populations are normally distributed.

(b) The variances of both populations are equal.

17. Small sample test about $(\mu_1 - \mu_2)$: Independent samples, unequal variances model

SMITH-SATTERTHWAITE TEST

One-tailed test

$H_o: (\mu_1 - \mu_2) = D_o$

$H_a: (\mu_1 - \mu_2) < D_o$

$(\mu_1 - \mu_2) > D_o$

Two-tailed test

$H_o: (\mu_1 - \mu_2) = D_o$

$H_a: (\mu_1 - \mu_2) \neq D_o$

Test statistic: $t = \dfrac{\overline{X}_1 - \overline{X}_2 - D_o}{\sqrt{\dfrac{S_1^2}{n_1} + \dfrac{S_2^2}{n_2}}}$

Rejection region

$t < -t_\alpha$

$t > t_\alpha$

and $\nu = \left(s_1^2 / n_1 + s_2^2 / n_2\right)^2 \Big/ \left(\frac{(s_1^2/n_1)^2}{n_1 - 1} + \frac{(s_2^2/n_2)^2}{n_2 - 1}\right)$ (round down to the nearest integer)

Rejection region

$|t| > t_{\alpha/2}$

Assumptions:

(a) Both populations are normally distributed.

(b) The variances of both populations are unequal.

18. Large sample test about $(\mu_1 - \mu_2)$: Dependent samples and matched pairs

One-tailed test

$H_o: (\mu_1 - \mu_2) = D_o$

Two-tailed test

$H_o: (\mu_1 - \mu_2) = D_o$

$H_a: (\mu_1 - \mu_2) < D_o$ $\qquad$ $H_a: (\mu_1 - \mu_2) \neq D_o$

$(\mu_1 - \mu_2) > D_o$

$$\textbf{Test statistic: } t = \frac{\bar{d} - D_o}{s_d/\sqrt{n}}$$

where d with the bar and s_d are the mean and standard deviation of differences respectively.

Rejection region $\qquad$ Rejection region

$Z < -z_\alpha$ $\qquad$ $|Z| > z_{\alpha/2}$

$Z > z_\alpha$

19. Small sample test about $(\mu_1 - \mu_2)$: Dependent samples and matched pairs

One-tailed test $\qquad$ Two-tailed test

$H_o: (\mu_1 - \mu_2) = D_o$ $\qquad$ $H_o: (\mu_1 - \mu_2) = D_o$

$H_a: (\mu_1 - \mu_2) < D_o$ $\qquad$ $H_a: (\mu_1 - \mu_2) \neq D_o$

$(\mu_1 - \mu_2) > D_o$

$$\textbf{Test statistic: } t = \frac{\bar{d} - D_o}{s_d/\sqrt{n}}$$

where d with the bar and s_d are the mean and standard deviation of differences respectively.

Rejection region $\qquad$ Rejection region

$t < -t_\alpha$ $\qquad$ $|t| > t_{\alpha/2}$

$t > t_\alpha$

ν = number of pairs – 1.

Assumption: The differences are normally distributed.

20. The hypotheses are tested using one of the following three methods:

(a) The rejection region is formed with reference to the alternative hypothesis and α. If the test statistic, when computed, falls in the rejection region, then the null hypothesis is rejected.

(b) If the p-value is less than α, then the null hypothesis is rejected.

(c) If a confidence interval of size $(1 - \alpha)$ is formed for the parameter and the value stated for the parameter in the null does not fall inside the confidence interval, then the null hypothesis is rejected.

CHAPTER 7

Inferences Concerning Variances

7.1 The Estimation of Variances

In this chapter we shall investigate how to make inferences about the variance of a single population and about the ratio of the variances of two populations. In particular, in this section we will see how to set a confidence interval on the variance of a single population or the ratio of two population variances. In Section 7.2, we shall investigate testing hypothesis about one population variance. In Section 7.3, we shall investigate testing hypothesis about two population variances.

The distribution of the variable $\frac{(n-1)S^2}{\sigma^2}$ is chi-square with $(n-1)$ degrees of freedom, σ^2 is the population variance and S^2 is the sample variance. The population from which the sample is taken has a normal distribution. The symbol χ^2_α represents the value on the horizontal axis of the chi-square curve with area α to its right. Recall that area represents probability.

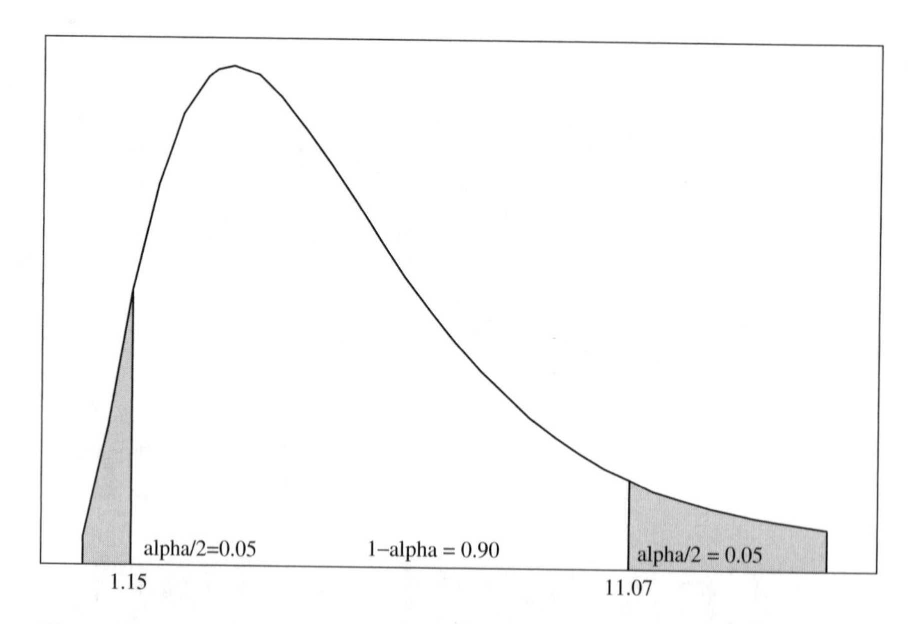

Figure 7-1 Chi-square distribution with 5 degrees of freedom, and $\chi^2_{0.05} = 11.07$ and $\chi^2_{0.95} = 1.15$.

Figure 7-1 shows a chi-square distribution with $\nu = 5$ degrees of freedom. The points $\chi^2_{0.05} = 11.07$ and $\chi^2_{0.95} = 1.15$ are also shown. They are found using EXCEL as follows: CHIINV(0.05,5) gives 11.0705 and CHIINV(0.95,5) gives 1.1455.

EXAMPLE

Take a random sample of size 10 from a normal population with unknown variance σ^2. Suppose we wish to construct a 95% confidence interval for σ^2. The variable $\frac{9S^2}{\sigma^2}$ will have a chi-square distribution with 9 degrees of freedom. Using EXCEL, the function =CHIINV(0.025,9) gives 19.02277. There is 0.025 area to the right of 19.02277 on the horizontal axis of the chi-square curve. There is 0.975 area to the right of CHIINV(0.975,9), which equals 2.70039. Now, we may write the following concerning the area under the curve:

$$P\left(2.700 < \frac{(n-1)S^2}{\sigma^2} < 19.023\right) = 0.95$$

Solving the inequality inside the probability statement for σ^2, we find

$$\frac{(10-1)S^2}{19.023} < \sigma^2 < \frac{(10-1)S^2}{2.0700}$$

Once the sample of size 10 is taken and S^2 is replaced by its value, we will have a 95% confidence interval for σ^2.

Generalizing from the example, we have the following $(1 - \alpha)$ confidence interval for σ^2:

$$\frac{(n-1)S^2}{\chi^2_{\alpha/2}} < \sigma^2 < \frac{(n-1)S^2}{\chi^2_{1-\alpha/2}}$$

It is interesting to contrast EXCEL and MINITAB as to how they find the values $\chi^2_{.025}$ and $\chi^2_{0.975}$. The value for $\chi^2_{.025}$ is 19.023. Using MINITAB, the area to the left of $\chi^2_{.025}$ is 0.975. Using the pull-down **Calc ⇒ Probability Distribution ⇒ Chi-Square** and choosing inverse cumulative probability gives the following:

Inverse Cumulative Distribution Function

```
Chi-Square with 9 DF

P(X <= x)          x
      0.975   19.0228
```

The area to the right of $\chi^2_{.025}$ is 0.025 and to the left of $\chi^2_{.025}$ is 0.975.

EXAMPLE

An industrial engineer is interested in the variation of content in containers labeled as 180 fluid ounces. Twenty are selected from production and the volumes in fluid ounces are shown in Figure 7-2. Set a 99% confidence interval on the variance in volume for this process. First, we need to confirm that the sample data comes from a normal population. The Kolmogorov-Smirnov test gives a p-value > 0.15. Normality is therefore a reasonable assumption.

Each bottle is labeled as 180 fluid ounces. The contents of each is measured to the nearest 10th to gain an idea of the variation in the contents.

181.5	180.8
179.1	182.4
178.7	178.5
183.9	182.2
179.7	180.9
180.6	181.4
180.4	181.4
178.5	180.6
178.8	180.1
181.3	182.2

Figure 7-2 Volume in fluid ounces.

Figure 7-3 shows the EXCEL computation of a 99% confidence interval for the population variance. The data is contained in A1:B10. The S^2 computation is in C1. The $19S^2$ computation is in C2. The chi-square values are in C3:C4. The lower limit of the 99% confidence interval is in C6 and the upper limit is in C7. The commands in column C are shown in column D. The population variance is between 1.061 and 5.980, with 99% confidence. The 99% confidence interval for the population standard deviation is found by taking square roots. The 99% confidence interval for σ is (1.030, 2.445).

If SAS is used to set a 99% confidence interval on the variance, the following is found:

```
                    One-Sample Chi-Square Test for  Variance
Sample Statistics for Volume
       N       Mean             SD        Variance
   -------------------------------------------------
     20     180.65            1.4677          2.1542
99% Confidence Interval for the Variance
                   Lower Limit    Upper Limit
                   -----------    -----------
                     1.06085        5.98045
```

A	B	C	D
181.5	180.8	2.154211	= VAR(A1:B10)
179.1	182.4	40.93	= 19*C1
178.7	178.5	38.58226	= CHIINV(0.005,19)
183.9	182.2	6.843971	= CHIINV(0.995,19)
179.7	180.9		
180.6	181.4	1.06085	= C2/C3
180.4	181.4	5.980446	= C2/C4
178.5	180.6		
178.8	180.1		
181.3	182.2		

Figure 7-3 EXCEL worksheet computation of 99% confidence interval for σ^2.

When the sample variances, S_1^2 and S_2^2, based on independent sample sizes n_1 and n_2, are taken from two normal populations, it can be shown that $\frac{S_1^2}{S_2^2}\frac{\sigma_2^2}{\sigma_1^2}$ has an F distribution with $\nu_1 = n_1 - 1$ and $\nu_2 = n_2 - 1$ degrees of freedom. Let $F_L = F_{1-\alpha/2}$ and $F_U = F_{\alpha/2}$, then we may write

$$P\left(F_L < \frac{S_1^2}{S_2^2}\frac{\sigma_2^2}{\sigma_1^2} < F_U\right) = 1 - \alpha$$

Now, we solve the inequality

$$F_L < \frac{S_1^2}{S_2^2}\frac{\sigma_2^2}{\sigma_1^2} < F_U \text{ for } \frac{\sigma_1^2}{\sigma_2^2}$$

and we will have a $(1 - \alpha)$ confidence interval for the ratio of population variances. After a bit of algebra, we have the following $(1 - \alpha)$ confidence interval for the ratio of population variances:

$$\frac{S_1^2}{S_2^2}\cdot\frac{1}{F_{\alpha/2(\nu_1,\nu_2)}} < \frac{\sigma_1^2}{\sigma_2^2} < \frac{S_1^2}{S_2^2}F_{\alpha/2(\nu_2,\nu_1)}$$

EXAMPLE

Two assembly lines using different methods of assembling PlayStation 3s (PS3s) were compared. The number of PS3s assembled per day was recorded for assembly line 1 for 13 days and the number assembled for assembly line 2 for 15 days. The engineer in charge of the process wanted a 95% confidence interval for the ratio of the variance of line 1 to the variance of line 2. The data for the two assembly lines is shown in Figure 7-4.

Line 1	Line 2
81	99
104	100
115	104
111	98
85	103
121	113
95	95
112	107
100	98
117	95
113	101
109	109
101	99
	93
	105

Figure 7-4 Number of PS3s assembled for two assembly lines.

A	B	C	D
Line 1	Line 2	148.5769231	= VAR(A2:A14)
81	99	31.06666667	= VAR(B2:B16)
104	100	3.050154789	= FINV(0.025,12,14)
115	104	3.2062117	= FINV(0.025,14,12)
111	98	1.567959436	= (C1/C2)/C3
85	103	15.33376832	= (C1/C2)*C4
121	113		
95	95	1.25218187	= SQRT(C5)
112	107	3.915835584	= SQRT(C6)
100	98		
117	95		
113	101		
109	109		
101	99		
	93		
	105		

Figure 7-5 EXCEL computation of 95% confidence interval for σ_1/σ_2.

Both sets of sample data pass the Kolmogorov-Smirnov normality test with p-values > 0.15.

Figure 7-5 gives the EXCEL computation of

$$\frac{S_1^2}{S_2^2}\cdot\frac{1}{F_{\alpha/2(v_1,v_2)}}<\frac{\sigma_1^2}{\sigma_2^2}<\frac{S_1^2}{S_2^2}F_{\alpha/2(v_2,v_1)}$$

i.e., the 95% confidence interval for $\frac{\sigma_1^2}{\sigma_2^2}$. The data is shown in A2:B16. The sample variance of assembly line 1 is shown in C1 as 148.58, and the sample variance of assembly line 2 is shown in C2 as 31.07. The commands given in column C are shown in column D. The 95% confidence interval for $\frac{\sigma_1^2}{\sigma_2^2}$ is (1.568, 15.334), and the 95% confidence interval for $\frac{\sigma_1}{\sigma_2}$ is (1.252, 3.916).

The SAS solution to the above example is as follows:

```
Two-Sample Test for Variances of Units within Line
Sample Statistics
        Line
        Group           N       Mean          SD        Variance
        ----------------------------------------------------------
        1              13    104.9231     12.189     148.5769
        2              15    101.2667     5.5737     31.06667
     95% Confidence Interval of the Ratio of Two Variances
              Lower Limit    Upper Limit
             -----------    -----------
                1.568          15.334
```

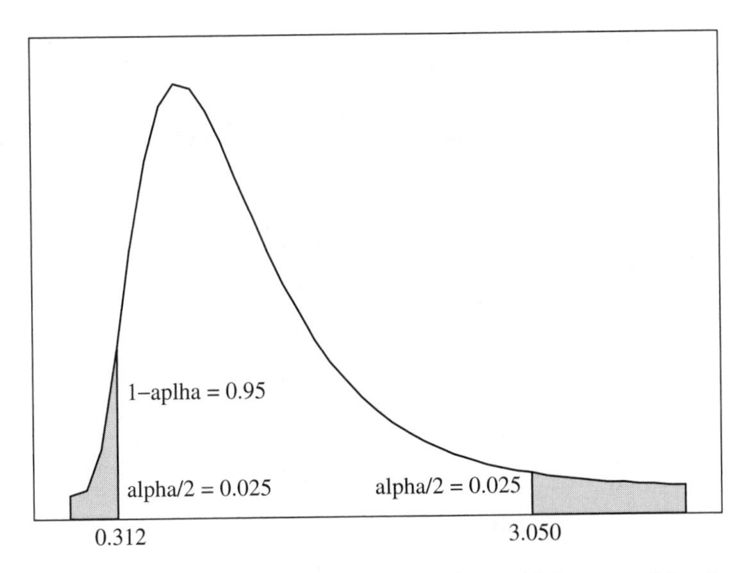

Figure 7-6 *F* distribution with $\nu_1 = 12$ and $\nu_2 = 14$ degrees of freedom.

EXAMPLE

Construct the *F* curve for $\nu_1 = 12$ and $\nu_2 = 14$ degrees of freedom and locate the points that give 0.025 area in the left-hand tail and 0.025 in the right-hand tail.

The two points shown on the graph in Figure 7-6 may be confirmed using MINITAB as follows:

Inverse Cumulative Distribution Function

```
F Distribution with 12 DF in Numerator and 14 DF in Denominator

  P(X <= x)        x
    0.025    0.311895
```

Inverse Cumulative Distribution Function

```
F Distribution with 12 DF in Numerator and 14 DF in Denominator

  P(X <= x)        x
    0.975    3.05015
```

7.2 Hypotheses Concerning One Variance

Testing a hypothesis about the variance of one population is the topic for discussion in this section. First, we shall outline the technique and then apply the technique to an example. We shall call upon some of the theory we have learned in Section 7.1 when estimating population variances.

One-tailed test	Two-tailed test
H_o: $\sigma^2 = \sigma_o^2$	H_o: $\sigma^2 = \sigma_o^2$
H_a: $\sigma^2 < \sigma_o^2$	H_a: $\sigma^2 \neq \sigma_o^2$
$\sigma^2 > \sigma_o^2$	

Test statistic: $\chi^2 = \frac{(n-1)S^2}{\sigma_o^2}$ has a chi-square distribution with $\nu = n - 1$.

Rejection region	Rejection region
$\chi^2 < \chi^2_{1-\alpha}$	$\chi^2 < \chi^2_{1-\alpha/2}$ or $\chi^2 > \chi^2_{\alpha/2}$
$\chi^2 > \chi^2_{\alpha}$	

Assumption: The population from which the sample is selected is normal.

EXAMPLE

An industrial engineer is interested in the variation of contents in containers labeled as 180 fluid ounces. Twenty are selected from production and the volumes in fluid ounces are shown in Figure 7-7. Test the null hypothesis that the standard deviation of the population is 1.5 versus the alternative that the standard deviation is not 1.5, at $\alpha = 0.05$. Test using EXCEL and the three methods that have been discussed in previous sections, and then follow this by SAS output that gives the solution.

First, we need to confirm that the sample data comes from a normal population. The Kolmogorov-Smirnov test gives a p-value > 0.15. Normality is therefore a reasonable assumption.

181.5	180.8
179.1	182.4
178.7	178.5
183.9	182.2
179.7	180.9
180.6	181.4
180.4	181.4
178.5	180.6
178.8	180.1
181.3	182.2

Figure 7-7 Volume in fluid ounces.

Method 1

The rejection region will be found first using EXCEL. The EXCEL solution is shown in Figure 7-8. The data is shown in A2:A21. The critical values are shown in B1:B2. The computed test statistic is given in B4. The EXCEL commands are given in C1, C2, and C4. The figure tells us that the rejection region is $\chi^2 < 8.906$ or $\chi^2 > 32.852$. The computed test statistic is 18.191. We are unable to reject the null hypothesis at $\alpha = 0.05$.

Figure 7-9 shows the chi-square distribution, the rejection region, and the computed test statistic.

Method 2

Compute the *p*-value, and then follow the rule that if the *p*-value is less than the preset α, reject the null. The *p*-value is given by 2*(1-CHIDIST(18.19,19)), which equals 0.98. Since 0.98 is not less than 0.05 the null is not rejected.

A	B	C
Contents	32.85233	= CHIINV(0.025,19)
181.5	8.906517	= CHIINV(0.975,19)
179.1		
178.7	18.19111	= 19*VAR(A2:A21)/2.25
183.9		
179.7		
180.6		
180.4		
178.5		
178.8		
181.3		
180.8		
182.4		
178.5		
182.2		
180.9		
181.4		
181.4		
180.6		
180.1		
182.2		

Figure 7-8 EXCEL solution to $H_o: \sigma^2 = 2.25$ versus $H_a: \sigma^2 \neq 2.25$.

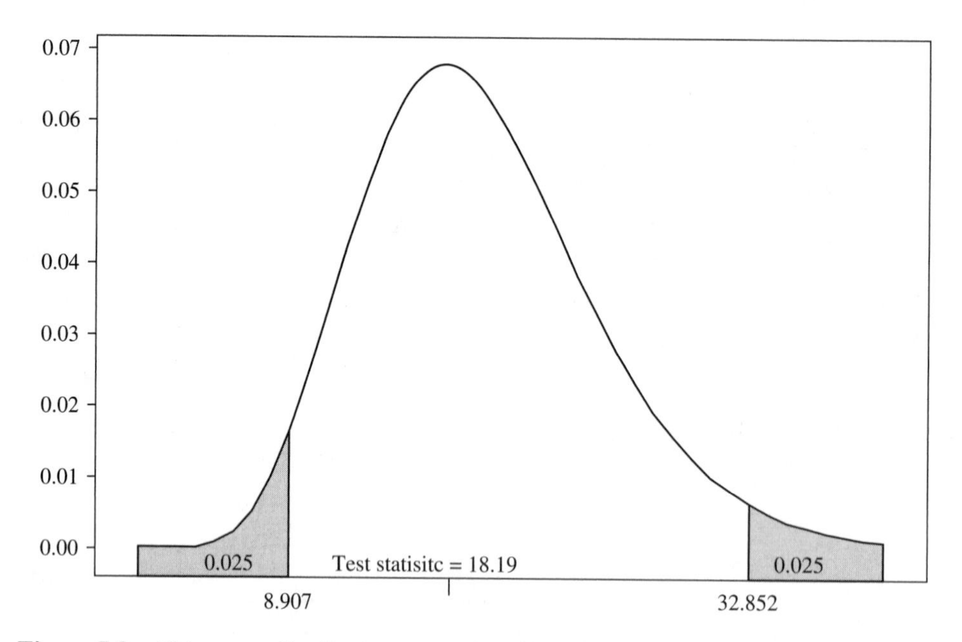

Figure 7-9 Chi-square distribution curve showing rejection region and test statistic for a two-tailed test.

Method 3

The 95% confidence interval is computed using

$$\frac{(n-1)S^2}{\chi^2_{0.025}} < \sigma^2 < \frac{(n-1)S^2}{\chi^2_{0.975}}$$

$$19*\text{VAR}(\text{A2:A21})/\text{CHIINV}(0.025,19) = 1.246$$

and

$$19*\text{VAR}(\text{A2:A21})/\text{CHIINV}(0.975,19) = 4.596$$

Since $\sigma_o^2 = 2.25$ is contained in the interval (1.246, 4.596), we do not reject the null hypothesis.

The SAS solution to the test of hypothesis is as follows:

```
                    One-Sample Chi-Square Test for a Variance
      Sample Statistics for Volume
           N        Mean          SD           Variance
        ---------------------------------------------
          20     180.65        1.4677          2.1542
   Hypothesis Test
        Null hypothesis:          Variance of volume = 2.25
```

181.5
179.1
178.7
183.9
179.7
180.6
180.4
178.5
178.8
181.3

Figure 7-10 Volume in fluid ounces.

```
Alternative:              Variance of volume ^= 2.25
      Chi-square          DF          Prob
      ---------------------------------
        18.191            19          0.9806
95% Confidence Interval for the Variance
                  Lower Limit    Upper Limit
                  -----------    -----------
                    1.2459          4.5955
```

The SAS output above contains the same information as the EXCEL output.

EXAMPLE

An industrial engineer is interested in the variation of contents in containers labeled as 180 fluid ounces. Ten are selected from production and the volumes in fluid ounces are shown in Figure 7-10. Test the null hypothesis that the standard deviation of the population is 1.0 versus the alternative that the standard deviation is greater than 1.0, at $\alpha = 0.05$. Test using EXCEL.

The EXCEL solution is shown in Figure 7-11. The critical value and the test statistic are shown. Since the computed test statistic falls in the rejection region,

A	B	C
Contents	16.91898	CHIINV(0.05, 9)
181.5		
179.1	25.325	9*VAR(A2:A11)/1
178.7		
183.9		
179.7		
180.6		
180.4		
178.5		
178.8		
181.3		

Figure 7-11 EXCEL solution to $H_o\!:\sigma^2 = 1.0$ versus $H_a\!:\sigma^2 > 1.0$.

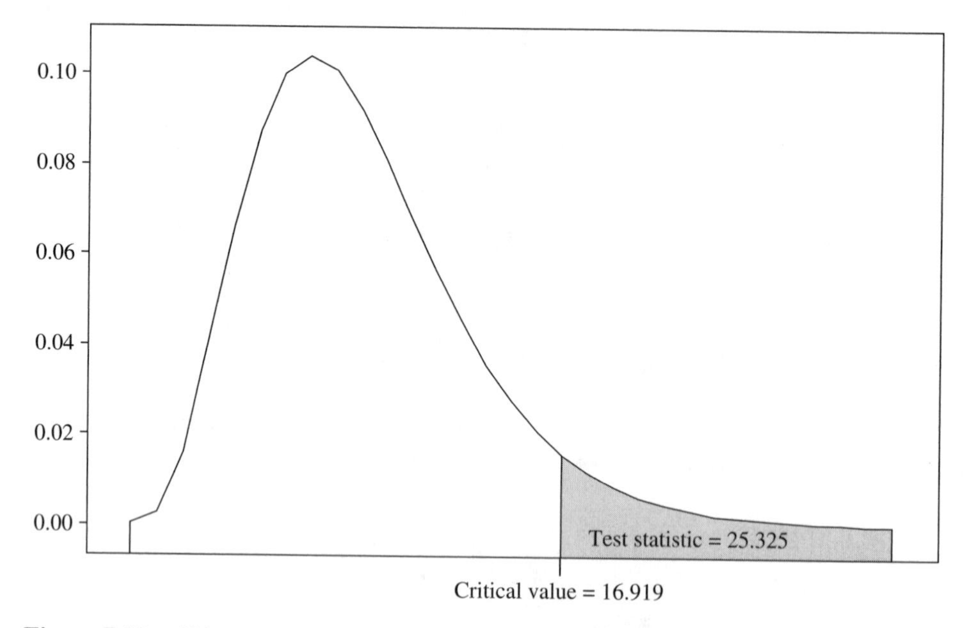

Figure 7-12 Chi-square distribution curve showing rejection region and test statistic for a one-tailed test (upper-tailed).

the null hypothesis is rejected. If the null hypothesis were true and the population variance were equal to 1.0, there would be less than a 5% chance of getting a sample variance equal to 2.814 or equivalently, a test statistic equal to 25.325.

The conclusion is that the population variance is most likely greater than 1.

EXAMPLE

An industrial engineer is interested in the variation of contents in containers labeled as 180 fluid ounces. Ten are selected from production and the volumes in fluid ounces are shown in Figure 7-13. Test the null hypothesis that the standard deviation of the

180.8
182.4
178.5
182.2
180.9
181.4
181.4
180.6
180.1
182.2

Figure 7-13 Volume in fluid ounces.

A	B	C
Contents	3.325113	= CHIINV(0.95,9)
180.8		
182.4	5.513333	= 9*VAR(A2:A11)/2.25
178.5		
182.2		
180.9		
181.4		
181.4		
180.6		
180.1		
182.2		

Figure 7-14 EXCEL solution to $H_o:\sigma^2 = 1.5$ versus $H_a:\sigma^2 < 1.5$.

population is 1.5 versus the alternative that the standard deviation is less than 1.5, at $\alpha = 0.05$. Test using EXCEL.

The EXCEL solution is shown in Figure 7-14. The critical value and the test statistic are shown in Figure 7-15. Since the computed test statistic does not fall in the rejection region, the null hypothesis is not rejected. It would not be considered unusual to get a sample variance equal to 1.378 if the population variance equaled 2.25.

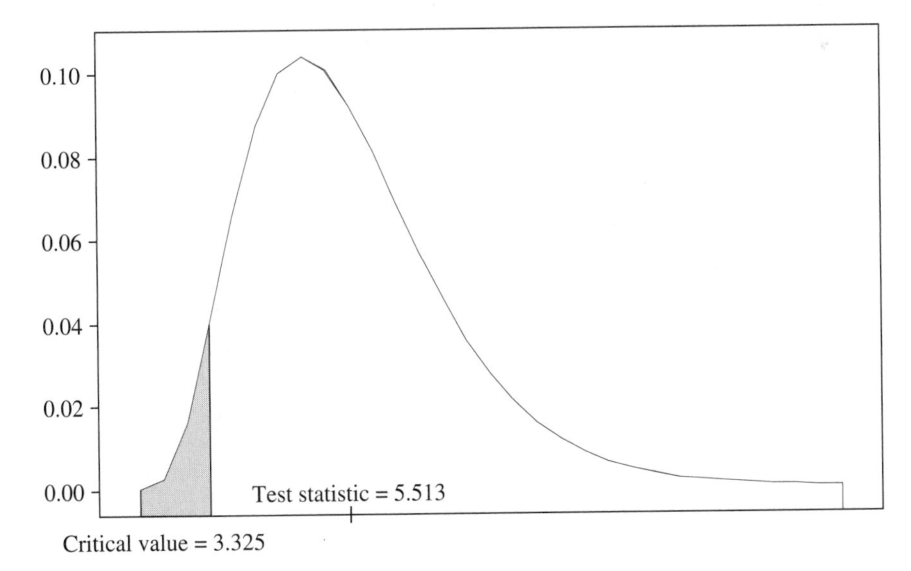

Figure 7-15 Chi-square distribution curve showing rejection region and test statistic for a one-tailed test (lower-tailed).

7.3 Hypotheses Concerning Two Variances

In this section we will start by giving an outline of the hypothesis test of the ratio of two variances.

One-tailed test

$H_o: \frac{\sigma_1^2}{\sigma_2^2} = 1$

$Ha: \frac{\sigma_1^2}{\sigma_2^2} < 1$

$H_a: \frac{\sigma_1^2}{\sigma_2^2} > 1$

Two-tailed test

$H_o: \frac{\sigma_1^2}{\sigma_2^2} = 1$

$H_a: \frac{\sigma_1^2}{\sigma_2^2} \neq 1$

Test statistic: $F = \frac{S_1^2}{S_2^2}$ has an F distribution with $\nu_1 = n_1 - 1$ and $\nu_2 = n_2 - 1$. ν_1 is called the numerator degrees of freedom and ν_2 the denominator degrees of freedom.

Rejection region (one-tailed)

$F < F_\alpha$

$F > F_\alpha$

Rejection region (two-tailed)

$F < F_{\alpha/2}$ or $F > F_{\alpha/2}$

Assumptions:

1. The samples are from normal populations.
2. The samples are selected independently of one another.

EXAMPLE

Two assembly lines using two different methods of assembling PS3s were compared. The number of PS3s assembled per day was recorded for assembly line 1 for 5 days and the number assembled for assembly line 2 for 7 days. The engineer in charge of the process wanted to test if the ratio of variances was equal to 1 versus not equal to 1. The data for the two assembly lines is shown in Figure 7-16.

Line 1	Line 2
81	99
104	100
115	104
111	98
85	103
	113
	95

Figure 7-16 Number of PS3s assembled for two assembly lines.

A	B	C	D
Line 1	Line 2	6.227161	= FINV(0.025, 4, 6)
81	99	0.108727	= FINV(0.975, 4, 6)
104	100	6.966573	= VAR(A2:A6)/VAR(B2:B8)
115	104		
111	98		
85	103		
	113		
	95		

Figure 7-17 EXCEL test $H_o: \frac{\sigma_1^2}{\sigma_2^2} = 1$ versus $H_a: \frac{\sigma_1^2}{\sigma_2^2} \neq 1$.

The EXCEL computation of the test statistic and the critical values are shown in Figure 7-17.

The rejection regions and the computed test statistic are shown with the *F* distribution in Figure 7-18. Since the computed test statistic falls within the rejection region, reject the null hypothesis and conclude that there is more variability in assembly line 1 than in assembly line 2.

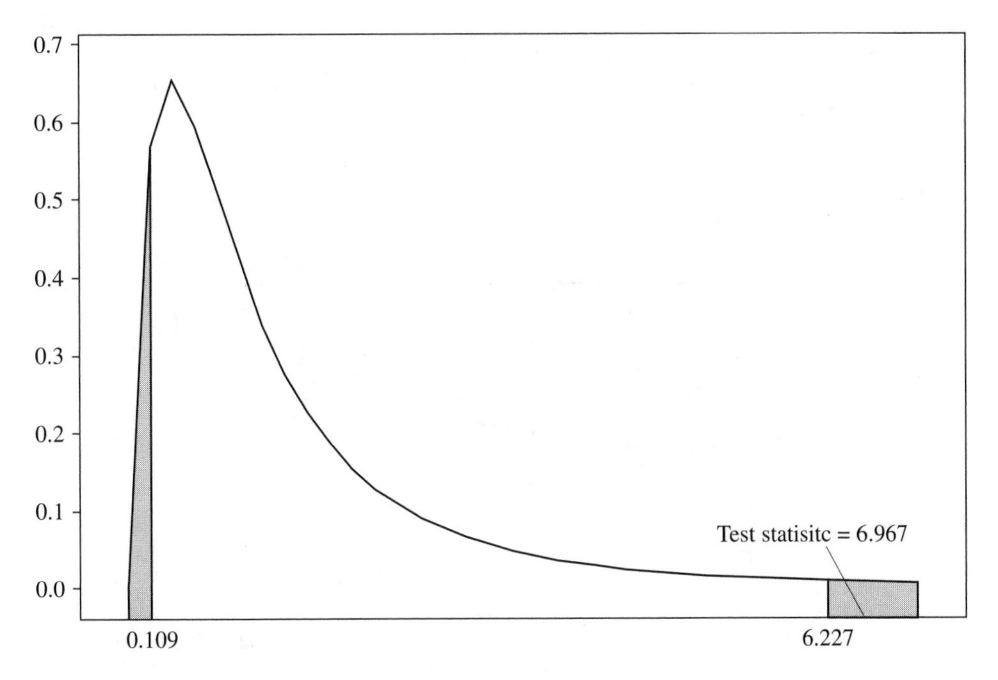

Figure 7-18 *F* distribution curve showing rejection region and test statistic for a two-tailed test.

Let us see how some of the software programs deal with this test of hypothesis. EXCEL has a built-in test of

$$H_o: \frac{\sigma_1^2}{\sigma_2^2} = 1 \text{ versus } H_a: \frac{\sigma_1^2}{\sigma_2^2} \neq 1$$

Use the EXCEL pull-down **Tools ⇒ Data Analysis ⇒ *F*-Test Two-Sample for Variances** to perform the test of hypothesis in Figure 7-17. The output for the test is shown in Figure 7-19.

The following are shown in Figure 7-19:

$$S_1^2 = 236.2,\ S_2^2 = 33.9,\ F = \frac{S_1^2}{S_2^2} = 6.97$$

The one-tailed *p*-value is

$$P(F > 6.967) = 0.019288$$

Since the test is two-tailed, *p*-value = 2(0.019288) = 0.039. The $\alpha = 0.05$ right-hand critical value is also given. A little work is saved if the built-in EXCEL routine in Figure 7-19 is used rather than the functions in Figure 7-17.

The MINITAB solution to the example uses the pull-down **Stat ⇒ Basic Statistics ⇒ Two Variances**. Figure 7-20 shows output graphic results.

The *F* test results are shown along with the two-tailed *p*-value in the upper right corner of Figure 7-20.

F-test two-sample for variances		
	Line 1	Line 2
Mean	99.2	101.7143
Variance	236.2	33.90476
Observations	5	7
df	4	6
F	6.966573	
P(<=*f*) one-tail	0.019288	
F Critical one-tail	4.533677	

Figure 7-19 EXCEL pull-down test $H_o: \frac{\sigma_1^2}{\sigma_2^2} = 1$ versus $H_a: \frac{\sigma_1^2}{\sigma_2^2} \neq 1$.

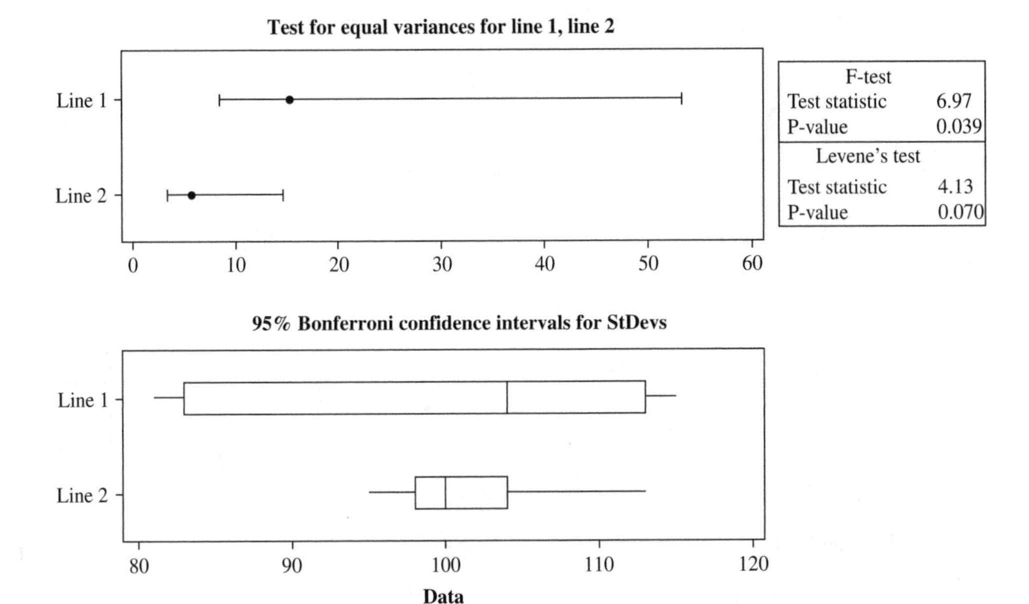

Figure 7-20 MINITAB pull-down test $H_o: \frac{\sigma_1^2}{\sigma_2^2} = 1$ versus $H_a: \frac{\sigma_1^2}{\sigma_2^2} \neq 1$.

EXAMPLE

Consider some of the changes that occur when the past example is one-tailed rather than two-tailed. Two assembly lines using two different methods of assembling PS3s were compared. The number of PS3s assembled per day was recorded for assembly line 1 for 5 days and the number assembled for assembly line 2 for 7 days. The engineer in charge of the process wanted to test if the ratio of variances was equal to 1 versus the ratio was greater than 1, at $\alpha = 0.05$. The data for the two assembly lines is shown in Figure 7-16. The computed test statistic $F = 6.97$ remains the same. Consider how the rejection region changes. It is now on the upper side of the F distribution curve.

The SAS output for the analysis of the hypothesis $H_o: \frac{\sigma_1^2}{\sigma_2^2} = 1$ versus $H_a: \frac{\sigma_1^2}{\sigma_2^2} > 1$ follows. Note the computed value of the test statistic under F in the SAS output. The one tailed p-value, listed as $\Pr > F$, is given as 0.0193. Because the p-value < the preset α, the null would be rejected. Also note the 95% confidence interval for the ratio of variances

$$1.1187 < \frac{\sigma_1^2}{\sigma_2^2} < 64.074$$

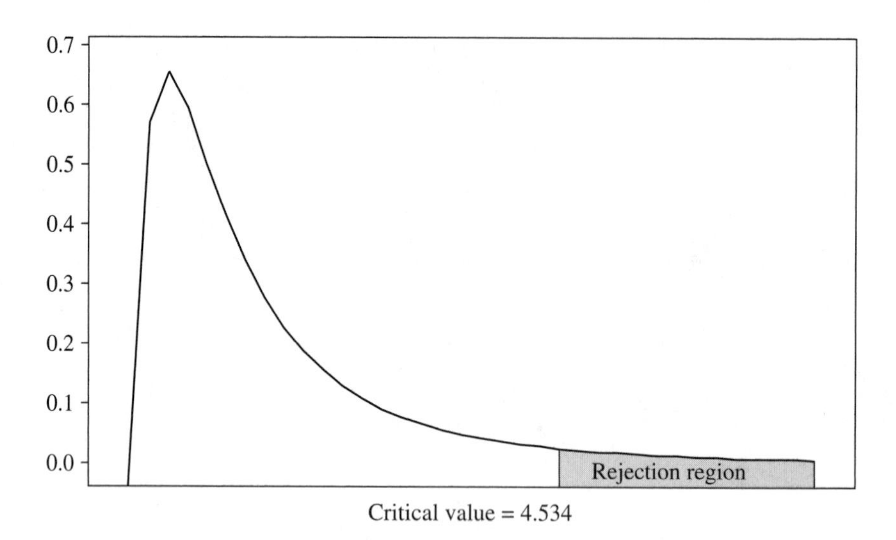

Figure 7-21 *F* distribution curve showing upper-tailed rejection region.

```
          Two-Sample Test for Variances of Units within Line
       Sample Statistics
          Line
          Group              N        Mean           SD       Variance
          ---------------------------------------------------------
          1                  5        99.2        15.369         236.2
          2                  7    101.7143        5.8228      33.90476
  Hypothesis Test
            Null hypothesis:     Variance 1 / Variance 2 <= 1
          Alternative:         Variance 1 / Variance 2 > 1

                      - Degrees of Freedom -
             F            Numer.    Denom.            Pr > F
       ----------------------------------------------
           6.97             4          6             0.0193

     95% Confidence Interval of the Ratio of Two Variances
                    Lower Limit     Upper Limit
                    -----------     -----------
                       1.1187          64.074
```

EXAMPLE

It is believed that the variability in female engineers' salaries is larger than the variability in male engineers' salaries. The salaries in Figure 7-22 were collected in a random sampling of males and females who had been working ten years or less.

Male	Female
61	65
73	67
70	70
71	70
71	51
71	75
65	72
65	58
71	67
71	83
67	73
68	53
65	70
77	47

Figure 7-22 Engineering salaries in thousands.

Identifying males as group 1 and females as group 2, we wish to test $H_o: \frac{\sigma_1^2}{\sigma_2^2} = 1$ versus $H_a: \frac{\sigma_1^2}{\sigma_2^2} < 1$ at preset α = 0.05. The critical value is =FINV(0.95,13,13), or 0.388. The rejection region is shown in Figure 7-23.

The variance of the 14 male salaries is $S_1^2 = 16.769$ and the variance of the 14 female salaries is $S_2^2 = 101.874$, and the computed test statistic is F = 16.769/ 101.874 = 0.165. This computed test statistic falls in the rejection region, and we would conclude that female salaries have a greater variance than male salaries.

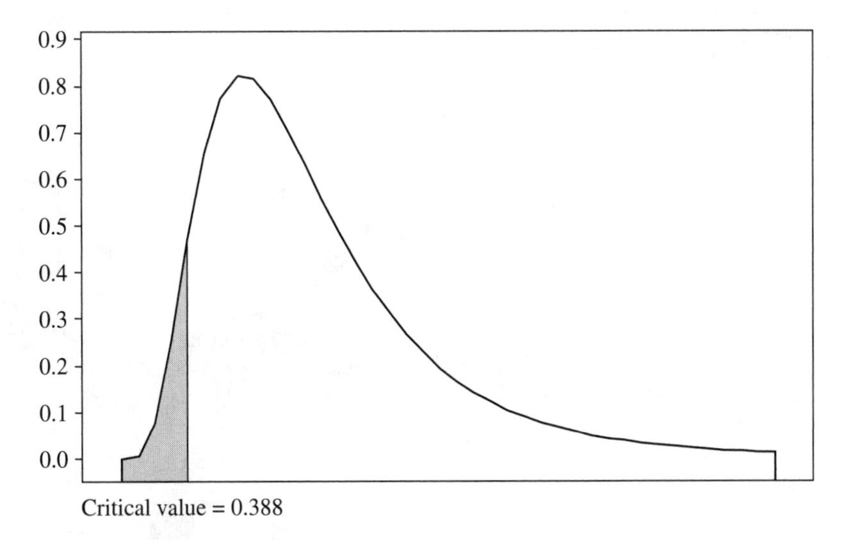

Figure 7-23 *F* distribution curve showing lower-tail rejection region.

The SAS output is as follows:

```
          Two-Sample Test for Variances of Salary within Sex
 Sample Statistics
        Sex
     Group              N       Mean          SD       Variance
     ----------------------------------------------------------
     1                 14         69         4.095     16.76923
     2                 14    65.78571       10.093     101.8736
   Hypothesis Test
     Null hypothesis:    Variance 1 / Variance 2 => 1
     Alternative:        Variance 1 / Variance 2 < 1

                   - Degrees of Freedom -
          F           Numer.     Denom.          Pr < F
     ------------------------------------------------------
        0.16            13         13            0.0013
95% Confidence Interval of the Ratio of Two Variances
               Lower Limit    Upper Limit
               -----------    -----------
                 0.0528          0.5128
```

The 95% confidence interval for $\frac{\sigma_1^2}{\sigma_2^2}$ is

$$0.0528 < \frac{\sigma_1^2}{\sigma_2^2} < 0.5128$$

Exercises

1.

The variance of the diameter of the pistons is of interest.

The diameter of a manufactured piston varies and the null hypothesis is that the variance is 0.01 centimeters. If the variance exceeds 0.01, the process must be stopped and repaired. The data is:

1.6, 1.5, 1.8, 1.5, 1.4, 1.5, 1.6, 1.7, 1.9, and 1.5

Use EXCEL to verify the following in the SAS output: standard deviation, variance, chi-square test statistic, probability, 90% lower limit, and 90% upper limit. What conclusion do you reach if $\alpha = 0.1$? Test using the three methods given in the book.

```
Sample Statistics for Diameter
      N       Mean          SD          Variance
    ---------------------------------------------
     10        1.6        0.1563         0.0244

Hypothesis Test
   Null hypothesis:          Variance of diameter <= 0.01
   Alternative:              Variance of diameter > 0.01
          Chi-square        DF          Prob
          ----------------------------------
            22.000           9          0.0089
90% Confidence Interval for the Variance
                 Lower Limit     Upper Limit
                 -----------     -----------
                   0.013           0.0662
```

2. Find $\chi^2_{.99}, \chi^2_{.95}, \chi^2_{.05}$ and $\chi^2_{.01}$ for a chi-square distribution having 10 degrees of freedom using EXCEL and MINITAB.

3. 

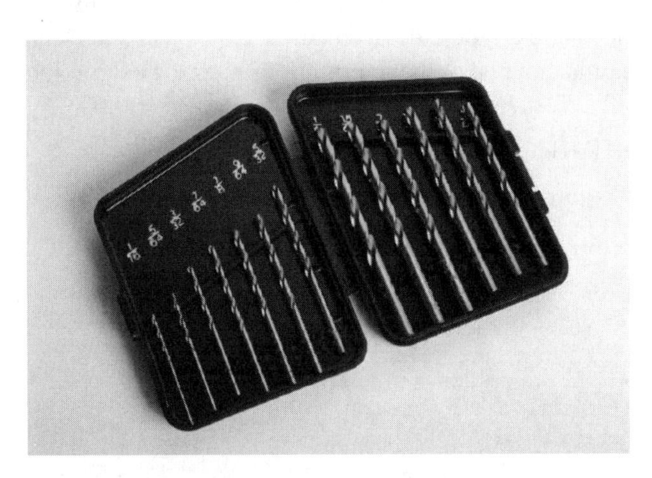

It was desired to compare the variation in the lifetimes of two types of drill bits.

Two types of drill bits were compared with respect to the variation in their lifetimes. Ten from company 1 were tested with the result that the sample variance equaled 3.78, and ten from company 2 were tested with the result that the sample variance equaled 1.76. Test $H_o: \frac{\sigma_1^2}{\sigma_2^2} = 1$ against $H_a: \frac{\sigma_1^2}{\sigma_2^2} \neq 1$ at $\alpha = 0.10$. Give your conclusion.

4. Do problem 3 of this chapter using MINITAB.
5. Draw the F distribution for problem 3 of this chapter using MINITAB. Show the left and right rejection regions.
6.

Two methods of controlling traffic were compared.

The number of accidents was measured at busy intersections being controlled by two different devices. A type 1 device is used at 10 intersections and a type 2 device is used at 12 other intersections. The numbers of accidents, at the intersections over a week period, are shown in Figure 7-24. Test $H_o: \frac{\sigma_1^2}{\sigma_2^2} = 1$ versus $H_a: \frac{\sigma_1^2}{\sigma_2^2} \neq 1$ at $\alpha = 0.05$.

The MINITAB analysis is as follows:

Test for Equal Variances: Type 1, Type 2

```
95% Bonferroni Confidence Intervals for Standard Deviations

         N     Lower       SD        Upper
Type 1   10   4.99309    7.6333     15.3707
Type 2   12   7.26499   10.7436     19.8596

F-test (normal distribution)
Test statistic = 0.50, p-value = 0.314
```

Type 1	Type 2
43	64
52	36
46	50
53	42
62	52
62	52
54	57
41	64
51	52
42	70
	68
	67

Figure 7-24 Comparison of the two devices used to control the traffic.

Use EXCEL to verify the *F*-test, test statistic and the *p*-value. What assumption concerning the data is made?

7. Find $F_{0.01}(4,7)$, $F_{0.05}(4,7)$, $F_{0.95}(4,7)$, and $F_{0.99}(4,7)$ using EXCEL and MINITAB.
8. Set a 95% confidence interval on $\frac{\sigma_1^2}{\sigma_2^2}$ using the data in Figure 7-24 and EXCEL.
9. The following is a SAS analysis of the data in Figure 7-24. Look over the output and identify the following: S_1^2, S_2^2, S_1, S_2, null hypothesis, alternative hypothesis, computed test statistic, two-tailed *p*-value, and 95% confidence interval for $\frac{\sigma_1^2}{\sigma_2^2}$, and compare the 95% confidence interval for $\frac{\sigma_1^2}{\sigma_2^2}$ with the one obtained using EXCEL in problem 8 of this chapter.

```
       Two-Sample Test for Variances of Accidents within Type
  Sample Statistics
      Type
      Group              N       Mean          SD       Variance
      ---------------------------------------------------------
      1                 10       50.6        7.6333     58.26667
      2                 12    56.16667       10.744     115.4242
Hypothesis Test
      Null hypothesis:      Variance 1 / Variance 2 = 1
      Alternative:          Variance 1 / Variance 2 ^= 1
                     - Degrees of Freedom -
           F         Numer.     Denom.          Pr > F
      -----------------------------------------------
         0.50            9         11          0.3142
    95% Confidence Interval of the Ratio of Two Variances
                   Lower Limit    Upper Limit
                   -----------    -----------
                   0.1407         1.9748
```

10. The following is the output from a MINITAB analysis of the data in Figure 7-24:

    Test for Equal Variances: Type 1, Type 2

    ```
    95% Bonferroni Confidence Intervals for Standard Deviations

              N     Lower       SD      Upper
    Type 1   10   4.99309    7.6333          15.3707
    Type 2   12   7.26499    10.7436         19.8596

    F-Test (normal distribution)
    Test statistic = 0.50, p-value = 0.314

    Levene's Test (any continuous distribution)
    Test statistic = 1.62, p-value = 0.217
    ```

    Look over the output and give the following: $S_1^2, S_2^2, S_1, S_2, F = \frac{S_1^2}{S_2^2}, p-value.$ If you do not have normality of the two populations, what procedure could you use?

Summary

1. The distribution of the variable $\frac{(n-1)S^2}{\sigma^2}$ is chi-square with $(n - 1)$ degrees of freedom, where σ^2 is the population variance and S^2 is the sample variance. The population from which the sample is taken has a normal distribution.
2. The symbol χ^2_α represents the value on the horizontal axis of the chi-square curve with area α to its right.
3. The following is a $(1 - \alpha)$ confidence interval for one population variance, σ^2:

$$\frac{(n-1)S^2}{\chi^2_{\alpha/2}} < \sigma^2 < \frac{(n-1)S^2}{\chi^2_{1-\alpha/2}}$$

4. The following is a $(1 - \alpha)$ confidence interval for the ratio of two population variances:

$$\frac{S_1^2}{S_2^2} \cdot \frac{1}{F_{\alpha/2(v_1,v_2)}} < \frac{\sigma_1^2}{\sigma_2^2} < \frac{S_1^2}{S_2^2} F_{\alpha/2(v_2,v_1)}$$

5. Outline for testing a hypothesis about a population variance.

One-tailed test

$H_o: \sigma^2 = \sigma_o^2$

$H_a: \sigma^2 < \sigma_o^2$

$\sigma^2 > \sigma_o^2$

Two-tailed test

$H_o: \sigma^2 = \sigma_o^2$

$H_a: \sigma^2 \neq \sigma_o^2$

Test statistic: $\chi^2 = \frac{(n-1)S^2}{\sigma_o^2}$ has a chi-square distribution with $\nu = n - 1$.

Rejection region

$\chi^2 < \chi^2_{1-\alpha}$

$\chi^2 > \chi^2_{\alpha}$

Rejection region

$\chi^2 < \chi^2_{1-\alpha/2}$ or $\chi^2 > \chi^2_{\alpha/2}$

Assumption: The population from which the sample is selected is normal.

6. Outline for testing a hypothesis about the ratio of two population variances.

One-tailed test

$H_o: \frac{\sigma_1^2}{\sigma_2^2} = 1$

$H_a: \frac{\sigma_1^2}{\sigma_2^2} < 1$

$H_a: \frac{\sigma_1^2}{\sigma_2^2} > 1$

Two-tailed test

$H_o: \frac{\sigma_1^2}{\sigma_2^2} = 1$

$H_a: \frac{\sigma_1^2}{\sigma_2^2} \neq 1$

Test statistic: $F = \frac{S_1^2}{S_2^2}$ has a F distribution with $\nu_1 = n_1 - 1$ and $\nu_2 = n_2 - 1$. ν_1 is called the numerator degrees of freedom and ν_2 the denominator degrees of freedom.

Rejection region

$F < F_{\alpha}$

$F > F_{\alpha}$

Rejection region

$F < F_{\alpha/2}$ or $F > F_{1\text{-}\alpha/2}$

Assumptions:

(a) The samples are from normal populations.

(b) The samples are selected independently of one another.

CHAPTER 8

Inferences Concerning Proportions

8.1 Estimation of Proportions

Engineers are often concerned with proportions or percents. As usual we can estimate the percent of defectives that a process produces when it is mass producing items such as cell phones, PlayStation 3s (PS3s), and automobiles. Or, we may be interested in testing hypotheses about the percent defectives that are produced when we are mass producing items. Suppose we are producing a product and are interested in the percent defectives that we are manufacturing, p. We need to keep up with the value of p so that we can remain competitive. The way we do this is to select a sample of size n and determine X, the number of defectives in the sample. The **sample proportion**, $\hat{p}$, is the proportion in the sample that has the characteristic of interest, such as being defective. Before the sample is taken, the sample proportion $\hat{p}=\frac{X}{n}$ is a random variable. In fact X is binomial with parameters n and p. The sample size, n, may be thought of as the number of trials, and p as the probability of success.

Before the sample is actually taken the probability that a given item will have the characteristic of interest (like being defective) is p. Remember, we do not know the value of p. The value of p is our main interest. We found when we studied the binomial, the mean equals np and the variance equals npq. Or,

$$E(X) = np \quad \text{and} \quad \text{Var}(X) = npq$$

Applying the properties of expectation and variance, we have, dividing both sides by n

$$E\left(\frac{X}{n}\right) = E(\hat{P}) = \frac{np}{n} = p$$

and we see that $\hat{p} = \frac{X}{n}$ is an unbiased estimator of p. Similarly,

$$\text{Var}\left(\frac{X}{n}\right) = \frac{pq}{n}$$

We also found when $np > 5$ and $nq > 5$, the distribution of X can be approximated by a normal distribution. The distribution of $\hat{p} = \frac{X}{n}$ can also be approximated by a normal distribution. The variable $Z = \frac{\hat{p} - p}{\sqrt{\frac{pq}{n}}}$ has a standard normal distribution. Recalling that

$$P(-1.96 < Z < 1.96) = 0.95$$

and substituting for Z, we have

$$P(-1.96 < \frac{\hat{p} - p}{\sqrt{\frac{pq}{n}}} < 1.96) = 0.95$$

Solving the inside part for p, we have $\hat{p} - 1.96\sqrt{\frac{pq}{n}}, \hat{p} + 1.96\sqrt{\frac{pq}{n}}$ as a 95% confidence interval for p. Note that we have a problem. The unknown p and q under the square root keeps us from being able to find a numerical value for our interval. We use the estimates for p and q obtained from the sample to finally arrive at our 95% confidence interval.

$$\left(\hat{p} - 1.96\sqrt{\frac{\hat{p}\hat{q}}{n}}, \hat{p} + 1.96\sqrt{\frac{\hat{p}\hat{q}}{n}}\right)$$

EXAMPLE

Cell phones are mass produced. A sample of 200 is taken and 10 have a small scratch on the body of the phone. Estimate the proportion of the day's production, p, that has a small scratch on the body with a 90% confidence interval. The point estimate of p is

$$\hat{p} = \frac{X}{n} = \frac{10}{200} = 0.05 \text{ or } 5\%$$

First question: Is the normal approximation valid? np is estimated by $n\hat{p} = 200(0.05) = 10$ and $n\hat{q} = 200(0.95) = 190$, and both exceed 5. So the normal approximation is valid. For 90% confidence we need to put 0.05 in the right and left tails of the standard normal curve. Using EXCEL to find the standard normal values, we find =NORMSINV(0.05) = −1.645 and =NORMSINV(0.95) = 1.645. The lower confidence limit is

$$0.05 - 1.645\sqrt{\frac{0.05(0.95)}{200}} = 0.025$$

and the upper confidence limit is

$$0.05 + 1.645\sqrt{\frac{0.05(0.95)}{200}} = 0.075$$

We are 90% confident that the percent with a small scratch is between 2.5% and 7.5%. Generalizing, a $(1 - \alpha)$ confidence interval for p is

$$\left(\hat{p} - z_{\alpha/2}\sqrt{\frac{\hat{p}\hat{q}}{n}}, \hat{p} + z_{\alpha/2}\sqrt{\frac{\hat{p}\hat{q}}{n}}\right)$$

The MINITAB solution to the above problem uses the pull-down menu **Stat ⇒ Basic Statistics ⇒ One Proportion**. The dialog box given in Figure 8-1 is filled in as shown. Under options, choose 90% confidence level, and choose use test and interval based on the normal distribution.

The following output is produced by the dialog box shown in Figure 8-1. *Since we are interested only in the 90% confidence interval at this point, any value may be placed in the test proportion value in the option box.*

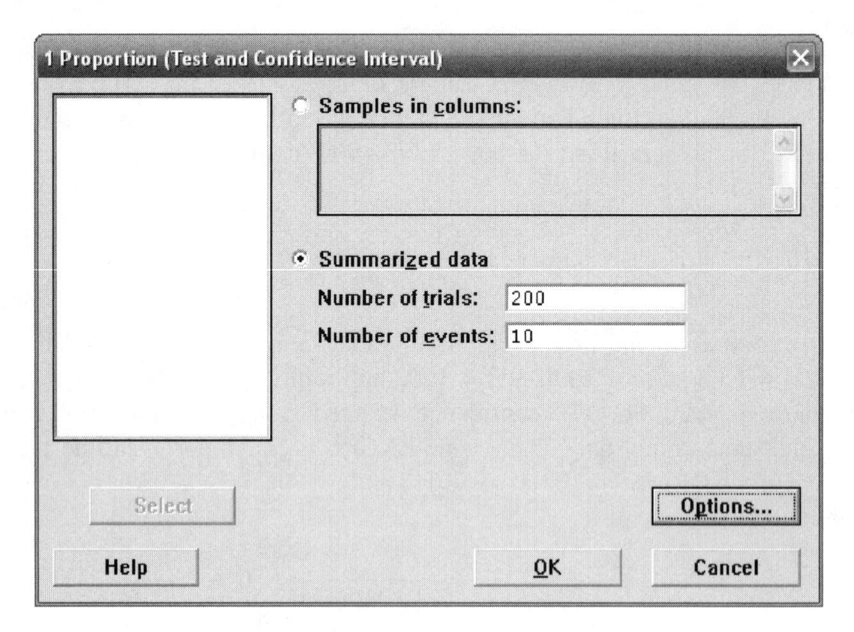

Figure 8-1 MINITAB dialog box for one proportion confidence interval.

Test and CI for One Proportion

```
Test of p = 0.5 vs p not = 0.5

Sample   X     N  Sample p           90% CI              z-Value  p-Value
1       10   200  0.050000   (0.024651, 0.075349)     -12.73    0.000
```

The 90% confidence interval is the same as was reached using EXCEL. The quantity $|\hat{p} - p|$ is the error made when p is estimated by $\hat{p}$. Another way of expressing the confidence interval is

$$|\hat{p} - p| < z_{\alpha/2}\sqrt{\frac{p(1-p)}{n}}$$

The quantity

$$E = z_{\alpha/2}\sqrt{\frac{p(1-p)}{n}}$$

is called the **maximum error of estimate**. If the error is specified, then the sample size needed to give that error is determinable. Simply solve the equation for n. When the equation is solved for n we get

$$n = p(1-p)[z_{\alpha/2}/E]^2$$

If a previous study involving p has been conducted, then p may be replaced by $\hat{p}$ to obtain

$$n = \hat{p}(1 - \hat{p})[z_{\alpha/2}/E]^2$$

as the needed sample size. If nothing is known concerning the possible size of p, then p may be replaced by $p = 0.5$ to obtain a conservative estimate of n. The reason for this is that $p(1 - p)$ is maximum when $p = 0.5$. The function

$$f(p) = p(1 - p) = p - p^2$$

is a quadratic function in p. It reaches its maximum value when $p = 0.5$. If the derivative is set equal to 0 and the resulting equation solved for p, the solution is $p = 0.5$. Therefore, if no estimate of p is available,

$$n = 0.25[z_{\alpha/2}/E]^2$$

is the sample size to use to give an error of estimate equal to E.

EXAMPLE

Surveys are conducted to estimate the proportion of teens who smoke.

The department of health wishes to estimate the percent of teens, p, who currently smoke. They wish to estimate p with a 95% maximum error of estimate equal to 3%.

a. If the estimate one year ago was 25%, what is the sample size needed today?
b. If no previous figure exists to use for p, what is the sample size?

a. Using last year's figure for p, the sample size is

$$n = \hat{p}(1-\hat{p})[z_{\alpha/2}/E]^2 = 25(75)[1.96/3]^2 = 800.33$$

Round up to 801.

b. No previous estimate exists. The sample size is

$$n = 0.25[z_{\alpha/2}/E]^2 = 0.25[1.96/0.03]^2 = 1067.11.$$

Round up to 1068.

NOTE *Use percent for both E and your estimate of p, or use decimal values for both. In part a, percents were used, and in part b, decimal values were used. Also, note that if no previous estimate for p is known, the sample size is considerably larger.*

Now we consider the case of two proportions, p_1 and p_2. Suppose we are comparing two machines with respect to the percent defectives that the two produce. Two samples of sizes n_1 and n_2 are taken. The binomial random variables X_1 and X_2 represent the number of defectives in the two samples. The sample proportions, $\hat{p}_1 = \frac{X_1}{n_1}$ and $\hat{p}_2 = \frac{X_2}{n_2}$, are random variables that represent the proportion of successes in each of the samples. The difference $\hat{p}_1 - \hat{p}_2$ is an unbiased point estimate of $p_1 - p_2$, that is

$$E(\hat{p}_1 - \hat{p}_2) = p_1 - p_2$$

It can be shown that the standard error of the difference in sample proportions is the following:

$$\sigma_{(\hat{p}_1 - \hat{p}_2)} = \sqrt{\frac{p_1 q_1}{n_1} + \frac{p_2 q_2}{n_2}}$$

This standard error is approximated by $\sqrt{\frac{\hat{p}_1 \hat{q}_1}{n_1} + \frac{\hat{p}_2 \hat{q}_2}{n_2}}$. When the sample sizes are large enough,

$$Z = \frac{\hat{p}_1 - \hat{p}_2 - (p_1 - p_2)}{\sqrt{\frac{\hat{p}_1 \hat{q}_1}{n_1} + \frac{\hat{p}_2 \hat{q}_2}{n_2}}}$$

has an approximate standard normal distribution. The probability is $1 - \alpha$ that

$$Z = \frac{\hat{p}_1 - \hat{p}_2 - (p_1 - p_2)}{\sqrt{\frac{\hat{p}_1\hat{q}_1}{n_1} + \frac{\hat{p}_2\hat{q}_2}{n_2}}}$$

is between $-z_{\alpha/2}$ and $z_{\alpha/2}$. If the resulting inequality is solved for $p_1 - p_2$, we obtain

$$(\hat{p}_1 - \hat{p}_2) - z_{\alpha/2}\sqrt{\frac{\hat{p}_1\hat{q}_1}{n_1} + \frac{\hat{p}_2\hat{q}_2}{n_2}} < p_1 - p_2 < (\hat{p}_1 - \hat{p}_2) + z_{\alpha/2}\sqrt{\frac{\hat{p}_1\hat{q}_1}{n_1} + \frac{\hat{p}_2\hat{q}_2}{n_2}}$$

as a $(1 - \alpha)$ confidence interval for $p_1 - p_2$.

EXAMPLE

Two machines are to be compared with respect to the percent of defectives that they produce. A sample of 200 is taken from each machine. The number defective from machine 1 is 20 and the number defective from machine 2 is 15. Use EXCEL to set a 99% confidence interval on $p_1 - p_2$.

Figure 8-2 gives an EXCEL method of setting a 99% confidence interval on $p_1 - p_2$.

Numerical value	EXCEL command	Book notation
A	B	C
0.1	= 20/200	$\hat{p}_1$
0.075	= 15/200	$\hat{p}_2$
0.9	= 1-A1	$1 - \hat{p}_1$
0.925	= 1-A2	$1 - \hat{p}_2$
0.028229	= SQRT((A1*A3/200) +(A2*A4/200)	$\sqrt{\frac{\hat{p}_1\hat{q}_1}{n_1} + \frac{\hat{p}_2\hat{q}_2}{n_2}}$
0.025	= A1-A2	$\hat{p}_1 - \hat{p}_2$
2.575829	= NORMSINV(0.995)	$Z_{0.005}$
−0.04771	= A6-A7*A5	$(\hat{p}_1 - \hat{p}_2) - z_{\alpha/2}\sqrt{\frac{\hat{p}_1\hat{q}_1}{n_1} + \frac{\hat{p}_2\hat{q}_2}{n_2}}$
0.097713	= A6+A7*A5	$(\hat{p}_1 - \hat{p}_2) + z_{\alpha/2}\sqrt{\frac{\hat{p}_1\hat{q}_1}{n_1} + \frac{\hat{p}_2\hat{q}_2}{n_2}}$

Figure 8-2 EXCEL routine for confidence interval for $p_1 - p_2$.

The MINITAB pull-down **Stat ⇒ Basic Statistics ⇒ Two Proportions** is used to set the confidence interval on $p_1 - p_2$. The results are

Test and CI for Two Proportions

```
Sample    X     N    Sample p
1        20   200    0.100000
2        15   200    0.075000

Difference = p (1) - p (2)
Estimate for difference:  0.025
99% CI for difference:  (-0.0477130, 0.0977130)
```

8.2 Hypotheses Concerning One Proportion

We have considered the estimation of a proportion in Section 8.1. Much of the same theory is used to test hypotheses about a proportion. An outline of the statistical test concerning a population proportion is given first, followed by an example. The test is based on the normal distribution. The sample size n will be assumed to be large enough so that $\hat{p} = \frac{X}{n}$ may be assumed to have an approximate normal distribution. The null hypothesis is H_o: $p = p_o$. When the null is true, the mean of the random variable $\hat{p} = \frac{X}{n}$ will be p_o, and the standard deviation of $\hat{p} = \frac{X}{n}$ will be $\sqrt{\frac{p_o q_o}{n}}$. *Remember, the null is assumed true until we get results that are not very likely to occur. When this happens, we reject the null hypothesis.* Assuming the null hypothesis is true, the variable

$$Z = \frac{\hat{p} - p_o}{\sqrt{\frac{p_o q_o}{n}}}$$

will have a standard normal distribution.

Large sample test of hypothesis about a population proportion or percent:

One-tailed test	Two-tailed test
H_o: $p = p_o$	H_o: $p = p_o$
H_a: $p < p_o$	H_a: $p \neq p_o$
H_a: $p > p_o$	

$$\textbf{Test statistic: } Z = \frac{\hat{p} - p_o}{\sqrt{\frac{p_o q_o}{n}}}$$

Rejection region	Rejection region
$Z < -z_{\alpha}$	$Z < -z_{\alpha/2}$ $\quad Z > z_{\alpha/2}$
$Z > z_{\alpha}$	

Assumption: The sample is large enough so that the normal approximation is valid.

EXAMPLE

A traffic engineer claims that over 20% of all trucks that cross a bridge exceed the weight limit for the bridge. One hundred trucks are stopped and weighed after crossing the bridge. Twenty-five of the 100 are over the weight limit. Test H_o: $p = 20\%$ versus H_a: $p > 20\%$ at $\alpha = 0.05$. Test using all three methods of testing hypotheses.

The MINITAB output is as follows:

Test and CI for One Proportion

```
Test of p = 0.2 vs p > 0.2
                                          95%
                                         Lower
Sample    X     N  Sample p          Bound    z-Value  p-Value
1        25   100  0.250000       0.178776       1.25    0.106
```

Method 1

The rejection region is $Z > 1.645$. Since the computed test statistic ($z = 1.25$) does not fall in the rejection region, we are unable to reject the null.

Method 2

The p-value is 0.106. Since the p-value is not less than α, we are unable to reject the null.

Method 3

The one sided confidence interval is $(0.179, \infty)$. The null value is in the interval. We are unable to reject the null.

Test the hypothesis using EXCEL.

From Figure 8-3, the rejection region is $Z > 1.645$, and the test statistic is 1.25, which does not exceed 1.645. Therefore, do not reject the null.

Value	EXCEL command	Book terminology
1.64485	=NORMSINV(0.95)	Critical value
1.25	=(0.25-0.2)/SQRT(0.2*0.8/100)	Test statistic
0.10565	=1-NORMSDIST(1.25)	p-value
0.17876	=0.25-1.645*SQRT(0.25*0.75/100)	95% lower bound

Figure 8-3 EXCEL solution to test of hypothesis about p.

The p-value is not less than $\alpha = 0.05$. Therefore, do not reject the null.

The 95% lower bound is $(0.179, \infty)$. The $p_o = 0.2$ is in this interval. Therefore, do not reject the null. The EXCEL and the MINITAB solutions are the same.

8.3 Hypotheses Concerning Two Proportions

When making decisions about two populations with respect to proportions, we use the random variables $\hat{p}_1 = \frac{X_1}{n_1}$ and $\hat{p}_2 = \frac{X_2}{n_2}$. Two cases present themselves and they require slightly different theory. The two cases are H_o: $(p_1 - p_2) = D_o$ and D_o is either 0 or not 0. In both the cases, the same test statistic will be used:

$$Z = \frac{\hat{p}_1 - \hat{p}_2 - D_o}{\sigma_{(\hat{p}_1 - \hat{p}_2)}}$$

However, our estimate of $\sigma_{(\hat{p}_1 - \hat{p}_2)}$, the standard error of the difference in sample proportions, will differ depending on the value of D_o. First, consider the case where the hypothesized difference in proportions is nonzero. An outline of this case is:

One-tailed test

H_o: $(p_1 - p_2) = D_o \neq 0$

H_a: $(p_1 - p_2) < D_o$

H_a: $(p_1 - p_2) > D_o$

Two-tailed test

H_o: $(p_1 - p_2) = D_o \neq 0$

H_a: $(p_1 - p_2) \neq D_o$

Test statistic: $Z = \frac{\hat{p}_1 - \hat{p}_2 - D_o}{\sigma_{(\hat{p}_1 - \hat{p}_2)}}$

where

$$\sigma_{(\hat{p}_1 - \hat{p}_2)} = \sqrt{\frac{\hat{p}_1\hat{q}_1}{n_1} + \frac{\hat{p}_2\hat{q}_2}{n_2}}$$

Rejection region

$Z < -z_\alpha$

$Z > z_\alpha$

Rejection region

$Z < -z_{\alpha/2}$ $\quad Z > z_{\alpha/2}$

Assumption: The samples are large enough so that the binomial distribution is approximated by the normal distribution.

$$n_1\hat{p}_1 > 5 \quad n_1\hat{q}_1 > 5 \quad n_2\hat{p}_2 > 5 \quad n_2\hat{q}_2 > 5$$

EXAMPLE

It is hypothesized that the percentage of women engineers who are considered to be at a healthy weight exceeds men engineers who are considered to be at a healthy weight by 10%. The alternative hypothesis is that the percentage is less than 10%. That is, the null hypothesis is H_o: $(p_1 - p_2) = 0.10$ and the alternative hypothesis is H_a: $(p_1 - p_2) < 0.10$, and $\alpha = 0.05$. The data in random samples was as follows:

$$\text{Women: } n_1 = 200, x_1 = 78 \qquad \text{Men: } n_2 = 200, x_2 = 63$$

Give an EXCEL solution followed by a MINITAB solution.

Figure 8-4 contains the EXCEL solution to the example. The number 0.39 is in cell A1.

The MINITAB solution is as follows:

```
Sample    X     N  Sample p
1        78   200  0.390000
2        63   200  0.315000

Difference = p (1) - p (2)
Estimate for difference:  0.075
95% upper bound for difference:  0.153340
Test for difference = 0.1 (vs < 0.1):  Z = -0.52  p-value = 0.300
```

More of the work is done for the engineer in MINITAB. A better understanding of how the test works is required for EXCEL.

Value	EXCEL command	Book notation
A	B	C
0.39	=78/200	$\hat{p}_1$
0.315	=63/200	$\hat{p}_2$
0.075	=78/200-63/200	$\hat{p}_1 - \hat{p}_2$
0.0476	=SQRT(A1*(1-A1)/200+A2*(1-A2)/200)	$\sigma_{(\hat{p}_1 - \hat{p}_2)}$
−0.5249	=(A3-0.1)/A4	Test statistic
−1.6448	=NORMSINV(0.05)	$Z < -z_\alpha$
0.2998	=NORMSDIST(A5)	p-value
0.1533	=A3+NORMSINV(0.95)*A4	Upper confidence bound

Figure 8-4 EXCEL solution to test of hypothesis about $(p_1 - p_2) = D_o$.

Now, consider the case where the hypothesized difference in proportions is zero. An outline of this case is:

One-tailed test	Two-tailed test
$H_o: (p_1 - p_2) = 0$	$H_o: (p_1 - p_2) = 0$
$H_a: (p_1 - p_2) < 0$	$H_a: (p_1 - p_2) \neq 0$
$H_a: (p_1 - p_2) > 0$	

$$\textbf{Test statistic: } Z = \frac{\hat{p}_1 - \hat{p}_2 - 0}{\sigma_{(\hat{p}_1 - \hat{p}_2)}}$$

where

$$\sigma_{(\hat{p}_1 - \hat{p}_2)} = \sqrt{\hat{p}\hat{q}\left(\frac{1}{n_1} + \frac{1}{n_2}\right)} \quad \text{and} \quad \hat{p} = \frac{x_1 + x_2}{n_1 + n_2}$$

Rejection region	Rejection region
$Z < -z_\alpha$	$Z < -z_{\alpha/2}$ $\quad Z > z_{\alpha/2}$
$Z > z_\alpha$	

Assumption: The samples are large enough so that the binomial distribution is approximated by the normal distribution.

$$n_1\hat{p}_1 > 5 \quad n_1\hat{q}_1 > 5 \quad n_2\hat{p}_2 > 5 \quad n_2\hat{q}_2 > 5$$

In this case, the null says the two proportions are equal. Assuming the null to be true, we pool our two samples to get an estimate of the common proportion, p. This pooled estimate

$$\hat{p} = \frac{x_1 + x_2}{n_1 + n_2}$$

is our best estimate of that common proportion. In the case where D_o was not 0, we did not have a common proportion in the two populations.

EXAMPLE

Engineers were interested in comparing the percentage of drivers who drive over the 75-mph limit on interstate 10 with the percentage of drivers who drive over the

75-mph limit on interstate 20 in 10 counties in West Texas. The null hypothesis is H_o: $(p_1 - p_2) = 0$ and the alternative hypothesis is H_a: $(p_1 - p_2) \neq 0$, with $\alpha = 0.10$. The data was:

$$n_1 = 400, x_1 = 235 \quad \text{and} \quad n_2 = 400, x_2 = 205$$

Solve using EXCEL and MINITAB.

The EXCEL solution is shown in Figure 8-5.

The MINITAB solution is as follows:

```
Sample     X     N  Sample p
1        235   400  0.587500
2        205   400  0.512500

Difference = p (1) - p (2)
Estimate for difference:  0.075
90% CI for difference:  (0.0173018, 0.132698)
Test for difference = 0 (vs not = 0):  Z = 2.13  p-value = 0.033
```

Value	EXCEL command	Book notation
A	B	C
0.5875	=235/400	$\hat{p}_1$
0.5125	=205/400	$\hat{p}_2$
0.55	=(235+205)/(400+400)	$\hat{p} = \frac{x_1 + x_2}{n_1 + n_2}$
0.035178	=SQRT(A3*(1-A3)*(1/400+1/400))	$\sigma_{(\hat{p}_1 - \hat{p}_2)}$
2.132014	=(A1-A2)/A4	Test statistic
−1.64485	=NORMSINV(0.05)	$Z < -z_{0.05}$
1.644854	=NORMSINV(0.95)	$Z > z_{0.05}$
0.033006	=2*(1-NORMSDIST(A5))	p-value
0.017137	=(A1-A2)-1.644854*A4	Lower limit
0.132863	=(A1-A2)+1.644854*A4	Upper limit

Figure 8-5 EXCEL solution to test of hypothesis about $(p_1 - p_2) = 0$.

8.4 Hypotheses Concerning Several Proportions

In this chapter we have been concerned with testing one and then two proportions. Our next step is testing several proportions. How do we decide whether it is reasonable to treat several proportions as if they have a common value? There are at least two techniques for testing hypotheses about several proportions that are equivalent. The second technique has certain advantages that we shall see in the next section entitled **the analysis of *r* × *c* tables**. We shall now discuss both techniques.

Technique 1 for testing $H_o: p_1 = p_2 = \cdots = p_k = p$ versus the alternative that not all the proportions are equal. Let X_i be the number of successes in samples of sizes n_i. Remember that even though the variables X_i are binomial, if the sample size is large enough, X_i are approximately normal with mean $n_i p_i$ and variance $n_i p_i q_i$, and the variable

$$Z_i = \frac{X_i - n_i p_i}{\sqrt{n_i p_i q_i}}$$

is approximately a standard normal. It can be shown that if a standard normal variable is squared, the result is a chi-square. If you add k chi-square variables, you obtain a chi-square variable having k degrees of freedom. However, you loose one degree of freedom when you estimate the common p-value. In conclusion, we have

$$\chi^2 = \sum_{i=1}^{k} \frac{(x_i - n_i p_i)^2}{n_i p_i q_i}$$

is chi-square with $k - 1$ degrees of freedom and the common p_i value is estimated by

$$\hat{p} = \frac{x_1 + x_2 + \cdot\cdot\cdot + x_k}{n_1 + n_2 + \cdot\cdot\cdot + n_k}$$

(This is an upper-tailed test only. The null hypothesis is rejected only if the observed values and their mean values differ which causes the test statistic to be large.)

EXAMPLE

Three assembly lines are assembling cell phones. Samples of sizes $n_1 = 100$, $n_2 = 80$, and $n_3 = 90$ are selected from the three lines. The number with a defect in the

Value	EXCEL command	Book notation
A	B	C
0.085185	=(7+10+6)/(100+80+90)	Pooled est. of p
0.295899	=(7-100*A1)^2/(100*A1*(1-A1))	First term of x^2
1.627354	=(10-80*A1)^2/(80*A1*(1-A1))	Second term of x^2
0.396057	=(6-90*A1)^2/(90*A1*(1-A1))	Third term of x^2
2.31931	=SUM(A2:A4)	Test statistic
5.991465	=CHIINV(0.05, 2)	Critical value
0.313594	=CHIDIST(A5, 2)	p-value

Figure 8-6 EXCEL test of H_o: $p_1 = p_2 = p_3 = p$ (Technique 1).

samples are $x_1 = 7$, $x_2 = 10$, and $x_3 = 6$. Use EXCEL to perform the test H_o: $p_1 = p_2 = p_3 = p$ for $\alpha = 0.05$. Calculate the test statistic, the critical value, and the p-value. (The excel output is shown in Figure 8-6.)

Technique 2 for testing H_o: $p_1 = p_2 = \cdots = p_k = p$ versus the alternative that not all the proportions are equal. Two tables are prepared called the observed and the expected tables. The test statistic is then computed using the following:

$$\chi^2 = \sum_{i=1}^{2}\sum_{j=1}^{k}\frac{(o_{ij} - e_{ij})^2}{e_{ij}}$$

This test statistic has $(k - 1)$ degrees of freedom.

EXAMPLE (alternative solution)

Observed table:

	Line 1	**Line 2**	**Line 3**
Defect	7	10	6
Nondefect	93	70	84

The estimated value of a defect is $p = 23/270 = 0.085185$.

Expected table:

	Line 1	**Line 2**	**Line 3**
Defect	0.085185(100)	0.085185(80)	0.085185(90)
Nondefect	(1–.085185)(100)	(1–.085185)(80)	(1–.085185)(90)

	Line 1	Line 2	Line 3
Defect	8.5185	6.548	7.66665
Nondefect	91.4815	73.452	82.33335

$$\chi^2 = \sum_{i=1}^{2}\sum_{j=1}^{k}\frac{(o_{ij}-e_{ij})^2}{e_{ij}}$$

$$\frac{(7-8.5185)^2}{8.5185}+\frac{(10-6.548)^2}{6.548}+\frac{(6-7.66665)^2}{7.66665}+\frac{(93-91.4815)^2}{91.4815}$$

$$+\frac{(70-73.452)^2}{73.452}+\frac{(84-82.33335)^2}{82.33335}=2.31931$$

Figure 8-7 contains the EXCEL test of H_o: $p_1 = p_2 = p_3 = p$ (Technique 2). The data is contained in A1:C2. The column totals are in A3:C3 and the row totals are in D1:D2. The grand total is in D3. The cell A5 contains the expression =A3*$D1/$D3. The $ symbol keeps D1 and D3 from changing when a click-and-drag is performed from left to right. The cell A6 contains the expression =A3*$D2/$D3. Again a click-and-drag is performed left to right. This saves a lot of keyboarding when entering the expected values in the space A5:C6. (Try it. You will like it.) Now, enter =(A1-A5)^2/A5 in cell A8 and perform a click-and-drag to the right and down until you cover the space A8:C9. In A11 enter =SUM(A8:C9). This will compute your test statistic,

$$\chi^2 = \sum_{i=1}^{2}\sum_{j=1}^{k}\frac{(o_{ij}-e_{ij})^2}{e_{ij}}$$

A	B	C	D	E
7	10	6	23	Observed
93	70	84	247	Data
100	80	90	270	
8.518519	6.814814815	7.666667		Expected
91.48148	73.18518519	82.33333		Data
0.270692	1.488727858	0.362319		
0.025206	0.138626481	0.033738		
2.31931				Test statistic
0.313594				*p*-value

Figure 8-7 EXCEL test of H_o: $p_1 = p_2 = p_3 = p$ (Technique 2).

In A12, the entry =CHIDIST(A11,2) will give the area to the right of 2.31931, which is the *p*-value. Go through the steps in the past paragraph. It will make a lot more sense if you do it while you have EXCEL in front of you.

The MINITAB solution to this example is as follows. Enter the data in the worksheet as follows:

Line 1	Line 2	Line 3
7	10	6
93	70	84

Execute the pull-down **Stat ⇒ Tables ⇒ Chi-Square Test**. The output is as follows:

Chi-Square Test: Line 1, Line 2, Line 3

```
Expected counts are printed below observed counts.
Chi-square contributions are printed below expected counts.

        Line 1  Line 2  Line 3  Total
    1        7      10       6     23
          8.52    6.81    7.67
         0.271   1.489   0.362

    2       93      70      84    247
         91.48   73.19   82.33
         0.025   0.139   0.034

Total      100      80      90    270

Chi-sq = 2.319, DF = 2, p-value = 0.314
```

Study and compare the two sets of output. The conclusion is that there is no difference in the assembly lines with respect to their proportion of defective cell phones output.

Technique 1 and Technique 2 are equivalent to the method given in Section 8.3 for testing the null hypothesis that p_1 and p_2 were equal. Recall the example we gave there.

EXAMPLE

Engineers were interested in comparing the percentage of drivers who drive over the 75-mph limit on interstate 10 with the percentage of drivers who drive over the 75-mph limit on interstate 20 in 10 counties in West Texas. The null hypothesis is H_o: $(p_1 - p_2) = 0$ and the alternative hypothesis is H_a: $(p_1 - p_2) \neq 0$, with $\alpha = 0.10$. The data was:

$$n_1 = 400, x_1 = 235 \quad \text{and} \quad n_2 = 400, x_2 = 205.$$

The test using the standard normal distribution in Figure 8-5 gave $Z = 2.132007$ and p-value = 0.033006. The test using the chi-square with 1 degree of freedom in Figure 8-8 gave $\chi^2 = 4.5454$ and p-value = 0.033006. Note that the square of the z-value, 2.132007^2, gives the chi-square value 4.5454. Both tests give the same p-value. This shows the equivalence of the two tests when only two proportions are involved.

The MINITAB solution is:

Chi-Square Test: Interstate 10, Interstate 20

```
Expected counts are printed below observed counts.
Chi-square contributions are printed below expected counts.

         Interstate 10  Interstate 20  Total
    1              235            205    440
                220.00         220.00
                 1.023          1.023

    2              165            195    360
                180.00         180.00
                 1.250          1.250

Total              400            400    800

Chi-sq = 4.545, DF = 1, p-value = 0.033
```

A	B	C	D
235	205	440	Observed
165	195	360	Data
400	400	800	
220	220		Expected
180	180		Data
1.022727273	1.022727		
1.25	1.25		
4.545454545			Test statistic
0.033006262			*p*-value

Figure 8-8 Showing the equivalence of tests for $p_1 = p_2$.

8.5 The Analysis of $r \times c$ Tables

A company assembles video game systems. The engineer in charge of the training program is concerned about the connection between the employees' performance in the training program and their success on the job. In particular, she is interested in whether or not the factors are independent of one another or whether there is a dependency. In statistics, this question is answered by performing a contingency table analysis. An outline of such a test is as follows:

H_o: The two classifications are independent of one another.

H_a: The two classifications are dependent upon one another.

$$\text{Test statistic: } \chi^2 = \sum_{j=1}^{c}\sum_{i=1}^{r}\frac{(o_{ij} - e_{ij})^2}{e_{ij}}$$

where $e_{ij} = \frac{n_i n_j}{n}$

n_i = total for row i

n_j = total for column j

n = observed counts

Rejection region: $\chi^2 > \chi^2_\alpha$ where χ^2_α has $(r - 1)(c - 1)$ degrees of freedom.

Assumptions:

1. The n observed counts are a random sample from the population of interest. This is a multinomial experiment with $r \times c$ possible outcomes.
2. For the χ^2 approximation to be valid, it is required that the estimated expected counts be ≥ 5 in all cells.

EXAMPLE

The company, discussed above, classified 200 employees according to their performance in the training program and success in the job. The results are shown in Figure 8-9. Test for independence at $\alpha = 0.05$.

		Performance in training program		
		Below avg.	Average	Above avg.
Success	Poor	25	15	10
In	Average	25	55	20
Job	Very good	10	25	15

Figure 8-9 Does success in the job depend on performance in training program?

Category	Row	Column
25	1	1
25	2	1
10	3	1
15	1	2
55	2	2
25	3	2
10	1	3
20	2	3
15	3	3

Figure 8-10 Data file for STATISTIX.

Figure 8-10 shows the file structure needed for analysis by STATISTIX.

The pull-down **Statistics ⇒ Association Tests ⇒ Chi-Square** is used in STATISTIX. The output is as follows:

```
Chi-Square Test for Heterogeneity or Independence
For category = row column

                                    Column
Row                       1           2            3
                    +-----------+-----------+-----------+
1       Observed    |   25      |   15      |   10      |    50
        Expected    |   15.00   |   23.75   |   11.25   |
    Cell Chi-Sq     |    6.67   |    3.22   |    0.14   |
                    +-----------+-----------+-----------+
2       Observed    |   25      |   55      |   20      |   100
        Expected    |   30.00   |   47.50   |   22.50   |
    Cell Chi-Sq     |    0.83   |    1.18   |    0.28   |
                    +-----------+-----------+-----------+
3       Observed    |   10      |   25      |   15      |    50
        Expected    |   15.00   |   23.75   |   11.25   |
    Cell Chi-Sq     |    1.67   |    0.07   |    1.25   |
                    +-----------+-----------+-----------+
                        60          95          45          200

Overall Chi-Square      15.31
p-Value                0.0041
Degrees of Freedom          4
```

The two variables are dependent since the p-value is less than α.

Exercises

1. The temperature of a process is important to the engineer in charge. The daily temperatures are determined, and it is found that in 8 days out of 35, the temperature is below a critical value. The proportion of days, p, that the temperature is below the critical value is to be estimated. It is hypothesized that the proportion is 0.1. The alternative is that the proportion is greater than 0.1. Refer to the following STATISTIX output, and answer the following questions:

```
Statistix 8.0
One-Sample Proportion Test

Sample size                35
Successes                   8
Proportion            0.22857

Null hypothesis: P = 0.1
Alternative hypothesis: P > 0.1

Difference            0.12857
Standard error        0.07098
Z (uncorrected)          2.54     P  0.0056
Z (corrected)            2.25     P  0.0121

                  95% Confidence Interval
Uncorrected          (0.08946, 0.36769)
Corrected            (0.07517, 0.38197)
```

(a) How is the sample proportion computed?

(b) What is the difference that is given?

(c) How is the standard error computed?

(d) How is Z (uncorrected) computed?

(e) How is the p-value that goes with Z (uncorrected) computed?

(f) How is the uncorrected 95% confidence interval for p computed?

2. The MINITAB solution to problem 1 of this chapter is given below.

```
Test of p = 0.1 vs p > 0.1
                                       95%
                                     Lower
Sample  X   N  Sample p              Bound  Z-Value  P-Value
1       8  35  0.228571           0.111823     2.54    0.006
```

Test the hypothesis using the three methods we have discussed. Refer to the MINITAB output to do the test and use $\alpha = 0.05$.

3.

Fuses from company A were compared with fuses from company B.

A sample of 100 fuses from company A had 10 defectives, and a sample of 125 fuses from company B had 8 defectives. Test that $p_1 = p_2$ versus $p_1 \neq p_2$ using the standard normal distribution.

4. In problem 3 of this chapter, test that $p_1 = p_2$ versus $p_1 \neq p_2$ using the chi-square distribution, and show that this test is equivalent to the one based on the standard normal.
5. Four age groups of engineers were compared with respect to their attitudes toward the Internet. The numbers in the four age groups with positive attitudes toward the Internet were: age group 1 : 68 out of 125, age group 2 : 75 out of 125, age group 3 : 80 out of 125, and age group 4 : 60 out of 125. The null hypothesis is that the four proportions are equal and the research hypothesis is that the proportions differ. Give the EXCEL solution and your conclusion at $\alpha = 0.05$.
6. The following is the MINITAB solution to problem 5 of this chapter:

Chi-Square Test: Group 1, Group 2, Group 3, Group 4

```
Expected counts are printed below observed counts
Chi-Square contributions are printed below expected counts

Group 1  Group 2  Group 3  Group 4  Total
    1       68       75       80       60    283
         70.75    70.75    70.75    70.75
         0.107    0.255    1.209    1.633
```

```
2          57      50      45      65     217
        54.25   54.25   54.25   54.25
        0.139   0.333   1.577   2.130

Total     125     125     125     125     500

Chi-sq = 7.385, DF = 3, p-value = 0.061
```

(a) In the formula

$$\chi^2 = \sum_{i=1}^{2}\sum_{j=1}^{k}\frac{(o_{ij}-e_{ij})^2}{e_{ij}}$$

what is the value of k?

(b) Identify o_{11}, e_{11}, and $(o_{11} - e_{11})^2/e_{11}$ in the output above.

(c) What is the value for

$$\chi^2 = \sum_{i=1}^{2}\sum_{j=1}^{k}\frac{(o_{ij}-e_{ij})^2}{e_{ij}}$$

in the output above?

(d) What is the area to the right of 7.385 under the chi-square curve with three degrees of freedom?

7. The following is the STATISTIX solution to problem 5 of this chapter:

```
Statistix 8
Chi-Square Test for Heterogeneity or Independence
For counts = row col
                                    Col.
Row                     1           2           3           4
                   +-----------+-----------+-----------+-----------+
1       Observed   | 68        | 75        | 80        | 60        |   283
        Expected   | 70.75     | 70.75     | 70.75     | 70.75     |
      Cell Chi-Sq  |  0.11     |  0.26     |  1.21     |  1.63     |
                   +-----------+-----------+-----------+-----------+
2       Observed   | 57        | 50        | 45        | 65        |   217
        Expected   | 54.25     | 54.25     | 54.25     | 54.25     |
      Cell Chi-Sq  |  0.14     |  0.33     |  1.58     |  2.13     |
                   +-----------+-----------+-----------+-----------+
                      125         125         125         125          500

Overall Chi-Square       7.38
P-Value                0.0606
Degrees of Freedom          3
```

Give the data file that STATISTIX requires.

	Appearance		
X-ray	Bad	Normal	Good
Bad	15	8	5
Normal	10	35	6
Good	8	13	25

Figure 8-11 Is the appearance of weld and x-ray classification dependent?

8.

Mechanical engineers test a new welding technique.

Mechanical engineers tested a new welding technique. They classified welds with respect to appearance and also by use of x-ray inspection. They were interested in finding out if the two classifications were independent of one another. The data is shown in Figure 8-11.

Use MINITAB to test for independence at $\alpha = 0.05$.

9. The SAS output for problem 8 is as follows:

```
The FREQ Procedure
              Frequency,
              Expected ,
              Deviation,
              Percent  ,
              Row Pct  ,
              Col Pct  ,        1,        2,        3,  Total
                     1 ,      15 ,       8 ,       5 ,     28
                       ,   7.392 ,  12.544 ,   8.064 ,
                       ,   7.608 ,  -4.544 ,  -3.064 ,
                       ,   12.00 ,    6.40 ,    4.00 ,  22.40
                       ,   53.57 ,   28.57 ,   17.86 ,
                       ,   45.45 ,   14.29 ,   13.89 ,
```

```
      2 ,      10 ,      35 ,       6 ,      51
        ,  13.464 ,  22.848 ,  14.688 ,
        ,  -3.464 ,  12.152 ,  -8.688 ,
        ,    8.00 ,   28.00 ,    4.80 ,   40.80
        ,   19.61 ,   68.63 ,   11.76 ,
        ,   30.30 ,   62.50 ,   16.67 ,

      3 ,       8 ,      13 ,      25 ,      46
        ,  12.144 ,  20.608 ,  13.248 ,
        ,  -4.144 ,  -7.608 ,  11.752 ,
        ,    6.40 ,   10.40 ,   20.00 ,   36.80
        ,   17.39 ,   28.26 ,   54.35 ,
        ,   24.24 ,   23.21 ,   69.44 ,

Total          33        56        36       125
            26.40     44.80     28.80    100.00
```

Each of the nine cells contains frequency, expected, deviation, percent, row pct, and col. pct. Explain what these terms are in cell (1, 1).

```
Statistic                        DF      Value       Prob

Chi-Square                        4    37.7817     <.0001
Likelihood Ratio Chi-Square       4    35.5092     <.0001
Mantel-Haenszel Chi-Square        1    17.7827     <.0001
Phi Coefficient                         0.5498
Contingency Coefficient                 0.4818
Cramer's V                              0.3888
```

What are the chi-square, DF, *p*-value, and probability?

10. Print the data as SAS must receive it.

Summary

1. The **sample proportion**, $\hat{p}$, is the proportion in the sample that has the characteristic of interest, such as being defective. Before the sample is taken, the sample proportion $\hat{p} = \frac{X}{n}$ is a random variable.
2. A **(1 – α) confidence interval** for p is

$$\left(\hat{p} - z_{\alpha/2}\sqrt{\frac{\hat{p}\hat{q}}{n}}, \hat{p} + z_{\alpha/2}\sqrt{\frac{\hat{p}\hat{q}}{n}}\right)$$

3. The quantity

$$E = z_{\alpha/2}\sqrt{\frac{p(1-p)}{n}}$$

is called the **maximum error of estimate.**

4.

$$(\hat{p}_1 - \hat{p}_2) - z_{\alpha/2}\sqrt{\frac{\hat{p}_1\hat{q}_1}{n_1} + \frac{\hat{p}_2\hat{q}_2}{n_2}} < p_1 - p_2 < (\hat{p}_1 - \hat{p}_2) + z_{\alpha/2}\sqrt{\frac{\hat{p}_1\hat{q}_1}{n_1} + \frac{\hat{p}_2\hat{q}_2}{n_2}}$$

is a $(1 - \alpha)$ confidence interval for $p_1 - p_2$.

5. Large sample test of hypotheses about a population proportion or percent:

One-tailed test

$H_o: p = p_o$
$H_a: p < p_o$
$H_a: p > p_o$

Two-tailed test

$H_o: p = p_o$
$H_a: p \neq p_o$

$$\textbf{Test statistic: } Z = \frac{\hat{p} - p_o}{\sqrt{\frac{p_o q_o}{n}}}$$

Rejection region (one-tailed test)

$Z < -z_\alpha$
$Z > z_\alpha$

Rejection region (two-tailed test)

$Z < -z_{\alpha/2} \quad Z > z_{\alpha/2}$

Assumption: The sample is large enough so that the normal approximation is valid.

6. Case where the hypothesized difference in two proportions is nonzero:

One-tailed test

$H_o: (p_1 - p_2) = D_o \neq 0$
$H_a: (p_1 - p_2) < D_o$
$H_a: (p_1 - p_2) > D_o$

Two-tailed test

$H_o: (p_1 - p_2) = D_o$
$H_a: (p_1 - p_2) \neq D_o$

$$\textbf{Test statistic: } Z = \frac{\hat{p}_1 - \hat{p}_2 - D_o}{\sigma_{(\hat{p}_1 - \hat{p}_2)}}$$

where

$$\sigma_{(\hat{p}_1 - \hat{p}_2)} = \sqrt{\frac{\hat{p}_1\hat{q}_1}{n_1} + \frac{\hat{p}_2\hat{q}_2}{n_2}}$$

Rejection region

$Z < -z_{\alpha}$

$Z > z_{\alpha}$

Rejection region

$Z < -z_{\alpha/2} \quad Z > z_{\alpha/2}$

Assumption: The samples are large enough so that the binomial distribution is approximated by the normal distribution.

$$n_1\hat{p}_1 > 5 \quad n_1\hat{q}_1 > 5 \quad n_2\hat{p}_2 > 5 \quad n_2\hat{q}_2 > 5$$

7. Case where the hypothesized difference in two proportions is zero.

One-tailed test

$H_o: (p_1 - p_2) = 0$

$H_a: (p_1 - p_2) < 0$

$H_a: (p_1 - p_2) > 0$

Two-tailed test

$H_o: (p_1 - p_2) = 0$

$H_a: (p_1 - p_2) \neq 0$

$$\textbf{Test statistic: } Z = \frac{\hat{p}_1 - \hat{p}_2 - 0}{\sigma_{(\hat{p}_1 - \hat{p}_2)}}$$

where

$$\sigma_{(\hat{p}_1 - \hat{p}_2)} = \sqrt{\hat{p}\hat{q}\left(\frac{1}{n_1} + \frac{1}{n_2}\right)} \quad \text{and} \quad \hat{p} = \frac{x_1 + x_2}{n_1 + n_2}$$

Rejection region

$Z < -z_{\alpha}$

$Z > z_{\alpha}$

Rejection region

$Z < -z_{\alpha/2} \quad Z > z_{\alpha/2}$

Assumption: The samples are large enough so that the binomial distribution is approximated by the normal distribution.

$$n_1\hat{p}_1 > 5 \quad n_1\hat{q}_1 > 5, \quad n_2\hat{p}_2 > 5 \quad n_2\hat{q}_2 > 5$$

8. The chi-square test of independency:

H_o: The two classifications are independent of one another.

H_a: The two classifications are dependent upon one another.

$$\textbf{Test statistic: } \chi^2 = \sum_{j=1}^{c}\sum_{i=1}^{r}\frac{(o_{ij} - e_{ij})^2}{e_{ij}}$$

where $e_{ij} = \frac{n_i n_j}{n}$

n_i = total for row i

n_j = total for column j

n = grand total

Rejection region: $\chi^2 > \chi^2_\alpha$ where χ^2_α has $(r - 1)(c - 1)$ degrees of freedom.

Assumptions:

(a) The n observed counts are a random sample from the population of interest. This is a multinomial experiment with $r \times c$ possible outcomes.

(b) For the χ^2 approximation to be valid, it is required that the estimated expected counts be ≥ 5 in all cells.

Final Examinations

It is time to find out what you have learned.

Final Examination One

A productivity study of industrial workers was made, and the numbers of acceptable pieces that the workers produced over an 8-hour shift were made. There were 80 workers in the study. The data is given in Figure Exam-1.

10	9	10	14	8	10	12	6
14	10	10	10	13	11	10	10
11	16	11	7	8	13	7	13
10	6	8	10	10	14	10	15
13	9	20	12	10	15	15	10
12	11	7	9	10	3	3	7
11	9	12	8	12	11	6	10
10	11	9	11	7	11	7	10
2	5	14	9	7	12	13	13
6	14	5	5	12	11	9	15

Figure Exam-1 Number of acceptable pieces produced per worker over 8 hours.

A MINITAB dot plot of the data is as follows: Refer to this graphic (Figure Exam-2) to answer problems 1 through 3.

1. The mode of the number of acceptable pieces produced is

   a. 9 b. 10 c. 11 d. 1

2. The range of the data is

   a. 18 b. 17 c. 19 d. 20

3. The mean is 10.113 and the standard deviation is 3.170. What percent of the data is between the (mean – 2 standard deviations) and the (mean + 2 standard deviations)?

   a. 90% b. 93% c. 95% d. 98%

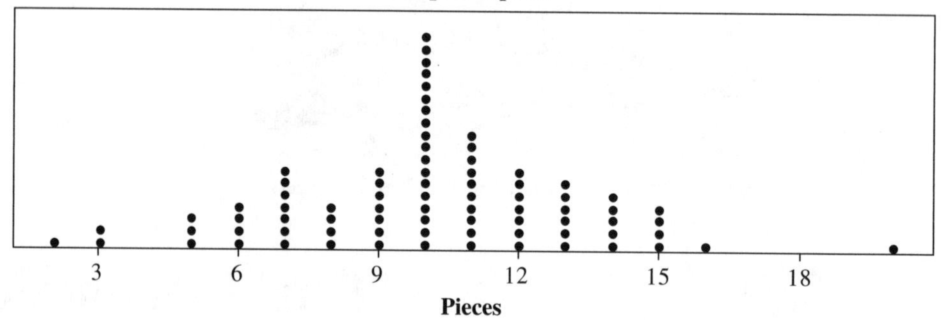

Figure Exam-2 MINITAB dot plot of data in Figure Exam-1.

A STATISTIX stem-and-leaf of the data follows:

```
Statistix 8.0
Stem-and-Leaf Plot of Pieces

  Leaf digit unit = 0.1                    Minimum   2.0000
  2  0  represents 2.0                     Median    10.000
                                           Maximum   20.000

      Stem   Leaves
       1      2  0
       3      3  00
       3      4
       6      5  000
      10      6  0000
      17      7  0000000
      21      8  0000
      28      9  0000000
     (18)    10  000000000000000000
      34     11  0000000000
      24     12  0000000
      17     13  000000
      11     14  00000
       6     15  0000
       2     16  0
       1     17
       1     18
       1     19
       1     20  0

80 cases included   0 missing cases
```

4. What basic shape does the stem-and-leaf show the data to have?

   a. Skewed to the right b. Skewed to the left

   c. Uniform d. Normal

5. A 95% confidence interval for the population mean would be (see problem 3 to get the mean and standard deviation of the variable):

   a. $10.113 \pm 1.96(3.170)$ b. $10.113 \pm 1.645(0.354)$

   c. $10.113 \pm 1.96(0.354)$ d. $10.113 \pm 2.575(0.354)$

6. When testing H_o: $\mu = 11.5$ versus H_a: $\mu \neq 11.5$, what is the computed test statistic? (See problem 3 to get the mean and standard deviation of the variable.)

   a. -3.92 b. -2.56 c. -1.99 d. -2.05

7. If one of the 80 numbers is selected randomly, what is the probability it is less than 7?

   a. 17/80 b. 6/80 c. 15/80 d. 10/80

8. If one of the 80 numbers is randomly selected, and you are told that it is less than the median, what is the probability it is less than 6?

   a. 10/46 b. 10/28 c. 6/46 d. 6/28

9. If one of the 80 numbers is randomly selected, what is the probability that the number is less than 8 or greater than 16?

   a. 17/80 b. 18/80 c. 19/80 d. 20/80

A random variable has the following probability density function:

$$f(x) = \begin{cases} 0.75(1-x^2), & -1 < x < 1 \\ 0, & \text{otherwise} \end{cases}$$

Problems 10 through 14 refer to this pdf:

10. The $P(X > 0)$ is

    a. 0.2 b. 0.4 c. 0.5 d. 0.6

11. The mean of X is

    a. –1 b. –0.5 c. 0 d. 0.5

12. The standard deviation of X is

    a. 0.227 b. 0.337 c. 0.447 d. 0.557

13. The median of X is

    a. –0.5 b. 0 c. 0.5 d. 1

14. The $P(X < 2)$ is

    a. 0 b. 0.5 c. 0.75 d. 1

Four plants manufacture electric tooth brushes. Four samples of size 200 are taken. The numbers of defective tooth brushes, found in samples from the four plants, are recorded with the following results: $x_1 = 12$, $x_2 = 14$, $x_3 = 11$, and $x_4 = 20$. An engineer used the data to test that the proportion of defectives are the same at the four plants versus they are different. The following MINITAB analysis resulted:

Chi-Square Test: Plant 1, Plant 2, Plant 3, Plant 4

```
Expected counts are printed below observed counts.
Chi-Square contributions are printed below expected counts.
        Plant 1  Plant 2  Plant 3  Plant 4  Total
   1       12       14       11       20     57
        14.25    14.25    14.25    14.25
        0.355    0.004    0.741    2.320
```

```
2         188       186       189       180        743
       185.75    185.75    185.75    185.75
        0.027     0.000     0.057     0.178

Total     200       200       200       200        800

Chi-sq = 3.684, DF = 3, p-value = 0.298
```

Refer to this output to answer 15 through 19.

15. What is the value of χ^2?

$$\chi^2 = \sum_{i=1}^{2}\sum_{j=1}^{k}\frac{(o_{ij} - e_{ij})^2}{e_{ij}}$$

a. 0.298 b. 185.75 c. 3 d. 3.684

16. The chi-square variable in problem 15 has how many degrees of freedom?

a. 4 b. 3 c. 2 d. 1

17. What would be the critical value that separates the rejection region from the nonrejection region if $\alpha = 0.05$? Give the EXCEL expression.

a. =CHIINV(0.05,3) b. =CHIINV(0.05,4)

c. =CHIINV(0.95,3) d. =CHIINV(0.95,4)

18. What would the expression for the p-value be equal? Give the EXCEL expression.

a. =CHIDIST(3.684,4) b. =CHIDIST(0.298,3)

c. =CHIDIST(4.536,3) d. =CHIDIST(3.684,3)

19. A 95% confidence interval for p_4 is

a. $0.1 \pm 1.96(0.0212)$ b. $0.2 \pm 1.96(0.0212)$

c. $0.1 \pm 1.96(0.1212)$ d. $0.1 \pm 1.96(0.2345)$

A product part has a maximum lifetime of 1 year. The lifetime, X, has the following density:

$$f(x) = 4x^3, \text{ where } 0 < x < 1$$

Refer to this scenario to answer 20 through 22.

20. The $P(X < 0.2)$ is

a. 0.16 b. 0.016 c. 0.0016 d. 0.26

21. Find the average lifetime of such product parts.

a. 0.6 b. 0.7 c. 0.8 d. 0.9

22. Find σ^2.

    a. 0.27 b. 0.37 c. 0.47 d. 0.027

23. Find the value of c that makes the following a pdf:

$$f(x) = c/x^2,\ 1 < x < 2 \quad \text{and} \quad f(x) \text{ is 0 otherwise}$$

    a. 2 b. 0.5 c. 1 d. 1.5

Four different brands of tires were driven, and their tread lives were measured. The results are shown below:

Miles	Brand A	Brand B	Brand C	Brand D
≤ 30,000 (1)	25	15	20	15
30 to 50 (2)	50	35	40	55
≥ 50,000 (3)	25	20	20	10

The null hypothesis is that tread life is independent of brand. Test at $\alpha = 0.05$. The SAS analysis follows:

```
The FREQ Procedure

                        Table of Mileage by Brand
          Mileage (mileage)     Brand (brand)
          Frequency,
          Expected ,
          Deviation,        1,         2,         3,         4,   Total

                 1 ,       25 ,       15 ,       20 ,       15 ,      75
                   ,   22.727 ,   15.909 ,   18.182 ,   18.182 ,
                   ,   2.2727 ,   -0.909 ,   1.8182 ,   -3.182 ,

                 2 ,       50 ,       35 ,       40 ,       55 ,     180
                   ,   54.545 ,   38.182 ,   43.636 ,   43.636 ,
                   ,   -4.545 ,   -3.182 ,   -3.636 ,   11.364 ,

                 3 ,       25 ,       20 ,       20 ,       10 ,      75
                   ,   22.727 ,   15.909 ,   18.182 ,   18.182 ,
                   ,   2.2727 ,   4.0909 ,   1.8182 ,   -8.182 ,

          Total           100         70         80         80      330

              Statistics for Table of Mileage by Brand
          Statistic                        DF       Value        Prob
          Chi-Square                        6     10.0670      0.1219
          Likelihood Ratio Chi-Square       6     10.5222      0.1043
          Mantel-Haenszel Chi-Square        1      0.4978      0.4805
          Phi Coefficient                          0.1747
          Contingency Coefficient                  0.1721
          Cramer's V                               0.1235
                              Sample Size = 330
```

Refer to the SAS output to answer 24 through 27.

24. The value of

$$\chi^2 = \sum_{j=1}^{c}\sum_{i=1}^{r}\frac{(o_{ij}-e_{ij})^2}{e_{ij}} \quad \text{is}$$

a. 10.067 b. 10.5222 c. 0.4978 d. 0.1747

25. How many degrees of freedom does the chi-square statistic have?

a. 3 b. 4 c. 5 d. 6

26. The critical value of the chi-square distribution for $\alpha = 0.05$ is

a. 12.59159 b. 10.23424 c. 11.78905 d. 13.09867

27. The *p*-value for the test is given by which of the following EXCEL expressions?

a. =CHIDIST(10.067,6) b. =CHIDIST(10.067,7)

c. =CHIDIST(10.067,5) d. =CHIDIST(10.067,4)

Use EXCEL to solve 28 through 30.

28. A shipment contains 15 PS3s. Two of them are defective. Two are selected for inspection. Find the probability both are defective.

a. 0.1233 b. 0.2354 c. 0.0095 d. 0.0135

29. A company produces a large number of PS3s. Ten percent have minor defects. One hundred are selected for inspection. What is the probability that the 100 contain 5 or fewer with minor defects?

a. 0.0987 b. 0.1234 c. 0.2335 d. 0.0576

30. A company produces a large number of PS3s. Ten percent have minor defects. One hundred are selected for inspection. What is the probability that the 100 contain between 5 and 10 with minor defects? That is, find $P(5 \le X \le 10)$, where X is the number in the 100 with minor defects.

a. 0.5594 b. 0.4567 c. 0.5564 d. 0.6578

In a torture test, a switch is turned on and off until it malfunctions. Assuming that the conditions for a geometric random variable hold and that the probability of malfunctioning on any one trial is 0.0005, answer 31 through 34.

31. On the average, how many turns on and off will be needed before it malfunctions?

a. 500 b. 1000 c. 1500 d. 2000

32. What is the probability that it malfunctions after being flipped 500 times?

    a. 0.7788 b. 0.5678 c. 0.6754 d. 0.8877

33. The standard deviation of the geometric random variable described above is

    a. 500 b. 999.5 c. 1999.5 d. 1500.5

34. According to Chebyshev's theorem, an interval in which the probability is at least 0.75 that the switch will malfunction is

    a. (500, 1000) b. (0, 5999) c. (500, 5999) d. (0, 10000)

The proportion of individuals who own a personal digital assistant (PDA) in a given year is a random variable having a beta distribution with $\alpha = 1$ and $\beta = 9$. The density is

$$f(x) = 9(1 - x)^8, 0 < x < 1$$

The use of MAPLE is assumed in this section of the test. However, most of the integrals are not too hard for those with three semesters of calculus. Problems 35 and 36 refer to this scenario.

35. What is the average proportion of individuals who own a PDA?

    a. 0.05 b. 0.10 c. 0.15 d. 0.20

36. What percent of the days will the ownership be less than 15%? (In the integral, make the substitution: $y = 1 - x$)

    a. 0.568 b. 0.668 c. 0.768 d. 0.868

The amount of water usage by a large city is uniformly distributed over the range 1 to 3 where the amount is measured in hundreds of thousands of gallons. Problems 37 and 38 refer to this scenario.

37. What is the standard deviation of the usage? (in hundreds of thousands of gallons)

    a. 0.577 gallons b. 0.167 gallons

    c. 0.344 gallons d. 0.444 gallons

38. What percent of the time is the usage over 250,000 gallons?

    a. 10% b. 15% c. 20% d. 25%

The weights of trucks traveling an interstate highway are normally distributed with mean equal to 15 tons and standard deviation equal to 2.5 tons. If 25 trucks are stopped and weighed, find the following probabilities: Problems 39 and 40 refer to this scenario.

39. What percent of the trucks are over 20 tons?

    a. 5.000% b. 2.275% c. 3.250% d. 4.355%

40. What is the probability that the mean of the sample is less than 14.5 tons?

    a. 0.1234 b. 0.1355 c. 0.1485 d. 0.1587

A study was conducted to compare the variability in price of PDAs in England and the United States. For a sample of 35 PDAs in the United States, it was found that S_1 = $5.50, and for 35 in England, it was found that S_2 = $8.35. The hypothesis of interest was $H_a: \frac{\sigma_1^2}{\sigma_2^2} \neq 1$ at $\alpha = 0.05$. Problems 41, 42, and 43 refer to this scenario.

41. Give the value of the test statistic that is used to test this hypothesis.

    a. 0.66 b. 1.52 c. 2.30 d. 0.43

42. What are the critical values for this test? What is your conclusion?

    a. 0.5048 and 1.9811, reject the null

    b. 0.2022 and 1.9811, do not reject the null

    c. 0.5048 and 1.9811, do not reject the null

    d. 0.2022 and 1.9811, reject the null

43. Find the *p*-value for the test.

    a. 0.080 b. 0.016 c. 0.008 d. 0.160

In order to compare the mean lifetimes of two different TV picture tubes, engineers tested each of the 35 and found the following results:

Tube 1 mean lifetime = 4.5 years and standard deviation = 1.5 years.

Tube 2 mean lifetime = 3.9 years and standard deviation = 2.8 years.

The null hypothesis H_o: $(\mu_1 - \mu_2) = 0$ versus the research hypothesis H_o: $(\mu_1 - \mu_2) \neq 0$ was tested at $\alpha = 0.01$. Problems 44, 45, and 46 refer to this scenario.

44. The test statistic is equal to:

    a. 2.12 b. 3.24 c. 0.98 d. 1.12

45. The *p*-value is equal to:

    a. 0.135 b. 0.269 c. 0.335 d. 0.013

46. Find a 99% confidence interval.

    a. (–0.8356, 2.0356) b. (–0.6677, 1.9800)

    c. (–0.5567, 1.5678) d. (–0.7786, 2.0356)

An engineering company was concerned about obesity among its employees. It required its employees to attend nutrition lectures. The data in Figure Exam-3 contains weights before the program started and weights of the same individuals after the program started: The difference is (after - before). The company wishes to test H_o: $\mu_d = 0$ versus H_a: $\mu_d < 0$ with $\alpha = 0.05$.

47. What assumption does this test make about the distribution of the differences?

    a. Uniform b. *t* c. Normal d. Log-normal

48. Which of the following are tests of normality? (Choose the best answer.)

    a. Anderson-Darling b. Kolmogorov-Smirnov

    c. Ryan-Joiner d. All three

49. The average difference is −5.524 and the standard deviation of the differences is 21.47. What is the computed test statistic?

    a. −1.34 b. −2.00 c. −1.56 d. −1.15

50. What is the EXCEL expression for the *p*-value?

    a. =TDIST(1.15,19,1) b. =TDIST(−1.15,19,1)

    c. =TDIST(1.15,19,2) d. =TDIST(1.15,20,1)

Before	After	Difference
204	170	−34
187	202	15
195	178	−17
163	148	−15
188	204	17
208	205	−4
155	180	25
177	172	−5
168	170	1
167	165	−1
154	176	22
185	193	7
219	172	−47
160	156	−4
203	154	−49
158	173	14
209	182	−27
158	172	13
166	158	−8
184	170	−14

Figure Exam-3 Weights before and after nutrition lectures.

Final Examination Two

There are 100 "fill in the blanks" worth 1 point each.

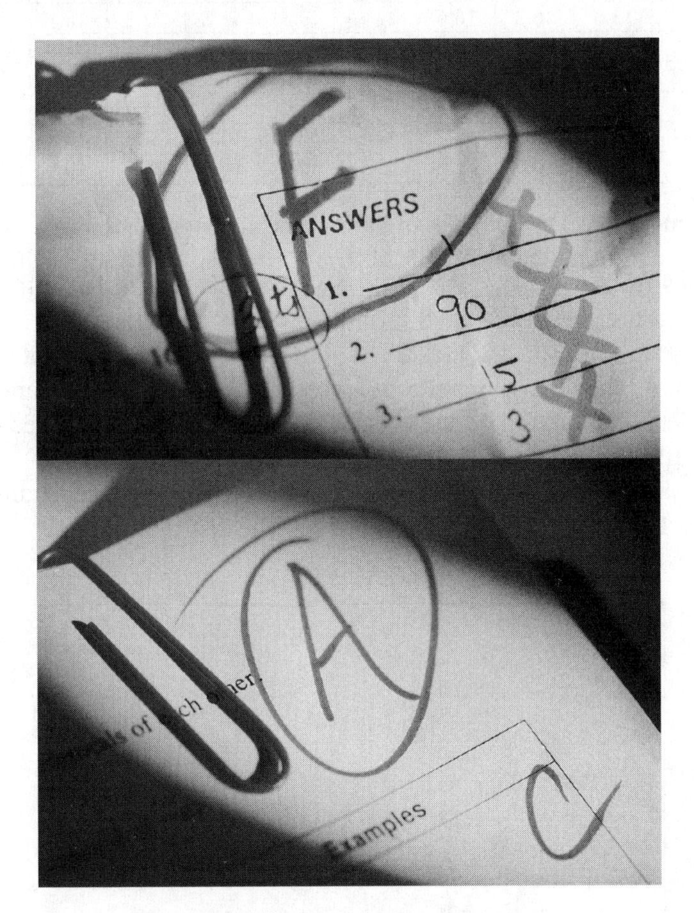

What have you learned from the book?

The weekly emissions of sulfur oxides (in metric tons) by an industrial manufacturer are shown in the following table:

An SAS box plot for the data in Figure Exam-4 is shown in Figure Exam-5. The lower whisker extends down to **1**_______. The upper whisker extends up to **2**_______. The distribution of sulfur oxide readings is skewed to the **3**_______. The box covers the middle **4**_______% of the data.

The SAS histogram in Figure Exam-6 reveals that **5**_______ of the sulfur oxide readings are between 0 and 12. This histogram consists of seven classes. The highest

27.1	42.2	4.5	5.6	0.8	31.1	3.6	1.1	24.7	44.7
10.7	4.5	22.8	12.6	4.8	11.0	35.9	34.9	1.0	8.8
6.7	5.4	23.5	15.2	24.6	43.1	15.4	29.0	61.1	26.5
0.8	11.8	1.9	10.9	2.8	3.7	11.5	32.7	7.8	4.2
16.8	13.6	46.2	16.6	21.5	9.8	41.2	22.2	18.1	54.6
42.0	11.8	9.2	15.0	3.4	0.5	44.2	10.8	33.8	20.7
8.2	39.5	18.3	5.2	3.1	6.6	2.2	78.0	11.3	35.5
50.3	43.2	15.4	17.7	27.1	55.2	4.9	5.6	4.5	0.4
75.2	6.9	9.5	5.7	4.6	2.6	40.6	7.6	0.8	35.9
26.7	11.0	10.6	1.6	15.3	0.2	10.0	8.1	10.1	13.2

Figure Exam-4 Weekly emissions of sulfur oxide in metric tons by Ace manufacturing.

class extends from **6** _______ to **7**_______. The histogram shows that the sulfur oxide readings are skewed to the **8**_______.

Figure Exam-7 gives a STATISTIX box and whisker plot for the data in Figure Exam-4. The three observations represented by the asterisks represent **outliers** in the data. By looking over the data in Figure Exam-4, the three outliers are **9**_______, **10**_______, and **11**_______.

Figure Exam-8 gives a MINITAB dot plot for the sulfur oxide readings.

There are **12**_______ values that exceed 60. The distribution is **13**_______ to the right.

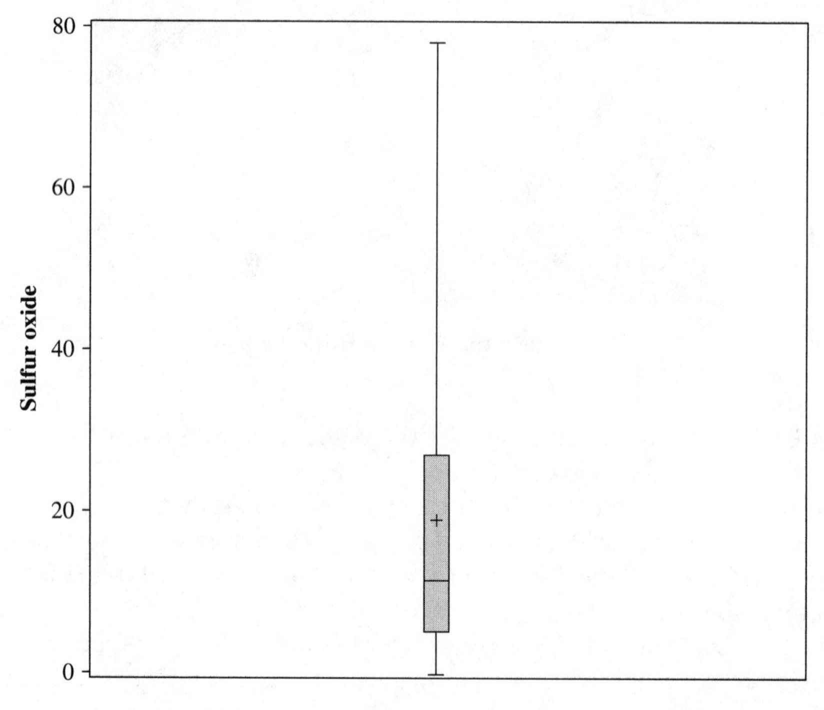

Figure Exam-5 SAS box plot for the sulfur oxide emissions.

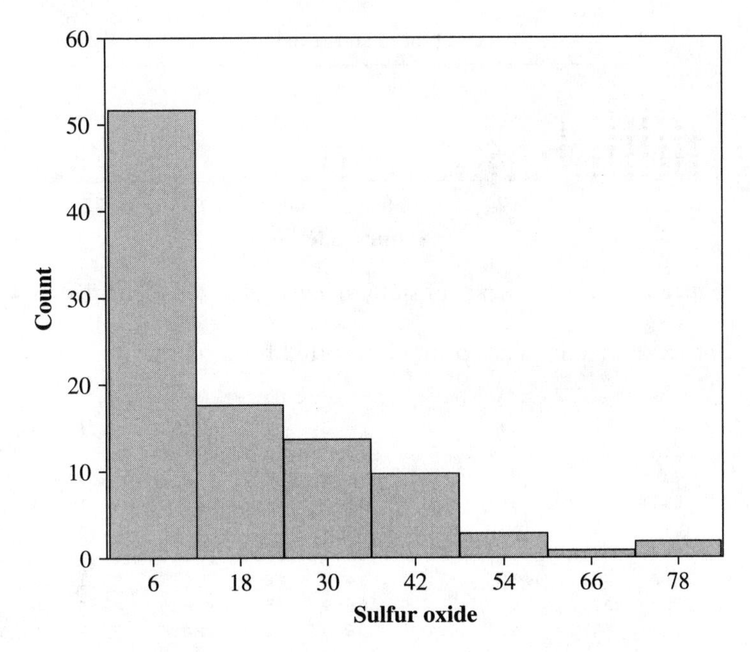

Figure Exam-6 SAS histogram for the data in Figure Exam-4.

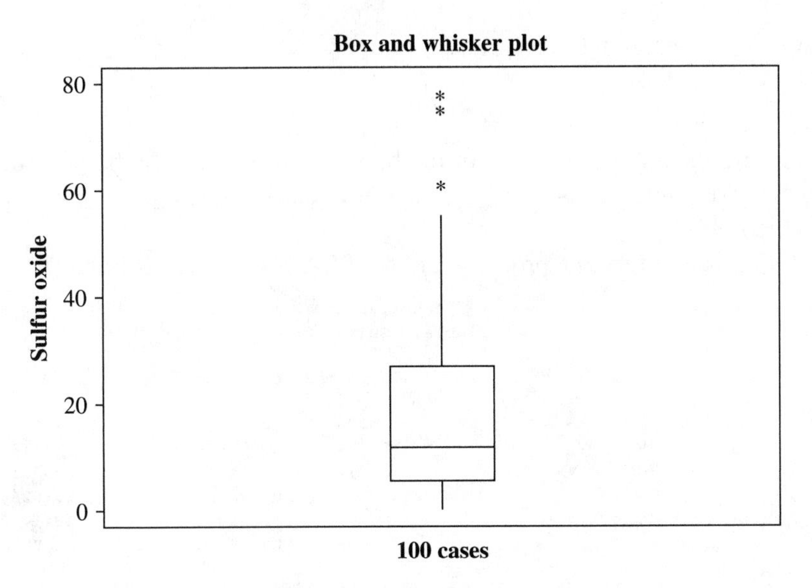

Figure Exam-7 A STATISTIX box and whisker plot for the data in Figure Exam-4.

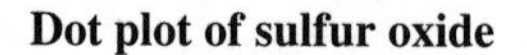

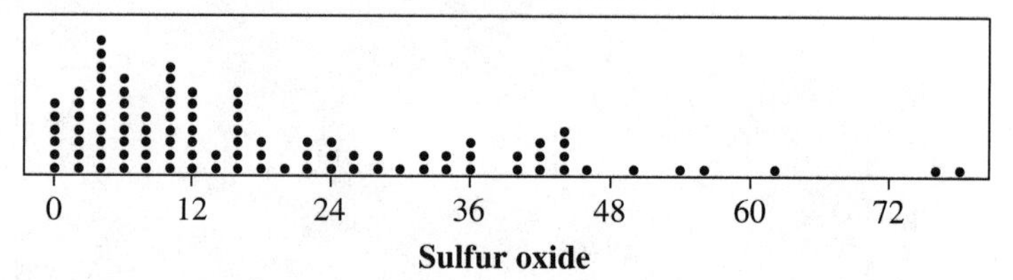

Figure Exam-8 Dot plot of sulfur oxide readings in Figure Exam-4.

The sulfur oxide readings are printed in sorted form below:

```
Sulfur oxide
   0.2    0.4    0.5    0.8    0.8    0.8    1.0    1.1    1.6    1.9    2.2
   2.6    2.8    3.1    3.4    3.6    3.7    4.2    4.5    4.5    4.5    4.6
   4.8    4.9    5.2    5.4    5.6    5.6    5.7    6.6    6.7    6.9    7.6
   7.8    8.1    8.2    8.8    9.2    9.5    9.8   10.0   10.1   10.6   10.7
  10.8   10.9   11.0   11.0   11.3   11.5   11.8   11.8   12.6   13.2   13.6
  15.0   15.2   15.3   15.4   15.4   16.6   16.8   17.7   18.1   18.3   20.7
  21.5   22.2   22.8   23.5   24.6   24.7   26.5   26.7   27.1   27.1   29.0
  31.1   32.7   33.8   34.9   35.5   35.9   35.9   39.5   40.6   41.2   42.0
  42.2   43.1   43.2   44.2   44.7   46.2   50.3   54.6   55.2   61.1   75.2
  78.0
```

The EXCEL solutions for the first quartile, the third quartile, and the 90th percentile are:

5.35	PERCENTILE(A2:A101,0.25)
27.1	PERCENTILE(A2:A101,0.75)
43.11	PERCENTILE(A2:A101,0.9)

If you use the algorithm given in the book where you figure the percentiles by hand, the first quartile is **14**_______, the third quartile is **15**_______, and the 90th percentile is **16**_______.

Part of the output from a proc univariate of SAS is given as follows:

```
                    The univariate Procedure
              Variable:  Sulfur Oxide  (Sulfur Oxide)

                              Moments

N                         100    Sum Weights                 100
Mean                   18.514    Sum Observations         1851.4
Std. Deviation     17.2380717     Variance            297.151115
Skewness           1.29877208    Kurtosis             1.36264216
Uncorrected SS       63694.78    Corrected SS         29417.9604
Coeff Variation    93.1083055    Std. Error Mean      1.72380717

                   Basic Statistical Measures

          Location                     Variability

    Mean     18.51400       Std. Deviation              17.23807
    Median   11.65000       Variance                   297.15112
```

```
Mode        0.80000     Range                       77.80000
                        Interquartile Range         21.80000

                  Quartiles (Definition 5)

                   Quantile          Estimate

                   100% Max             78.00
                   99%                  76.60
                   95%                  52.45
                   90%                  43.15
                   75% Q3               27.10
                   50% Median           11.65
                   25% Q1                5.30
                   10%                   2.05
                   5%                    0.80
                   1%                    0.30
                   0% Min                0.20
```

Under quartiles in the above SAS output, it is shown that the 95th percentile is 52.45, the median or the 50th percentile is 11.65, and the 10th percentile is 2.05. If you use the algorithm given in the book where you figure the percentiles by hand, the 95th percentile is **17**_______, the median is **18**_______, and the 10th percentile is **19**_______.

The numbers of engineering degrees, Midwestern University conferred, are shown in the table for 2006: An EXCEL pie chart for the data is given in Figure Exam-9.

Engineering area	Count
Mechanical	13
Chemical	5
Computer	5
Industrial	10
Civil	17

Fill in the angle for each of the categories in the following table:

Category	Angle of the slice
Mechanical	**20**
Chemical	**21**
Computer	**22**
Industrial	**23**
Civil	**24**

Suppose one of the 50 engineering graduates was randomly chosen to be interviewed about his or her experience at Midwestern University. Answer the following questions by referring to Figure Exam-9.

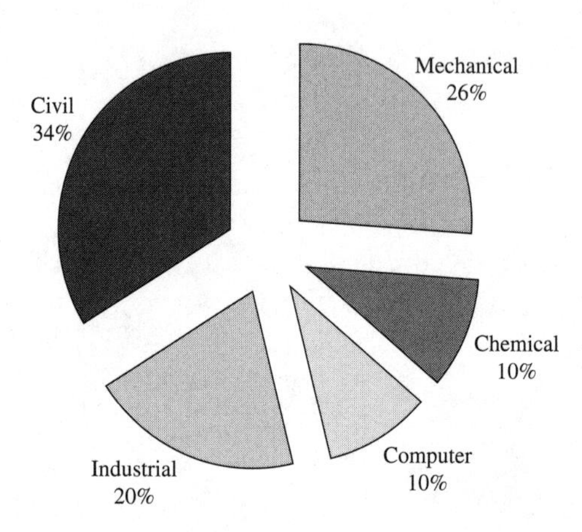

Figure Exam-9 Engineering graduates from Midwestern University.

The probability that the chosen student was a mechanical or a civil engineering graduate is **25**_______. The probability that the chosen student was not a civil engineering graduate is **26**_______.

Three of the engineering graduates were randomly selected to appear on University TV to talk about the engineering programs at Midwestern University. The probability that all three were civil engineering graduates is **27**_______. The probability that all three were chemical engineering graduates is **28**_______. The probability that one was a computer engineering graduate, one was an industrial engineering graduate, and one was a mechanical engineering graduate is **29**_______.

If it is known that the first one selected was a civil engineering graduate, what is the probability that the other two are computer engineering graduates? **30**_______ If the first two selected were mechanical engineering graduates, what is the probability that the third is an industrial engineering graduate? **31**_______.

The number of defectives in a sample of size 50 is described by the following probability function:

$$p(x) = \frac{4-x}{10}, x = 0, 1, 2, 3$$

A quality control engineer selects a sample of size 50. What is the probability that the sample has: no defectives **32**_______, at least two defectives **33**_______, and at most one defective **34**_______?

The mean number of defectives is **35**_______ and the standard deviation of the number of defectives is **36**_______.

An engineer makes measurements that have measurement errors with the following density:

$$f(x) = \begin{cases} 80x^4, & -0.5 < x < 0.5 \\ 0, & \text{otherwise} \end{cases}$$

Find the probability that the measurement error is at least 0.3. **37**_______ is between –0.2 and 0.2. **38**_______.
is either < –0.3 or >0.4. **39**_______. The average measurement error is **40**_______ and the standard deviation of measurement error is **41**_______.

The probability that a DVD has a flaw is 0.05. In the process of producing the DVDs, the probability that the third such flaw occurs on the 50th DVD produced is **42**_______. On the average, the third such flaw occurs on the **43**_______ trial. The probability that the first such flaw occurs after 30 DVDs have been manufactured is **44**_______.

In a stress test performed on an iPod nano, it was found that the height that it could be dropped from and would no longer function properly was normally distributed with a mean of 8 feet and a standard deviation of 2 feet. Use NORMDIST and NORMINV of EXCEL to answer the following questions: The probability that the height is less than 5 feet is **45**_______. The first quartile of heights is **46**_______.

An environmental engineer has found that the logarithm of natural gas use (in therms) for Colorado in summer per month has a normal distribution with mean $\alpha = 2$ and standard deviation $\beta = 1$. The mean of the summer natural gas use per month is **47**_______ therms, and the standard deviation of natural gas use per month is **48**_______ therms. Use EXCEL and the function NORMDIST to find the probability that the natural gas use per month is between 5 therms and 15 therms **49**_______.

An environmental engineer compared the monthly use of electricity of two groups of Colorado residents. One group is called the energy-conscious group and the other group is called the non-energy-conscious group. Summary statistics for the two groups are given in the accompanying table. The response variable is electricity use in kilowatts.

Group	*N*	Sample mean	Sample SD
Energy-conscious	35	185	95
Non-energy-conscious	35	257	105

The hypothesis to be tested is H_o: $(\mu_1 - \mu_2) = 0$ versus H_a: $(\mu_1 - \mu_2) \neq 0$ at $\alpha = 0.01$.

The value of the test statistic used to perform the test is **50**_______.

The critical values for performing the test are **51**_______ and _______.

The *p*-value for performing the test is **52**_______.

The 99% confidence interval for $\mu_1 - \mu_2$ is **53**_______.

Since the 99% confidence interval does not contain 0, the null hypothesis is **54**_______.

Four age groups of engineers were compared with respect to the proportion who had a statistics course as part of their degree program. The results of the study are as follows:

Years engineering degree completed	Sample size	Yes
0 to < 10	200	165
10 to < 20	200	159
20 to < 30	200	132
Thirty or more	200	124

The null hypothesis is $H_o: p_1 = p_2 = \cdots = p_k = p$ versus the alternative that not all the proportions are equal. α is 0.05. The MINITAB analysis of the data is as follows:

Chi-Square Test: Group 1, Group 2, Group 3, Group 4

```
Expected counts are printed below observed counts.
Chi-square contributions are printed below expected counts.

        Group 1  Group 2  Group 3  Group 4  Total
    1       165      159      132      124    580
         145.00   145.00   145.00   145.00
          2.759    1.352    1.166    3.041

    2        35       41       68       76    220
          55.00    55.00    55.00    55.00
          7.273    3.564    3.073    8.018

Total       200      200      200      200    800

Chi-sq = 30.245, DF = 3, p-value = 0.000
```

The test statistic has a **55**_______ distribution with **56**_______ degrees of freedom. The EXCEL expression for the *p*-value is **57**_______.

A large computer company hires people with disabilities. The firm wishes to test whether the performance of its employees is independent of the disability type. The firm gathers the following data to test if performance is independent of the disability type:

	Performance		
	Above average	Average	Below average
Seeing impaired	15	35	17
Hearing impaired	13	38	16
No disability	26	55	24

58 Find the expected number in the nine cells assuming the null is true_______.
59 Find the value of the chi-square test statistic_______.
60 Find the critical value for the test of independence, assuming $\alpha = 0.05$ _______.
61 Use EXCEL to find the p-value for the test _______.
62 At $\alpha = 0.05$, are the type of disability and the level of performance dependent _______?

Three dice are rolled, and the sum on the three dice, X, is recorded. Give the probability distribution for X **63**_______. Find the mean of X **64**_______.
Find the variance of X **65**_______. Find the probability, $P(8 \le X \le 11)$ **66**_______.
An experiment consists of rolling three dice, and the sum on the three dice, X, is recorded. Samples of size 36 are taken and the sample means are computed for these samples. The central limit theorem assures us that this sample mean will have a **67**_______ distribution. The mean of this sample mean will equal **68**_______. The standard error of this sample mean will equal **69**_______.
Alloy 600 is a product used in steam generators, chemical processing, food processing, superheaters, jet engines, and electronic parts. Determination of tensile strength was made on 35 Alloy 600 products, and the mean was 551.6 mega pascals and the standard deviation was 5.7 mega pascals. The hypothesis $H_o: \mu = 555.0$ versus $H_a: \mu \neq 555.0$ was tested at $\alpha = 0.1$. The value of the test statistic is **70**_______. The critical values are **71**_______ and **72**_______. The p-value is **73**_______. Suppose the variability is tested by testing $H_o: \sigma = 3$ versus $H_a: \sigma \neq 3$ at $\alpha = 0.05$. The value of the test statistic is **74** _______. The critical values are **75**_______ and **76**_______. Use EXCEL to determine the p-value **77**_______. A 95% confidence interval for σ is **78**_______.
Three random variables are jointly distributed with the following joint probability density function:

$$f(x) = \frac{c}{xyz}, 1 < x < 2, 2 < y < 3, 3 < z < 4 \text{ and 0 otherwise}$$

The value of c is **79** _______. The marginal density of X is **80** _______. The marginal density of Y is **81** _______. The marginal density of Z is **82** _______. The probability that $1 < X < 1.5, 2 < Y < 2.5, 3 < Z < 3.5$ is **83** _______.
Using the sulfur oxide data in Figure Exam-4 and the Anderson-Darling test of normality, the null hypothesis would reject normality because the p-value is **84** _______. Using the sulfur oxide data in Figure Exam-4 and the Ryan-Joiner test of normality, the null hypothesis would reject normality because the p-value is **85** _______. Using the sulfur oxide data in Figure Exam-4, and the Kolmogorov-Smirnov test of normality, the null hypothesis would reject normality because the p-value is **86** _______. Use MINITAB to answer 84, 85, and 86.
The proportion of engineers at Allstate Engineering who utilize the Internet in their planning and operations daily has a beta distribution with $\alpha = 3$ and $\beta = 2$.

The density for this proportion is **87** _______. The average value for this random variable is **88**_______. The probability that the proportion is greater than 0.50 is **89**_______.

The basic assumption when small samples are used to test means is the **90** _______ assumption.

A traffic engineer measures the number of accidents a week before and a week after the installation of a traffic control device at 10 intersections. The data is shown in the following table. The engineer wishes to test H_o: $\mu_d = 0$ versus H_a: $\mu_d > 0$ for $\alpha = 0.05$.

Intersection	Before	After	Difference
1	10	8	2
2	13	12	1
3	15	14	1
4	20	16	4
5	16	14	2
6	17	19	–2
7	19	14	5
8	25	20	5
9	18	16	2
10	23	24	–1

The MINITAB output is

```
Paired t for before - after

                N      Mean       SD    SE Mean
Before         10   17.6000   4.4771     1.4158
After          10   15.7000   4.4734     1.4146
Difference     10   1.90000  2.33095    0.73711

95% lower bound for mean difference: 0.54879
t-test of mean difference = 0 (vs > 0): t-value = 2.58 p-value = 0.015
```

The computed test statistic is **91** _______. The EXCEL expression to compute the p-value is **92**_______. The conclusion at $\alpha = 0.05$ is **93**_______. A lower one-sided 95% confidence interval for μ_d is **94**________.

A random variable has a uniform density function over the interval [10, 12]. The mean of the random variable is **95** _______ and the standard deviation of the random variable is **96**_______.

It is hypothesized that 30% of all engineers have a degree beyond the bachelor's degree, and the research hypothesis is that the figure is greater than 30%. α is chosen to be 0.1. In a random selection across the United States, it is found that 375 out of 1000 have an advanced degree. To test $H_o: p = 0.30$ versus $H_a: p > 0.30$, what is the value of the test statistic? **97** _______. What is the p-value? **98** _______. What is your conclusion at $\alpha = 0.1$? **99** _______. Give a 90% two-sided confidence interval for p. **100**_______.

Solutions to Chapter Exercises and Final Exams

Chapter 1

1.

The data in Table 1-5 is entered into the EXCEL worksheet. The label lifetime is entered into cell A1, and the data itself is entered into A2:A76. A sort is performed using the sort ascending icon. The data is converted to two dimensional data by adding a second number to each value as follows: If a number, such as 1470, is repeated three times, then the numbers 1, 2, and 3 are paired with 1470 as follows:

1470 1
1470 2
1470 3

If a number is not repeated, such as 1403, it is paired with 1. In this way the data is converted to 75 pairs and then the scatter plot from the chart wizard is used to plot the two dimensional data. The dot plot shown in Figure 1-15 is the result. The dot plot reveals a set of data that is not normally distributed. The data has a tail to the right due to a few bulbs that have long lifetimes. Such a set of data is said to be **skewed to the right**.

2.

Each class is 25 hours wide. The frequency for the class 1700–1725 is one. The data has a tail to the right, or it is **skewed to the right**.

3.

(a) The four outliers are 1625, 1660, 1670, and 1705.

(b) 1500.5

(c) 1403

(d) 1600

4.

There are 7 numbers equal to 1470. There are 26 numbers that are 1520 or larger. There are 9 numbers that are equal to 1450 or smaller.

5.

(a) (1457.156, 1561.991) 78.7%

(b) (1404.738, 1614.409) 93.3%

(c) (1352.320, 1666.827) 97.3%

(d) The 68%, 95%, and 99.7% hold only for a normal distribution.

6.

(a) $Z = \frac{1403 - 1509.6}{52.418} = -2.03$

(b) $Z = \frac{1705 - 1509.6}{52.418} = 3.73$

7.

Lower outer fence = 1301

Lower inner fence = 1388

Upper inner fence = 1620

Upper outer fence = 1707

There are no outliers on the lower side. The mild outliers are 1625, 1660, 1670, and 1705.

8.

Defect	Percent	Angle
Dent	40	144
Other	14	50.4
Scratch	20	72
Smudge	8	28.8
Spec	18	64.8

9.

Dent is the most frequently occurring. Dent and scratch are the two that occur the most often. Dent, scratch, and specifications are the three that occur the most often.

10.

Percentile	Value			
5	1448		1448.7	PERCENTILE(A1:A75,0.05)
25	1475		1475	PERCENTILE(A1:A75,0.25)
50	1502		1502	PERCENTILE(A1:A75,0.5)
75	1533		1531	PERCENTILE(A1:A75,0.75)
95	1625		1607.5	PERCENTILE(A1:A75,0.95)

The technique given in the book gave the values on the left side in the table above. The built-in EXCEL function gave the values on the right side.

Chapter 2

1. (a)

$$S = \{CCC, CNC, CCN, CNN, NCC, NNC, NCN, NNN\}$$

$$P(CCC) = P(C)P(C)P(C) = (0.97)^3 = 0.912673$$

The events are independent. So, the probabilities can be multiplied without conditioning. Since we are selecting from a large batch, the 0.97 figure will not change significantly from one selection to the next. *CCC* stands for conforming on the first draw, the second draw, and the third draw.

$$P(CNC) = (.97)(.03)(.97) = 0.028227$$

$$P(CCN) = (.97)(.97)(.03) = 0.028227$$

$$P(CNN) = (.97)(.03)(.03) = 0.000873$$

$$P(NCC) = (.03)(.97)(.97) = 0.028227$$

$$P(NNC) = (.03)(.03)(.97) = 0.000873$$

$$P(NCN) = (.03)(.97)(.03) = 0.000873$$

$$P(NNN) = (.03)(.03)(.03) = 0.000027$$

Note that $P(S) = 1$.

$A = \{CCC, CNC, CCN, NCC\}$ $B = \{NNN, CNN, NCN, NNC, CNC, CCN, NCC\}$

(b) $P(A) = P(CCC) + P(CNC) + P(CCN) + P(NCC) = 0.997354$

(c) $P(B) = 1 - P(CCC) = 0.087327$

(d) $P(A \cap B) = P(CNC) + P(CCN) + P(NCC) = 0.084681$

(e) $A \cup B = S$, so $P(A \cup B) = 1$

(f) $P(A^c) = 1 - P(A) = 1 - 0.997354 = 0.002646$

(g) $P(B^c) = 1 - P(B) = 1 - 0.087327 = 0.912673$

2.

C	*C*	*C*	*C*	0.6561
C	*C*	*C*	*N*	0.0729
C	*C*	*N*	*C*	0.0729
C	*C*	*N*	*N*	0.0081
C	*N*	*C*	*C*	0.0729
C	*N*	*C*	*N*	0.0081
C	*N*	*N*	*C*	0.0081
C	*N*	*N*	*N*	0.0009
N	*C*	*C*	*C*	0.0729
N	*C*	*C*	*N*	0.0081
N	*C*	*N*	*C*	0.0081
N	*C*	*N*	*N*	0.0009
N	*N*	*C*	*C*	0.0081
N	*N*	*C*	*N*	0.0009
N	*N*	*N*	*C*	0.0009
N	*N*	*N*	*N*	0.0001
				1

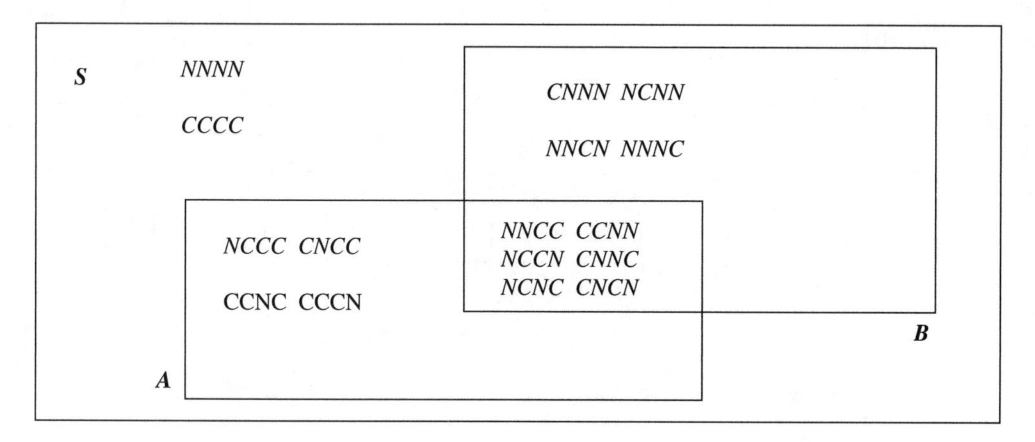

(a) $A \cap B^c = \{NCCC, CNCC, CCNC, CCCN\}$

(b) $B \cap A^c = \{CNNN, NCNN, NNCN, NNNC\}$

(c) $A \cap B = \{NNCC, CCNN, NCCN, CNNC, NCNC, CNCN\}$

(d) $(A \cup B)^c = \{NNNN, CCCC\}$

(e) $P(A) = 0.3402$

(f) $P(B) = 0.0522$

(g) $P(A \cap B) = 0.0486$

(h) $P(A \mid B) = 0.9310$

(i) $P(B \mid A) = 0.1429$

3. (a)

$A^c \cup B^c$

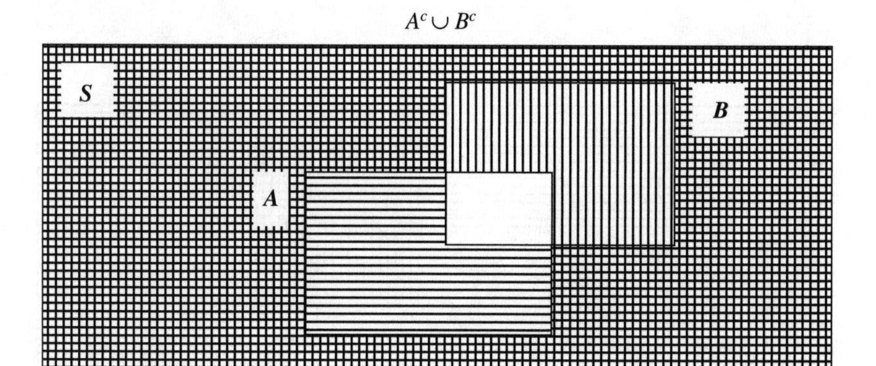

A^c is shaded vertically and B^c is shaded horizontally. $A^c \cup B^c$ will have either type of shading. $(A^c \cup B^c)^c$ will have neither type of shading. We see that $(A^c \cup B^c)^c$ is $A \cap B$. Hence, $(A^c \cup B^c)^c = A \cap B$.

(b)

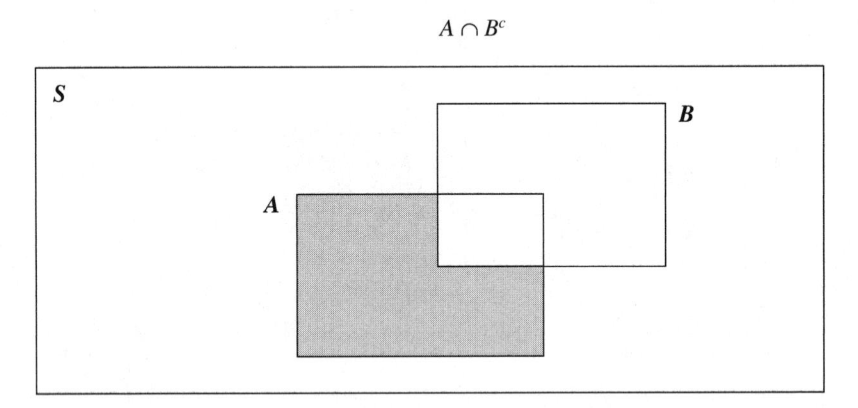

The shaded portion is $A \cap B^c$. The complement of $A \cap B^c$ is shown in the following graphic:

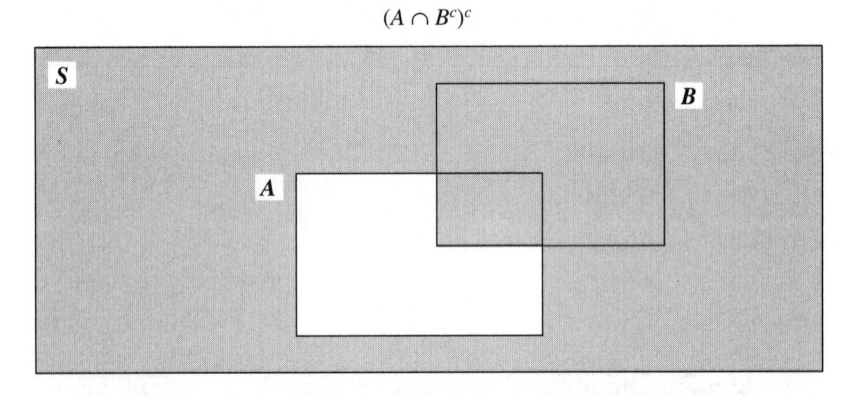

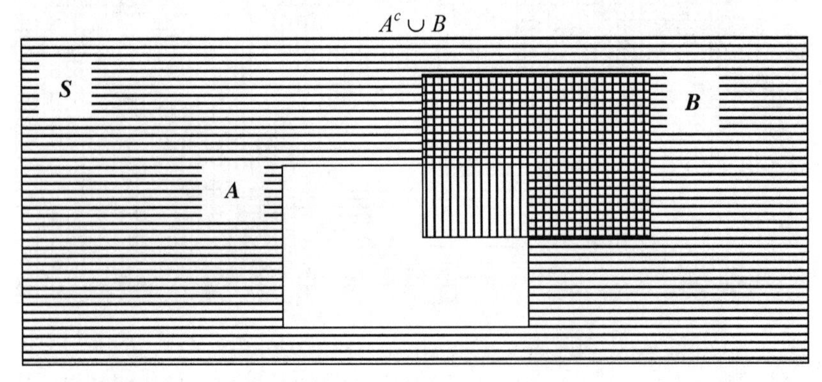

In this figure, the complement of A is shaded horizontally and B is shaded vertically. The union is any shaded portion: horizontal, vertical, or both. Comparing the last two figures, we see that

$$(A \cap B^c)^c = A^c \cup B$$

4.

(a) Order is important. So, the number of permutations is P_4^{10}. Using EXCEL, the answer is =PERMUT(10,4) = 5040.

(b) In a sample, order is ordinarily unimportant. The answer is $\binom{50}{10}$. Using EXCEL, the answer is =COMBIN(50,10) = 10,272,278,170.

(c) Using the multiplicative rule, the answer is $2^2 \times 3^8 = 26,244$.

(d) $3(4)(3) = 36$.

5.

$$P(A) = 0.35,\ P(B) = 0.45, \text{ and } P(A \text{ and } B) = 0.25$$

(a) $P(A \text{ or } B) = P(A) + P(B) - P(A \text{ and } B) = 0.35 + 0.45 - 0.25 = 0.55$.

(b) $P(\text{not } A) = 1 - P(A) = 1 - 0.35 = 0.65$.

(c) $P(A \cup B)^c = (A \text{ or } B)^c = 1 - 0.55 = 0.45$.

(d) $P(B \mid A) = P(A \text{ and } B)/P(A) = 0.25/0.35 = 0.714$.

(e) $P(A \cap B)^c = 1 - P(A \text{ and } B) = 1 - 0.25 = 0.75$.

6.

(a) .6(.005) + .15(.01) + .25(.0075) = 0.006375, or 0.6375 per 100 items.

(b) Plant A 0.003/0.006375 = 0.47 Plant B 0.0015/0.006375 = 0.24
Plant C 0.001875/0.006375 = 0.29

7.

$$P(0) = \frac{\binom{5}{0}\binom{25}{4}}{\binom{30}{4}} = 0.461595$$

The probabilities of 0, 1, 2, 3, and 4 with a minor discoloration are as follows:

0	0.461595
1	0.419631
2	0.109469
3	0.009122
4	0.000182

The expected number with a minor discoloration is

$$0(0.461595) + 1(0.419631) + 2(0.109469) + 3(0.009122) + 4(0.000182) = 0.67$$

8.

The part selected from box 2 is nonconforming if a nonconforming part is moved from box 1 **and** a nonconforming part is selected from box 2 **or** a conforming part is moved from box 1 **and** a nonconforming part is selected from box 2. (Remember **multiply** for **and** and **add** for **or**.)

$$\frac{4}{40}\frac{9}{41} + \frac{36}{40}\frac{8}{41} = \frac{324}{1640} = 0.198$$

9.

(a) P(nondefective **and** nondefective **and** nondefective **and** nondefective **and** nondefective) = $(0.85)^5 = 0.4437$

(b) P(defective **and** nondefective **and** nondefective **and** nondefective **and** nondefective **or** nondefective **and** defective **and** nondefective **and** nondefective **and** nondefective **or** nondefective **and** nondefective **and** defective **and** nondefective **and** nondefective **or** nondefective **and** nondefective **and** nondefective **and** defective **and** nondefective **or** nondefective **and** nondefective **and** nondefective **and** nondefective **and** defective) = $(0.15)(0.85)^4 + (0.85)(0.15)(0.85)^3 + (0.85)^2(0.15)(0.85)^2 + (0.85)^3(0.15)(0.85) + (0.85)^4(0.15) = 0.3915$.

Similarly, using **and**'s as well as **or**'s, we find the answers to the other parts to be

2	0.138178
3	0.024384
4	0.002152
5	7.59E-05

10.

Twenty-two are nondiscolored and three are discolored. Draw the three without replacement.

$$P(\text{no discolored in the three drawn}) = \frac{22}{25}\frac{21}{24}\frac{20}{23} = 0.6696$$

$$P(\text{one discolored in the three drawn}) = \frac{3}{25}\frac{22}{24}\frac{21}{23} + \frac{22}{25}\frac{3}{24}\frac{21}{23} + \frac{22}{25}\frac{21}{24}\frac{3}{23} = 0.3013$$

$$P(\text{two discolored in the three drawn}) = \frac{3}{25}\frac{2}{24}\frac{22}{23} + \frac{3}{25}\frac{22}{24}\frac{2}{23} + \frac{22}{25}\frac{3}{24}\frac{2}{23} = 0.0287$$

$$P(\text{three discolored in the three drawn}) = \frac{3}{25}\frac{2}{24}\frac{1}{23} = 0.0004$$

Chapter 3

1.

	Outcome		Sum		*x*	*p*(*x*)		*xp*(*x*)		*x*^2**p*(*x*)
1	1	1	3		3	0.0156		0.0469		0.1406
1	1	2	4		4	0.0469		0.1875		0.75
1	1	3	5		5	0.0938		0.4688		2.3438
1	1	4	6		6	0.1563		0.9375		5.625
1	2	1	4		7	0.1875		1.3125		9.1875
1	2	2	5		8	0.1875		1.5		12
1	2	3	6		9	0.1563		1.4063		12.656
1	2	4	7		10	0.0938		0.9375		9.375
1	3	1	5		11	0.0469		0.5156		5.6719
1	3	2	6		12	0.0156		0.1875		2.25
1	3	3	7			1		7.5		60
1	3	4	8							
1	4	1	6							
1	4	2	7			Mean	7.5			
1	4	3	8			Variance	3.75			
1	4	4	9		Standard	deviation	1.9365			

The above worksheet shows 16 of the 64 outcomes, the probability distribution, the mean, and the standard deviation of the distribution.

2.

(a) The scatterplot is symmetrical about the mean. Seven and 8 are the most likely values for the variable and 3 and 12 are the least likely values. The only values that the variable can take on are 3, 4, 5, 6, 7, 8, 9, 10, 11, and 12. The scatterplot is shown in Figure Solutions-1.

(b) $$(\mu - 2\sigma, \mu + 2\sigma) = (3.627, 11.373)$$

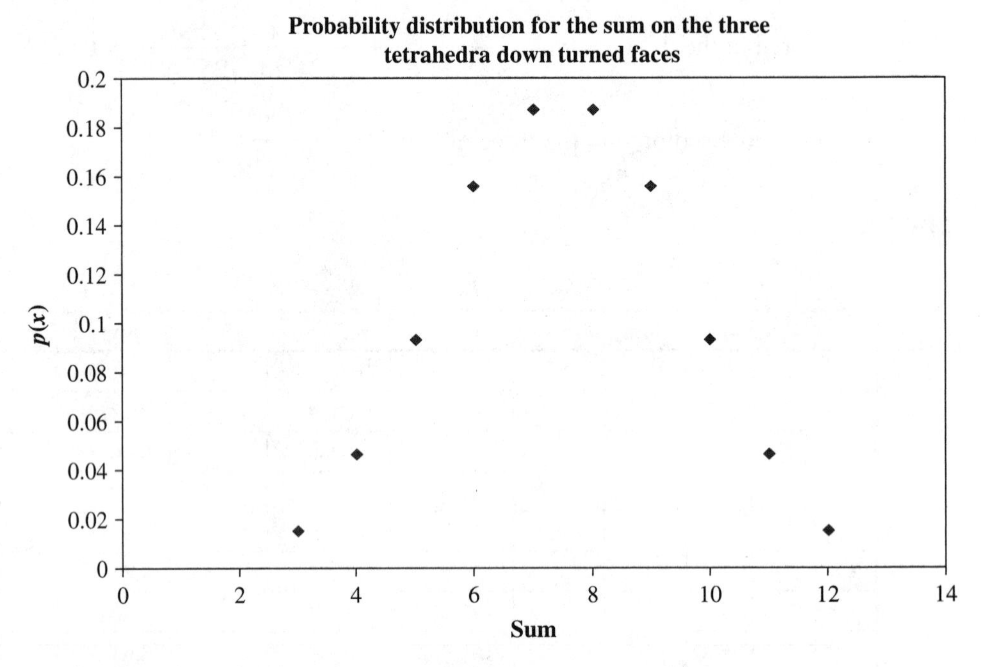

Figure Solutions-1 Scatter plot of the probability distribution of sum on the three tetrahedra.

This interval captures the numbers 4, 5, 6, 7, 8, 9, 10, and 11. The probability corresponding to these numbers is 0.9688. Chebyshev's theorem states there is at least 75% in this interval. $(\mu - 3\sigma, \mu + 3\sigma)$ captures all the probability. Chebyshev's theorem states there is at least 89% in the interval.

(c)

x	$F(x)$
3	0.0156
4	0.0625
5	0.1563
6	0.3126
7	0.5001
8	0.6876
9	0.8439
10	0.9377
11	0.9846
12	1.0000

(d) First we build the following table:

x	$p(x)$	$F(x)$	Range
3	0.0156	0.0156	0000 - 0155
4	0.0469	0.0625	0156 – 0624
5	0.0938	0.1563	0625 – 1562
6	0.1563	0.3126	1563 – 3125
7	0.1875	0.5001	3126 – 5000
8	0.1875	0.6876	5000 – 6875
9	0.1563	0.8439	6876 – 8438
10	0.0938	0.9377	8439 – 9376
11	0.0469	0.9846	9377 – 9845
12	0.0156	1.0000	9846 – 9999

Now, select 20 random numbers between 0 and 9999.

```
MTB > print c1
```

Data Display

```
C1
  5362.67   8388.85   9574.04   6921.12   9457.83   4380.39   9925.58
   278.86   8684.35   4708.12   2278.93   6236.93   9231.74   2473.96
  6505.28   2024.75   5332.86   4717.10   1476.33   7988.55
```

Using these random numbers and the above table, our random numbers are

8, 9, 11, 9, 11, 7, 12

4, 10, 7, 6, 8, 10, 6

8, 6, 8, 7, 5, 9

If we wish to view this as simulating what would happen if we tossed the three tetrahedra, each second for 20 seconds, we would get the following simulated time series shown in Figure Solutions-2:

3.

(a)

i. $p(x) \geq 0$ for every x.

ii. $\Sigma\, p(x) = 0.05 + 0.10 + 0.20 + 0.25 + 0.40 = 1$.

Since the two properties are satisfied, $p(x)$ is a probability distribution function.

(b) Figure Solutions-3 shows the scatterplot for $p(x)$.

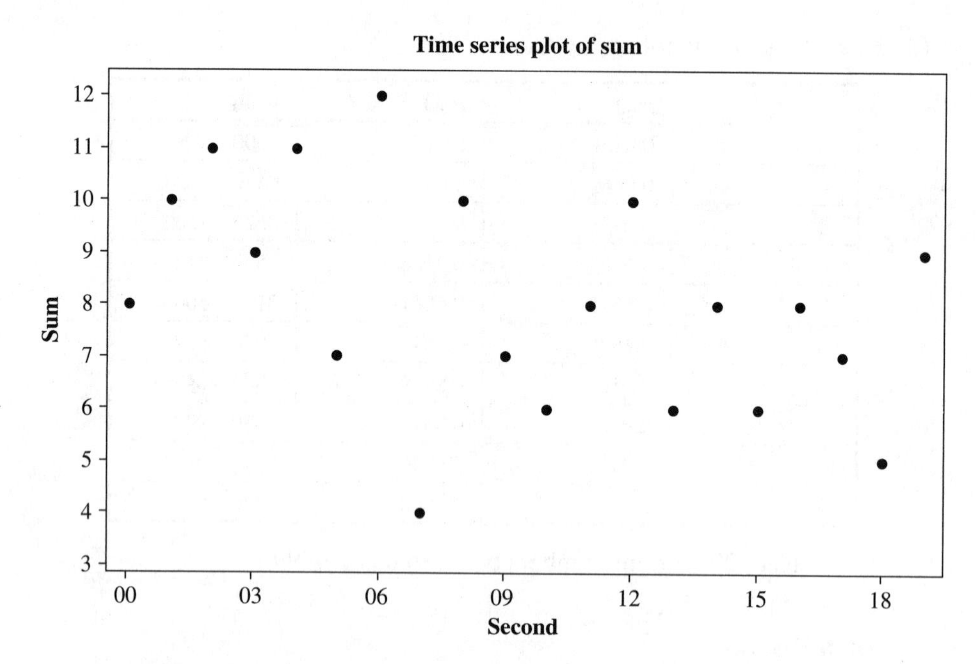

Figure Solutions-2 Viewing the random sample as a time series.

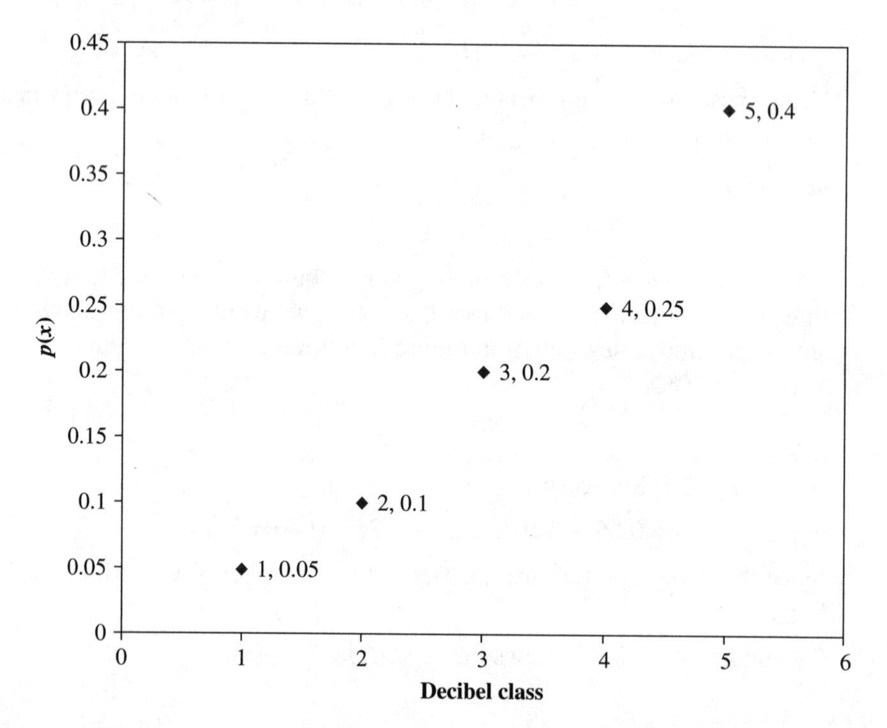

Figure Solutions-3 Probability distribution for sound levels at which iPods are played.

(c)

x	$p(x)$		$xp(x)$		$x^{2}p(x)$
1	0.05		0.05		0.05
2	0.1		0.2		0.4
3	0.2		0.6		1.8
4	0.25		1		4
5	0.4		2		10
		Mean =	3.85		16.25
		Variance =	1.4275		
		SD =	1.19478		

(d)

x	$p(x)$	$F(x)$
1	0.05	0.05
2	0.1	0.15
3	0.2	0.35
4	0.25	0.6
5	0.4	1

(e) A graph of $F(x)$ is shown in Figure Solutions-4.

Note the following:

$F(2.5) = P(X \leq 2.5) = 0.15$, $F(3.8) = P(X \leq 3.8) = 0.35$, $F(5.3) = P(X \leq 5.3) = 1$, and so forth.

(f)

x	$p(x)$	Cumulative	Range
1	0.05	0.05	00–04
2	0.1	0.15	05–14
3	0.2	0.35	15–34
4	0.25	0.6	35–59
5	0.4	1	60–99

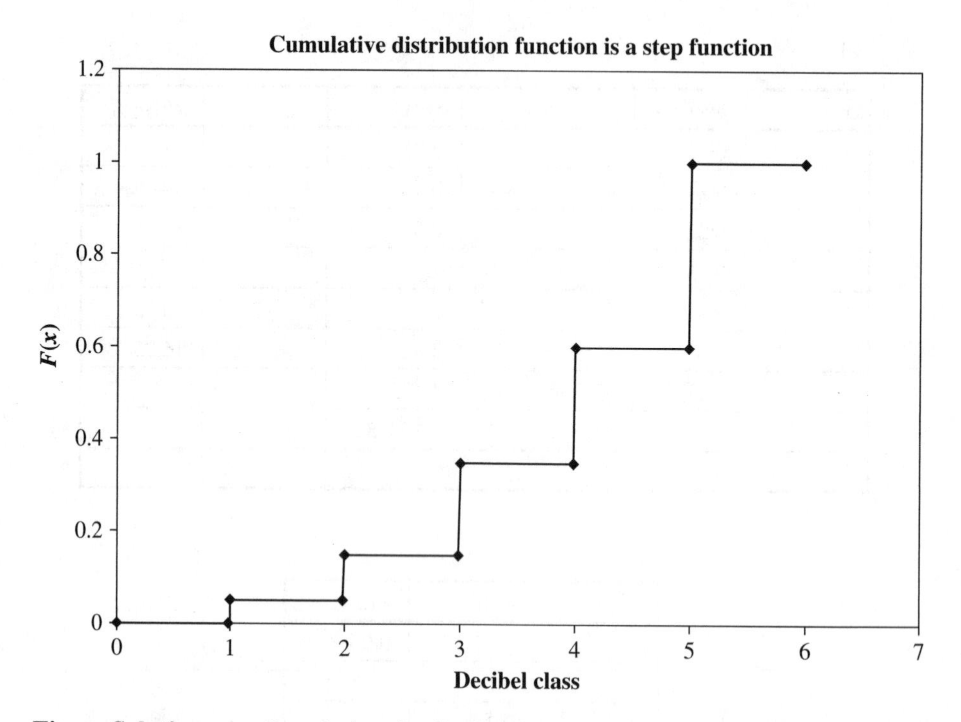

Figure Solutions-4 Cumulative distribution functions are step functions for discrete random variables.

(g) Use =RANDBETWEEN(0,99) of EXCEL to choose 50 random numbers.

67	82	53	81	40
8	93	22	55	38
88	76	53	37	92
73	71	52	88	77
22	47	51	17	51
80	94	14	20	78
75	26	26	41	73
28	64	81	18	15
56	42	77	18	61
32	90	83	51	88

(h) Use the range values in (f) to convert the random numbers in (g) to sample values. The sample mean is 4.16 and the standard deviation is 0.912. The EXCEL commands are shown in the upper right corner of the worksheet.

5	5	4	5	4				
2	5	3	4	4		4.16	AVERAGE(A1:E10)	
5	5	4	4	5		0.911603	STDEV(A1:E10)	
5	5	4	5	5				
3	4	4	3	4				
5	5	2	3	5				
5	3	3	4	5				
3	5	5	3	3				
4	4	5	3	5				
3	5	5	4	5				

The mean of the probability distribution, called the population mean, is 3.85, and the population standard deviation is 1.19. Note that the population mean, μ, is constant and the simulated sample mean, $\overline{X}$, is variable; and the population standard deviation, σ, is constant and the simulated sample standard deviation, S, is variable. In other words if another simulation were performed, a different sample mean and a different sample standard deviation would result.

4.

(a) The mean of a binomial probability distribution is

$$\mu = np = 30(0.7) = 21$$

(b) The standard deviation of a binomial distribution is

$$\mu = \sqrt{npq} = \sqrt{30(0.7)(0.3)} = 2.51$$

(c) Using the software package STATISTIX, fill in the dialog box as shown in Figure Solutions-5. The cumulative probability $P(X \le 15) =$ 0.01694. The dialog box is shown in Figure Solutions-5.

(d) Using EXCEL.

$P(25 \text{ or more}) = 1 - P(24 \text{ or less}) = 1 - B(24) = 1 - 0.923405 = 0.0.076595$

The cumulative function $B(24)$ is given by =BINOMDIST(24,30,0.7,1)

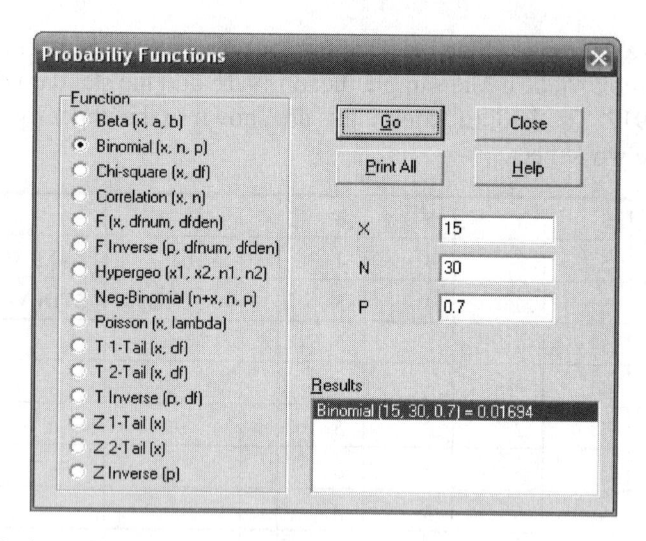

Figure Solutions-5 Cumulative binomial distribution function from STATISTIX.

5.

(a)

0	0.310563	COMBIN(10,A1)*COMBIN(40,5-A1)/COMBIN(50,5)
1	0.431337	COMBIN(10,A2)*COMBIN(40,5-A2)/COMBIN(50,5)
2	0.20984	COMBIN(10,A3)*COMBIN(40,5-A3)/COMBIN(50,5)
3	0.044177	COMBIN(10,A4)*COMBIN(40,5-A4)/COMBIN(50,5)
4	0.003965	COMBIN(10,A5)*COMBIN(40,5-A5)/COMBIN(50,5)
5	0.000119	COMBIN(10,A6)*COMBIN(40,5-A6)/COMBIN(50,5)

(b) **Output Produced by the Dialog Box in Figure Solutions-6**

Probability Density Function

```
Hypergeometric with N = 50, M = 10, and n = 5

x  P(X = x)
0     0.310563
1     0.431337
2     0.209840
3     0.044177
4     0.003965
5     0.000119
```

In 5a, the computations using the COMBIN function are shown. In 5b, the computations using the built-in function of MINITAB are shown.

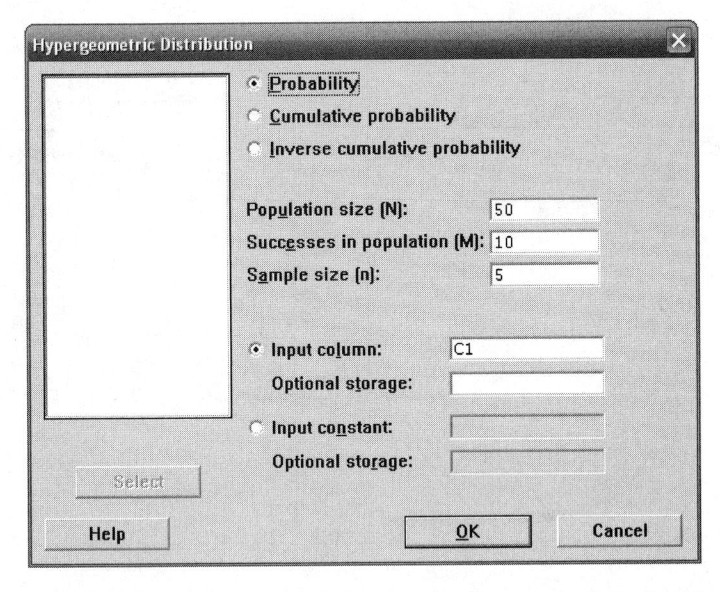

Figure Solutions-6 MINITAB dialog box for hypergeometric distribution.

The dialog box is produced by the pull-down **Calc ⇒ Probability Distributions ⇒ Hypergeometric**.

6.

(a) Figure Solutions-7 is a MINITAB plot of the probability function and the cumulative probability function for the random variable X.

Note that the graph stops at $x = 30$. This is because the $p(x)$ values are already close to 0 at this point or even before. The function $F(x)$ is actually a step function. However, the only $F(x)$ values plotted are for integral values of x. $F(30) = .999381$, indicating that there is no need to plot $F(x)$ any further.

(b) $\mu = \frac{1}{p} = \frac{1}{0.25} = 4$ and $\sigma^2 = \frac{1-p}{p^2} = \frac{1-0.25}{0.0625} = 12$

(c) It is skewed to the right.

(d) $F(5) = 0.762695$ $F(10) = 0.943686$ $F(15) = 0.986637$

(e) The probability that it takes more than 15 tosses with a balanced regular tetrahedron is 0.013363. Even though this is a possible event, you might begin to believe that the tetrahedron is not balanced if such an outcome occurred.

7.

(a) $p(x) = NB(x) = \binom{x-1}{2}(0.45)^3(0.55)^{(x-3)}$, $X = 3, 4, \ldots$

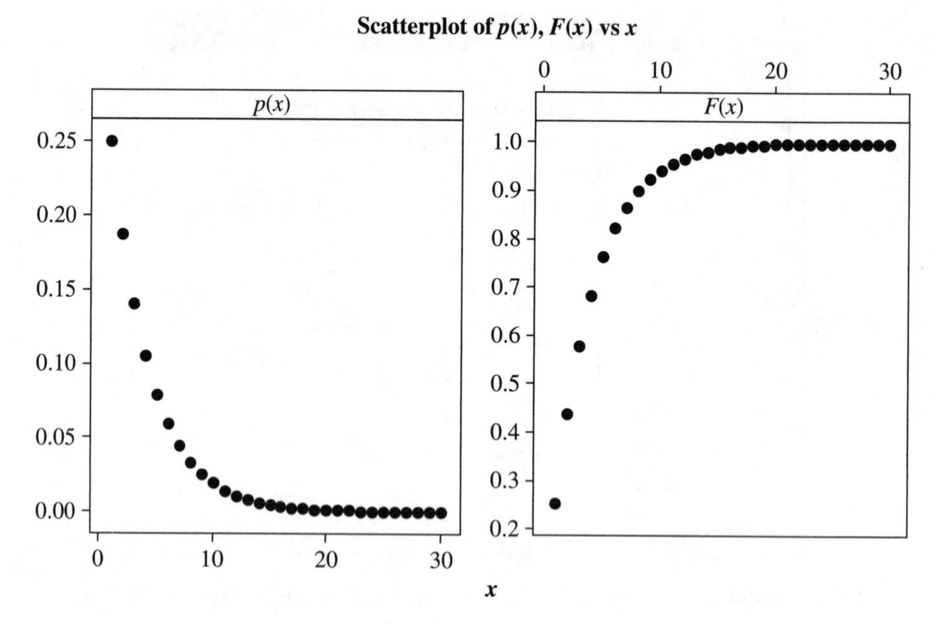

Figure Solutions-7 Plot of the probability function and the cumulative probability function for $p(x) = (0.75)\text{^}(x{-}1)*(0.25)$.

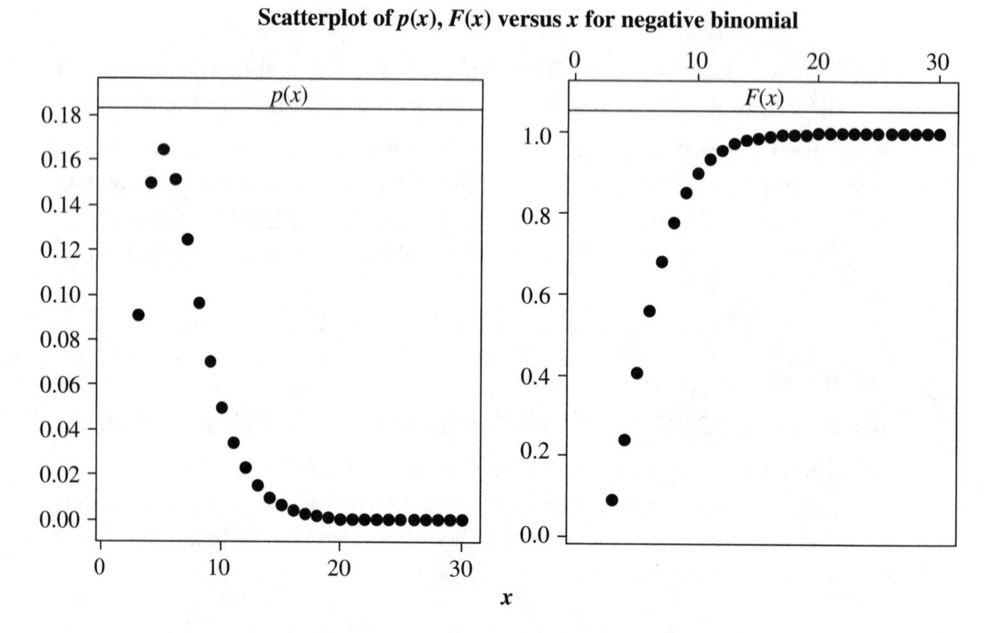

Figure Solutions-8 Plot of the probability function and the cumulative probability function for $NB(x) = \binom{x-1}{2}(0.45)^3(0.55)^{(x-3)}$.

(b)

Note that the graph stops at $x = 30$. This is because the $p(x)$ values are already close to 0 at this point or even before. The function $F(x)$ is actually a step function. However, the only $F(x)$ values that are plotted are for integral values of x. $F(30) = .999995$, indicating that there is no need to plot $F(x)$ any further as it gets closer and closer to 1 as x gets larger.

(c) The mean is

$$\mu = \frac{r}{p} = \frac{3}{0.45} = 6.67$$

and the variance is

$$\sigma^2 = \frac{r(1-p)}{p^2} = \frac{3(0.55)}{0.2025} = 8.15$$

(d) The graph in Figure Solutions-8 is skewed to the right.

(e) $F(5) = 0.406873$ $F(10) = 0.90044$ $F(15) = 0.989384$

8.

(a) $p(x) = P_o(x) = \frac{e^{-15}15^x}{15!}$, $x = 0, 1, 2, \ldots$

(b) The numbers 0 through 30 are entered into column C1. The pull-down **Calc ⇒ Probability Distributions ⇒ Poisson** is used to calculate both the Poisson probabilities as well as the cumulative Poisson probabilities using MINITAB. Note again the probabilities can be calculated from 0 to any large number. Thirty is sufficient to obtain the probabilities that are useful.

x	$P_o(x)$	$F(x)$
0	0.000000	0.000000
1	0.000005	0.000005
2	0.000034	0.000039
3	0.000172	0.000211
4	0.000645	0.000857
5	0.001936	0.002792
6	0.004839	0.007632
7	0.010370	0.018002
8	0.019444	0.037446
9	0.032407	0.069854
10	0.048611	0.118464

11	0.066287	0.184752
12	0.082859	0.267611
13	0.095607	0.363218
14	0.102436	0.465654
15	0.102436	0.568090
16	0.096034	0.664123
17	0.084736	0.748859
18	0.070613	0.819472
19	0.055747	0.875219
20	0.041810	0.917029
21	0.029865	0.946894
22	0.020362	0.967256
23	0.013280	0.980535
24	0.008300	0.988835
25	0.004980	0.993815
26	0.002873	0.996688
27	0.001596	0.998284
28	0.000855	0.999139
29	0.000442	0.999582
30	0.000221	0.999803

(c) $P(X > 30) = 1 - P(X \leq 30) = 1 - 0.999803 = 0.000197$

(d) $P(5 \leq X \leq 15) = F(15) - F(4) = 0.568090 - 0.000857 = 0.567233$

(e) The Poisson plot is shown in Figure Solutions-9.

9.

(a) $P(X_1 = 0, X_2 = 25, X_3 = 0, X_4 = 0) = 2.10297\text{E-}05$ as given by the following EXCEL statement =MULTINOMIAL(0,25,0,0)*0.15^0*0.65^25*0.15^0*0.05^0, or the probability of selecting 25 two car accidents is $(0.65)^{25} = 0.00002103$. Because of the independence we simply multiply the probabilities.

(b) $P($ all 25 involve the same number of cars$) = P(X_1 = 25, X_2 = 0, X_3 = 0, X_4 = 0) + P(X_1 = 0, X_2 = 25, X_3 = 0, X_4 = 0) + P(X_1 = 0, X_2 = 0, X_3 = 25, X_4 = 0) + P(X_1 = 0, X_2 = 0, X_3 = 0, X_4 = 25) = (.15)^{25} + (.65)^{25} + (.15)^{25} + (.05)^{25} = 2.525 \times 10^{-21} + 2.102 \times 10^{-5} + 2.525 \times 10^{-21} + 2.980 \times 10^{-33}$, which is approximately $2.102 \times 10^{-5} = 0.00002103$.

(c) The EXCEL expression is =MULTINOMIAL(6,6,6,7)*0.15^6*0.65^6*0.15^6*0.05^7, which equals 6.30352E-08.

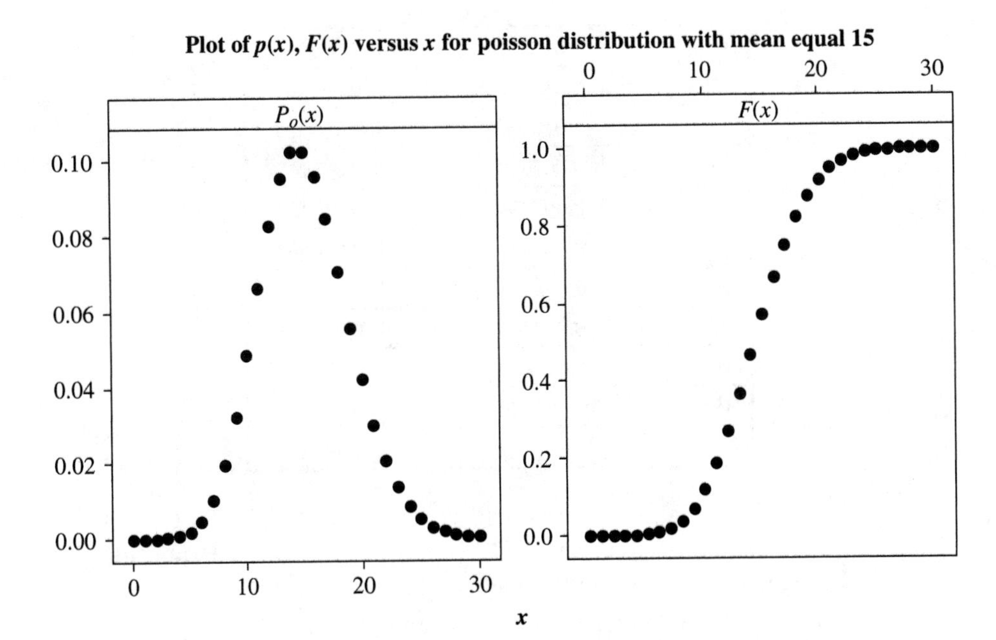

Figure Solutions-9 Plot of the probability function and the cumulative probability function for $P_o(x) = \frac{e^{-15}15^x}{15!}$.

10.

(a) =MULTINOMIAL(5,0,0,0)*0.25^5*0.25^0*0.25^0*0.25^0 = 0.000977. There would be four terms like this, and they would be added to give 4(0.000977) = 0.003908.

(b) The following is an EXCEL solution. In B2, enter =0.75^(A2-1)*0.25 and perform a click-and-drag. In C2, enter = B2. In C3, enter =C2+B3 and perform a click-and-drag.

A	B	C
x	$p(x)$	$F(x)$
1	0.25	0.25
2	0.1875	0.4375
3	0.140625	0.578125
4	0.105469	0.683594
5	0.079102	0.762695
6	0.059326	0.822021
7	0.044495	0.866516
8	0.033371	0.899887
9	0.025028	0.924915

10	0.018771	0.943686
11	0.014078	0.957765
12	0.010559	0.968324
13	0.007919	0.976243
14	0.005939	0.982182
15	0.004454	0.986637
16	0.003341	0.989977
17	0.002506	0.992483
18	0.001879	0.994362
19	0.001409	0.995772
20	0.001057	0.996829

(c) X has a binomial distribution with $n = 6$ trials, $p = 12/52 = 0.23$, and $q = 1 - 0.23 = 0.77$. The MINITAB pull-down **Calc ⇒ Probability Distributions ⇒ Binomial** leads to the binomial probability and cumulative probability function. The output is:

Row	x	$p(x)$	$F(x)$
1	0	0.208422	0.20842
2	1	0.373536	0.58196
3	2	0.278939	0.86090
4	3	0.111093	0.97199
5	4	0.024888	0.99688
6	5	0.002974	0.99985
7	6	0.000148	1.00000

(d) The MINITAB pull-down **Calc ⇒ Probability Distributions ⇒ Hypergeometric** is given after the numbers 0 through 4 are entered into C1. The dialog box in Figure Solutions-10 is produced, which is filled in as shown. After going through this again and choosing the cumulative probability function, the following output is given:

x	$p(x)$	$F(x)$
0	0.658842	0.65884
1	0.299474	0.95832
2	0.039930	0.99825
3	0.001736	0.99998
4	0.000018	1.00000

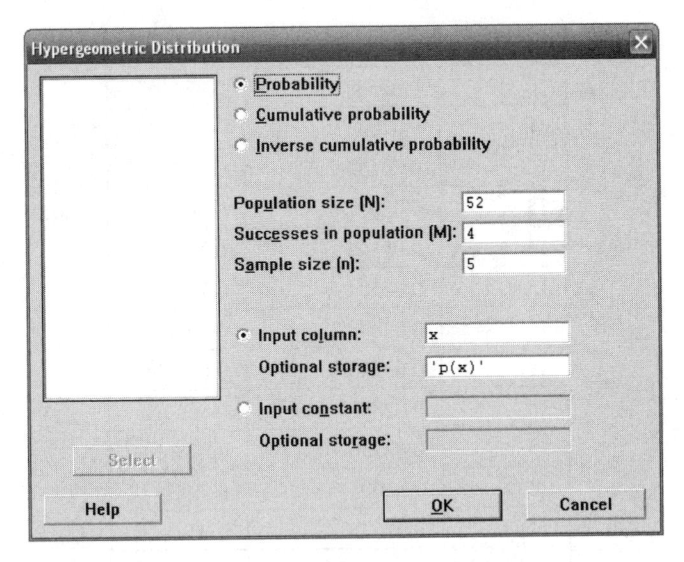

Figure Solutions-10 Dialog box for hypergeometric probability function for the number of Aces dealt in a 5 card hand (without replacement).

The deck of cards is dichotomized as follows: 4 Aces and 48 Nonaces. R = 5 cards are selected from the 52 without replacement. The hypergeometric probabilities are calculated for the numbers in the output column. The most likely occurrence is no Aces in the hand of 5, and the least likely thing to occur is 4 Aces in the hand of 5. Do not count on being dealt 4 Aces very often.

(e) The EXCEL computations are shown below. The expression =COMBIN(A2-1,2)*(0.25)^3*(0.75)^(A2-3) is entered into B2 and a click-and-drag is performed to generate the $p(x)$ values for $x = 3$ through 25. The expression =B2 is entered in C2 and the expression =B3+C2 is entered into C3, and then a click-and-drag is performed. The results in A, B, and C are copied from EXCEL into MINITAB, and then **Graph ⇒ Scatterplot** is used to graph $p(x)$ and $F(x)$ in separate panels of the same graph. The results are shown in Figure Solutions-11.

A	*B*	*C*
x	*p*(*x*)	*F*(*x*)
3	0.015625	0.015625
4	0.035156	0.050781
5	0.052734	0.103516
6	0.065918	0.169434
7	0.074158	0.243591
8	0.077866	0.321457

9	0.077866	0.399323
10	0.075085	0.474407
11	0.070392	0.544799
12	0.064526	0.609325
13	0.058073	0.667398
14	0.051474	0.718872
15	0.04504	0.763912
16	0.038977	0.802889
17	0.033409	0.836298
18	0.028397	0.864695
19	0.02396	0.888655
20	0.020084	0.90874
21	0.016737	0.925477
22	0.013874	0.939351
23	0.011446	0.950797
24	0.009402	0.960199
25	0.007693	0.967891

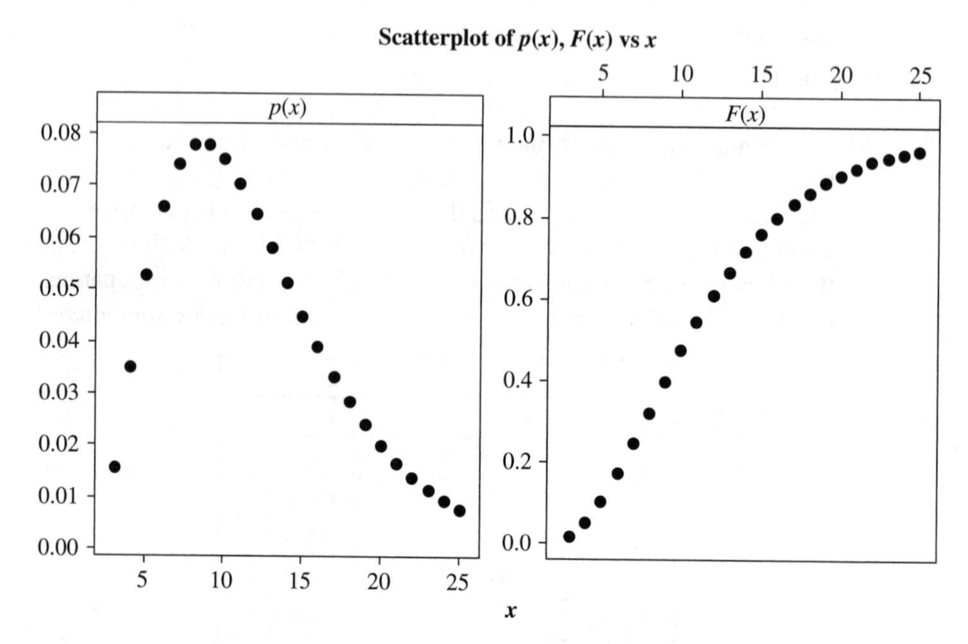

Figure Solutions-11 $p(x)$ and $F(x)$ for a negative binomial distribution.

Chapter 4

1.

(a)

0	0
0.1	0.004
0.2	0.031949
0.3	0.107129
0.4	0.24953
0.5	0.469707
0.6	0.758978
0.7	1.079146
0.8	1.359699
0.9	1.513028
1	1.471518
1.1	1.231367
1.2	0.869062
1.3	0.505244
1.4	0.235537
1.5	0.085451
1.6	0.023347
1.7	0.004636
1.8	0.000644
1.9	6.01E-05
2	3.6E-06

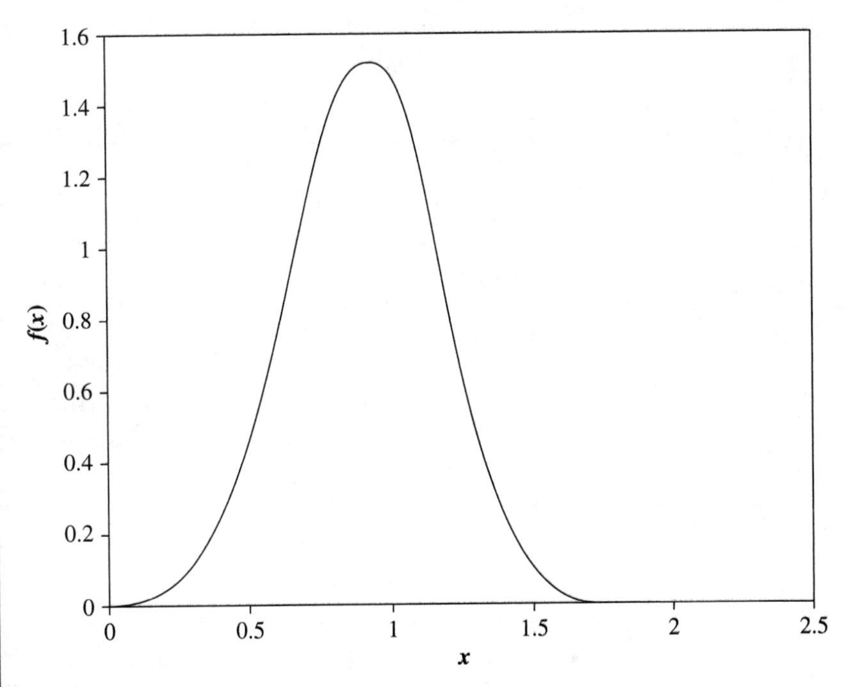

Even though the function is defined from 0 to infinity, it is sufficient to go from 0 to 2. At $x = 2$, the $f(x)$ value is 3.6E-06, and the $f(x)$ values decrease from 2 to infinity.

(b) The antiderivative of $f(x) = 4x^3e^{-x^4}$ is $-e^{-x^4}$. The antiderivative at the upper limit of integration (∞) is 0. Evaluation at the lower limit of integration gives -1. The value of the integral is $0 - (-1) = 1$.

(c)

```
> evalf(int(4*x^3*exp(-x^4),x=0..infinity));
```

1.

(d) The interval from 0 to 2 is divided into 25 equal intervals. This makes each interval 0.08 wide. Evaluating $f(x) = 4x^3e^{-x^4}$ at the interval midpoints and multiplying by 0.08 gives rectangle areas which are added up to give 0.999999, or approximately 1. The details are as follows: This is the approximate evaluation of an integral as a limiting sum of rectangles.

Midpoint	Height	Area of rectangle
0.04	0.000256	2.04799E-05
0.12	0.006911	0.000552845
0.2	0.031949	0.002555907
0.28	0.08727	0.006981595
0.36	0.183516	0.014681249
0.44	0.328201	0.026256101
0.52	0.522777	0.041822122
0.6	0.758978	0.060718239
0.68	1.015614	0.081249104
0.76	1.257797	0.100623789
0.84	1.441037	0.115282942
0.92	1.521591	0.121727289
1	1.471518	0.117721421
1.08	1.292642	0.103411395
1.16	1.021135	0.081690821
1.24	0.717068	0.057365402
1.32	0.441857	0.035348588
1.4	0.235537	0.018842929
1.48	0.106946	0.008555678
1.56	0.040678	0.003254275
1.64	0.012733	0.001018633
1.72	0.003218	0.000257474
1.8	0.000644	5.15131E-05
1.88	9.98E-05	7.98774E-06
1.96	1.17E-05	9.38961E-07
	Sum =	0.999998718

We shall use MAPLE to evaluate **e** through **h**. The integrals involved could be evaluated using integration by parts or as the limiting areas of rectangles.

(e)

```
> evalf(int(4*x*x^3*exp(-x^4),x=0..infinity)); 0.9064024772
```

The mean is 0.9064.

The variance is given by

```
> evalf(int(4*x^2*x^3*exp(-x^4),x=0..infinity))- 0.906402^2;
0.0646623399
```

The standard deviation is 0.254.

(f) $P(X < 1)$ is

```
> evalf(int(4*x^3*exp(-x^4),x=0..1)); 0.6321205588
```

(g) $P(X > 2)$ is given by

```
> evalf(int(4*x^3*exp(-x^4),x=2..infinity)); 0.1125351747 10^-6
```

(h) $P(0.5 < X < 1.5)$ is given by

```
> evalf(int(4*x^3*exp(-x^4),x=(0.5)..(1.5))); 0.9330833474
```

2.

(a) =NORMDIST(105,100,5,1)-NORMDIST(90,100,5,1)   0.818595

(b) **Cumulative Distribution Function**

```
Normal with mean = 100 and standard deviation = 5

  x      P(X <= x)
105       0.841345
```

Cumulative Distribution Function

```
Normal with mean = 100 and standard deviation = 5

 x     P(X <= x)
90     0.0227501
```

$0.841345 - 0.0227501 = 0.818595$

(c)

```
> 1/(2*3.14159*25)^(0.5); 0.07978848978
> evalf(int(0.0797885*exp(-(x-100)^2/50),x=90..105));
0.8185950647
```

(d) The region from $x = 90$ to $x = 105$ is divided into 30 equal intervals. The intervals are 90 to 90.5, 90.5 to 91, … , and 104.5 to 105. The midpoints of the intervals are 90.25, 90.75, and so on. These are entered into the A column of the EXCEL worksheet. The normal pdf =0.079788*EXP(-((A1-100)^2/50)) is entered into B1 and a click-and-drag is performed from B1 to B30. In cell C1, the area =B1*(0.5) is calculated and another click-and-drag is performed from C1 to C30. When =SUM(C1:C30) is executed, it gives the sum of the rectangular areas that approximate

$$\int_{90}^{105} \frac{1}{\sqrt{2\pi}\sigma} e^{-\frac{(x-100)^2}{2\sigma^2}} dx$$

The answer given is 0.818735793. This is a close approximation to the true value 0.818595.

Midpoint	Height	Area of Rectangle
90.25	0.011919	0.005959437
90.75	0.014413	0.007206446
91.25	0.017255	0.008627683
91.75	0.020453	0.010226434
92.25	0.024002	0.012000831
92.75	0.027886	0.013942977
93.25	0.032076	0.016038241
93.75	0.03653	0.018264804
94.25	0.041187	0.020593509
94.75	0.045976	0.022988083
95.25	0.050812	0.02540576
95.75	0.055597	0.02779833
96.25	0.060227	0.030113571
96.75	0.064594	0.032297051
97.25	0.068588	0.034294189
97.75	0.072105	0.03605249
98.25	0.075048	0.03752382
98.75	0.077333	0.038666591
99.25	0.078895	0.039447708
99.75	0.079688	0.039844164
100.25	0.079688	0.039844164
100.75	0.078895	0.039447708
101.25	0.077333	0.038666591

101.75	0.075048	0.03752382
102.25	0.072105	0.03605249
102.75	0.068588	0.034294189
103.25	0.064594	0.032297051
103.75	0.060227	0.030113571
104.25	0.055597	0.02779833
104.75	0.050812	0.02540576
		0.818735793

3.

X, the number in the survey to answer yes has a binomial distribution with $n = 500$ and $p = 0.15$. Fit a normal curve to this distribution that has mean = $np = 500(0.15) = 75$ and standard deviation equal to the square root of npq, or 7.98.

(a) The area under the normal curve to the right of 99.5 will approximate this. The normal approximation is given by =1-NORMDIST(99.5,75,7.98,1) = 0.00107. It is not very likely that 100 or more will answer yes.

(b) The area under the normal curve between 49.5 and 80.5 will approximate the binomial probabilities of 50 or 51 or …or 80. The normal approximation is given by =NORMDIST(80.5,75,7.98,1)-NORMDIST(49.5,75,7.98,1) = 0.753961.

(c) The area under the normal curve to left of 59.5 will approximate the binomial probabilities of 59, 58, 57, ⋯ ,0. The normal approximation is given by =NORMDIST(59.5,75,7.98,1) = 0.026047.

4.

(a) The density is

$$f(x) = \begin{cases} \dfrac{\Gamma(20)}{\Gamma(1)\Gamma(19)} x^0 (1-x)^{18}, & 0 < x < 1 \\ 0, & \text{otherwise} \end{cases}$$

The density reduces to $19(1-x)^{18}$, $0 < x < 1$. The MINITAB plot of the graph is shown in Figure Solutions-12.

(b)

```
> evalf(int(19*(1-x)^18,x=0..1));          1.
```

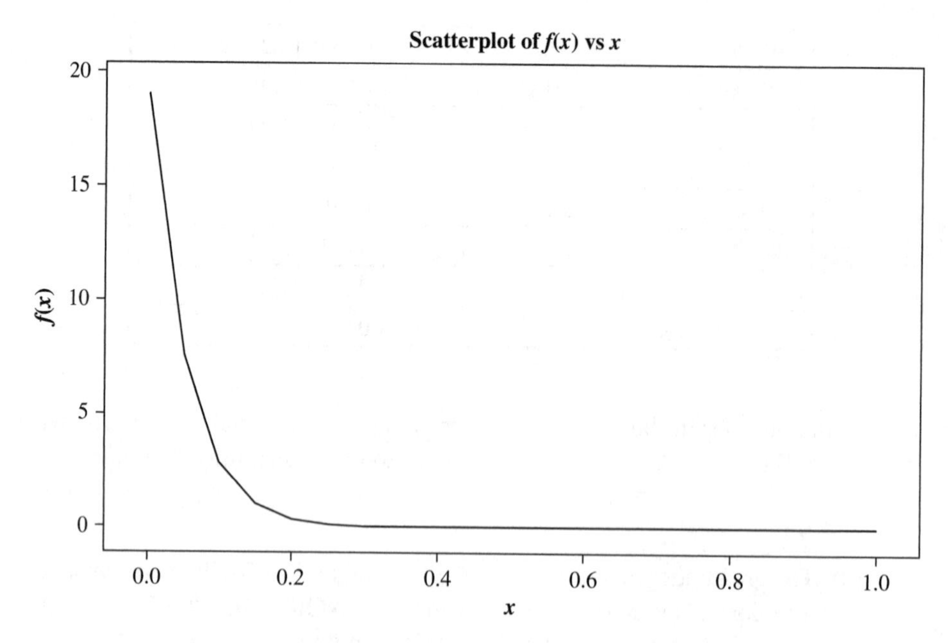

Figure Solutions-12 Beta density with $\alpha = 1$ and $\beta = 19$.

(c) The mean is

$$\mu = \frac{\alpha}{\alpha + \beta} = \frac{1}{20} = 0.05$$

The variance is

$$\sigma^2 = \frac{\alpha\beta}{(\alpha + \beta)^2(\alpha + \beta + 1)} = \frac{19}{8400} = 0.00226$$

The standard deviation is 0.049.

(d)

```
> evalf(int(19*(1-x)^18,x=0..(.1)));
```

0.8649148282

(e)

Data Display

0.048	0.029	0.043	0.045	0.070	0.035	0.040	0.020	0.005	0.057
4.8%	2.9%	4.3%	4.5%	7.0%	3.5%	4.0%	2.0%	0.5%	5.7%

5.

(a) We need to find c such that

$$\int_0^\infty \int_0^2 \int_0^1 cxy^2 e^{-z} dx\, dy\, dz = 1$$

The value of the integral is $\frac{4}{3}c$. Set this equal to 1, and solving for c gives $c = 0.75$.

(b)

```
> evalf(int(int((3/4)*x*y^2*exp(-z),y=0..2),z=0..infinity));
```

$2.x$

The marginal of X is $f_1(x) = 2x,\ 0 < x < 1$.

```
> evalf(int(int((3/4)*x*y^2*exp(-z),x=0..1),z=0..infinity));
```

$0.3750000000y^2$

The marginal of Y is $f_2(y) = 0.375y^2,\ 0 < y < 2$

```
> evalf(int(int((3/4)*x*y^2*exp(-z),x=0..1),y=0..2));
```

$e^{(-1.z)}$

The marginal of Z is $f_3(z) = e^{-z},\ Z > 0$

Since $f(x, y, z) = f_1(x)\, f_2(y)\, f_3(z)$, X, Y, and Z are independent.

(c)

```
> evalf(int(int(int((3/4)*x*y^2*exp(-z),x=0..1),y=0..1),z=0..1));
```

0.07901506985

$$P(X < 1, Y < 1, Z < 1) = 0.07902$$

6.

(a)

i. Anderson-Darling p-value < 0.005

ii. Ryan-Joiner p-value < 0.01, and

iii. Kolmogorov-Smirnov p-value < 0.010 (all three from MINITAB)

iv. Shapiro-Wilk p-value $= 0.0008$ (from STATISTIX)

The tests indicate that we may not make the normality assumption.

(b)

1.93	4.50	1.32	2.24	3.37
3.66	2.03	2.76	1.63	1.73
2.26	3.48	2.47	3.70	2.72

i. Anderson-Darling p-value $= 0.542$

ii. Ryan-Joiner p-value > 0.10

iii. Kolmogorov-Smirnov p-value > 0.15 (all three from MINITAB)

iv. Shapiro-Wilk p-value $= 0.6243$ (from STATISTIX)

The tests indicate that we may make the normality assumption on the transformed data.

7.

(a) The mean is given by

$$\mu = \int_{-1}^{1} xf(x)\,dx$$

```
> evalf(int(-(3/4)*x*(x^2-1),x=-1..1));
```
0.

(b) The standard deviation is given by

$$\sigma = \sqrt{\int_{-1}^{1} x^2 f(x)\,dx - \mu^2} = \sqrt{\int_{-1}^{1} x^2 f(x)\,dx}, \text{ since } \mu = 0$$

```
> evalf(sqrt(int(-(3/4)*x^2*(x^2-1),x=-1..1)));
```
0.4472135954

(c) $P(-0.5 < X < 0.5)$ is given by

```
> evalf(int(-(3/4)*(x^2-1),x=-(0.5)..(0.5)));
```
0.6875000000

8.

(a) The EXCEL plot is shown in Figure Solutions-13.

(b)
```
> evalf(int(2*t*exp(-t^2),t=0..x));
```
$1. - 1.e^{(-1.x^2)}$

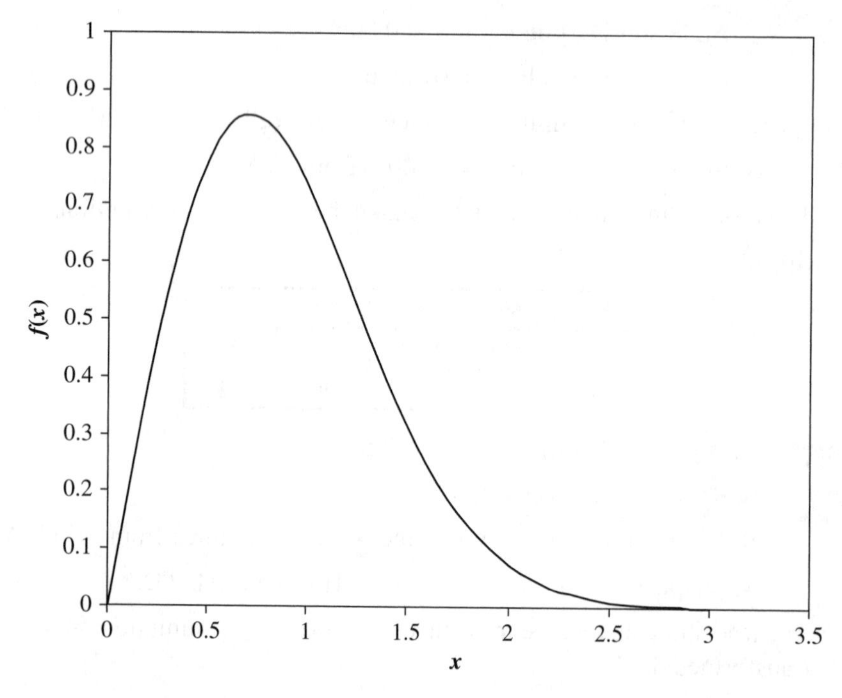

Figure Solutions-13 EXCEL plot of $f(x) = 2xe^{-x^2}$.

The cumulative distribution function is

$$F(x) = 1 - e^{-x^2}$$

Now, select a uniform value between 0 and 1, and call it u. Set $u = F(x)$ and solve for x that gives $x = \sqrt{-\ln(1-u)}$. Do this 20 times.

```
MTB > print c1
```

Data Display

```
C1
  0.714885   0.097200   0.463522   0.981267   0.034246   0.080840   0.415917
  0.774619   0.445700   0.174317   0.828826   0.192090   0.170119   0.320307
  0.441448   0.812440   0.148943   0.139102   0.226046   0.749110
```

```
MTB > print c2
```

Data Display

```
C2
  1.12021   0.31977   0.78913   1.99436   0.18667   0.29034   0.73329
  1.22064   0.76815   0.43766   1.32856   0.46185   0.43182   0.62138
  0.76316   1.29370   0.40159   0.38701   0.50620   1.17590
```

The 20 numbers in C1 are uniform from (0, 1). The 20 numbers from C2 are given by $C2 = \sqrt{-\ln(1-C1)}$ and represent a random sample from the density

$$f(x) = 2xe^{-x^2}$$

9.

(a) The gamma distribution is

$$f(x) = \frac{1}{\beta^{\alpha}\Gamma(\alpha)} x^{\alpha-1} e^{-x/\beta}, x > 0$$

Substitute $\alpha = \nu/2$ and $\beta = 2$.

The chi-square distribution is

$$f(x) = \frac{1}{2^{\nu/2}\Gamma(\nu/2)} x^{\nu/2-1} e^{-x/2}, x > 0$$

(b) The mean of the gamma distribution is $\mu = \alpha\beta$. Substitute $\alpha = \nu/2$ and $\beta = 2$.

The chi-square mean is

$$\mu = (\nu/2)(2) = \nu$$

(c) The variance of the gamma distribution is $\alpha\beta^2$. Substitute $\alpha = \nu/2$ and $\beta = 2$.

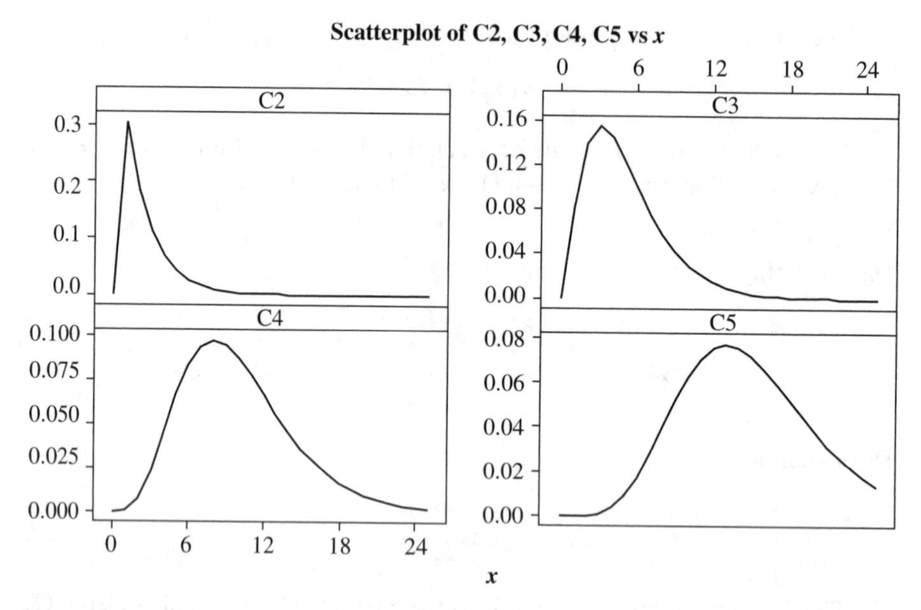

Figure Solutions-14 Chi-square distributions with 2, 5, 10, and 15 degrees of freedom.

The chi-square variance is

$$\sigma^2 = (\nu/2)(4) = 2\nu$$

(d) Figure Solutions-14 shows 4 chi-square distributions. C2 is a chi-square with 2 degrees of freedom. C3 is a chi-square with 5 degrees of freedom. C4 is a chi-square with 10 degrees of freedom. C5 is a chi-square with 15 degrees of freedom.

10.

(a) In MINITAB, α is called the location parameter and β is called the shape parameter in the dialog box for the log-normal distribution. The log-normal is shown in Figure Solutions-15.

(b)

$$\mu = e^{\alpha+\beta^2/2} = e^{3.5+1.125} = e^{4.625} = 102$$

$$\sigma^2 = e^{2\alpha+\beta^2}(e^{\beta^2} - 1) = e^{9.25}(e^{2.25} - 1) = 88311 \text{ and } \sigma = 297$$

(c) Log-normal with location = 3.5 and scale = 1.5

x	$P(X <= x)$
150	0.843054

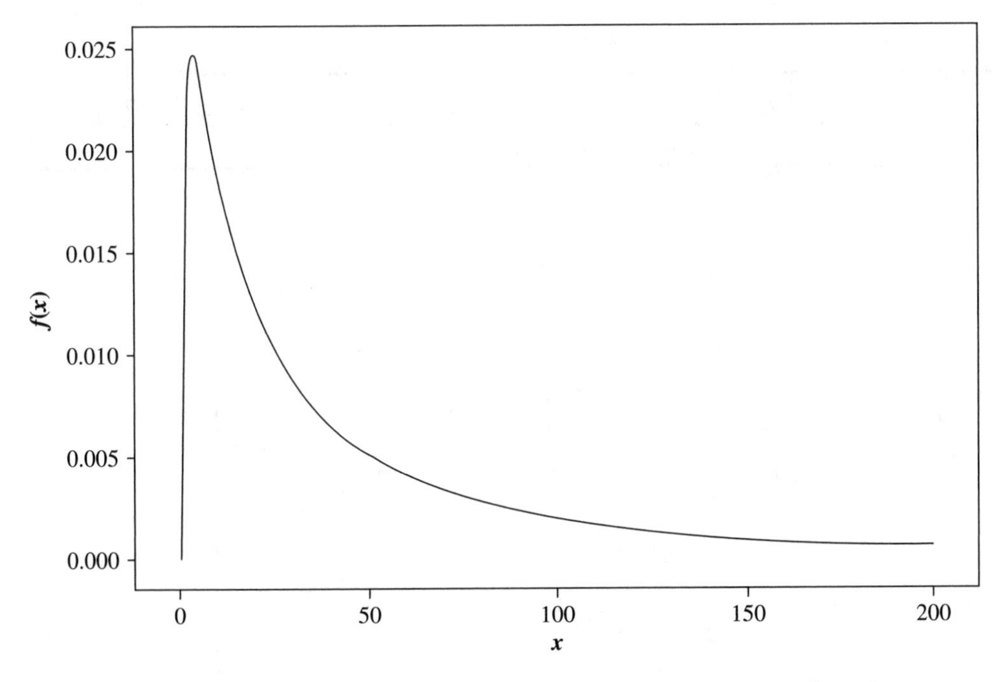

Figure Solutions-15 Log-normal distribution with $\alpha = 3.5$ and $\beta = 1.5$.

Chapter 5

1.

(a) $\mu = 1(0.2) + 2(0.2) + 3(0.2) + 4(0.2) + 5(0.2) = 3$

$\sigma^2 = 1^2(0.2) + 2^2(0.2) + 3^2(0.2) + 4^2(0.2) + 5^2(0.2) - 3^2 = 2$ and $\sigma = 1.414$

(b) Uniform

(c)

$\bar{x}$	1	1.5	2	2.5	3	3.5	4	4.5	5
$p(\bar{x})$	0.04	0.08	0.12	0.16	0.20	0.16	0.12	0.08	0.04

$$\mu_{\bar{x}} = 3 \qquad \sigma_{\bar{x}} = \frac{1.414}{\sqrt{2}} = 1$$

(d)

$\bar{x}$	1	1.33	1.67	2	2.33	2.67	3	3.33	3.67	4	4.33	4.67	5
$P(\bar{x})$	.008	.024	.048	.080	.120	.144	.152	.144	.120	.080	.048	.024	.008

$$\mu_{\bar{x}} = 3 \qquad \sigma_{\bar{x}} = \frac{1.414}{\sqrt{3}} = 0.816$$

(e)

$\bar{x}$	Count	$p(\bar{x})$
1.00	1	0.0016
1.25	4	0.0064
1.50	10	0.0160
1.75	20	0.0320
2.00	35	0.0560
2.25	52	0.0832
2.50	68	0.1088
2.75	80	0.1280
3.00	85	0.1360
3.25	80	0.1280
3.50	68	0.1088
3.75	52	0.0832
4.00	35	0.0560
4.25	20	0.0320
4.50	10	0.0160
4.75	4	0.0064
5	1	0.0016

$$\mu_{\bar{x}} = 3 \qquad \sigma_{\bar{x}} = \frac{1.414}{\sqrt{4}} = 0.707$$

2.

Figures Solutions-16 through Solutions-19 show distributions of $\bar{X}$.

The plots show that as the sample size (n) increases for a particular distribution, the curve approaches a normal curve and the standard deviation of the curve is decreasing. The same phenomenon happens no matter what the original population distribution is. It happens at different rates depending on the shape of the original distribution.

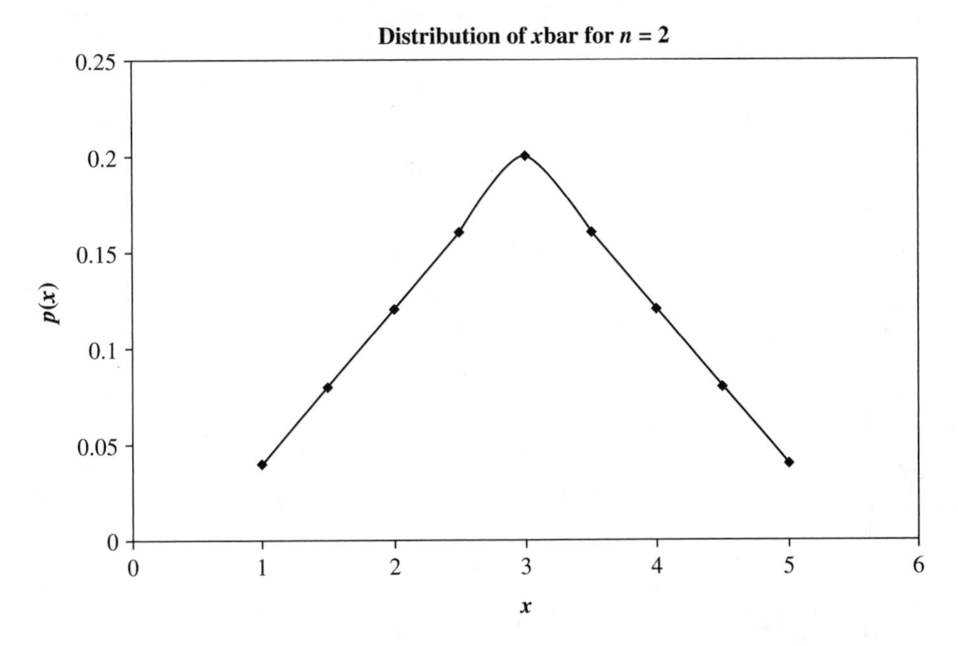

Figure Solutions-16 Distribution of $\overline{X}$ for sample size 2.

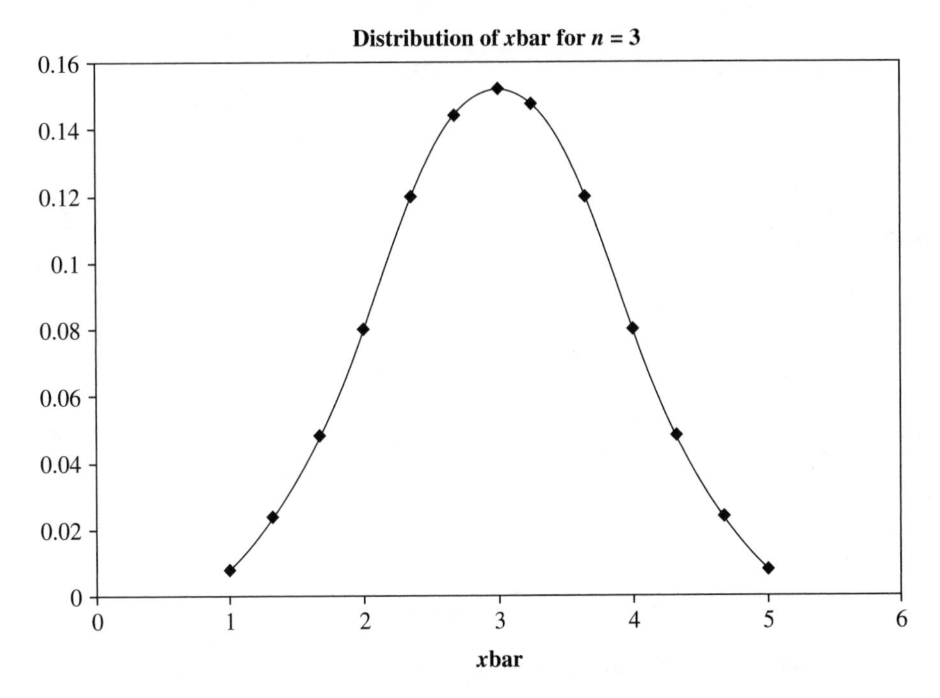

Figure Solutions-17 Distribution of $\overline{X}$ for sample size 3.

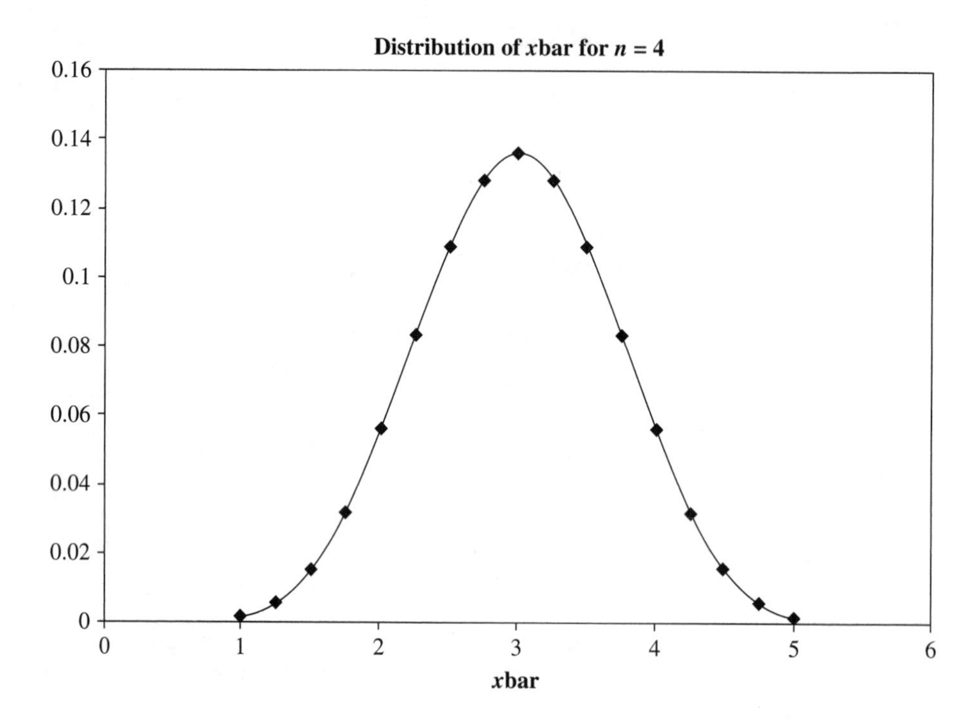

Figure Solutions-18 Distribution of $\bar{X}$ for sample size 4.

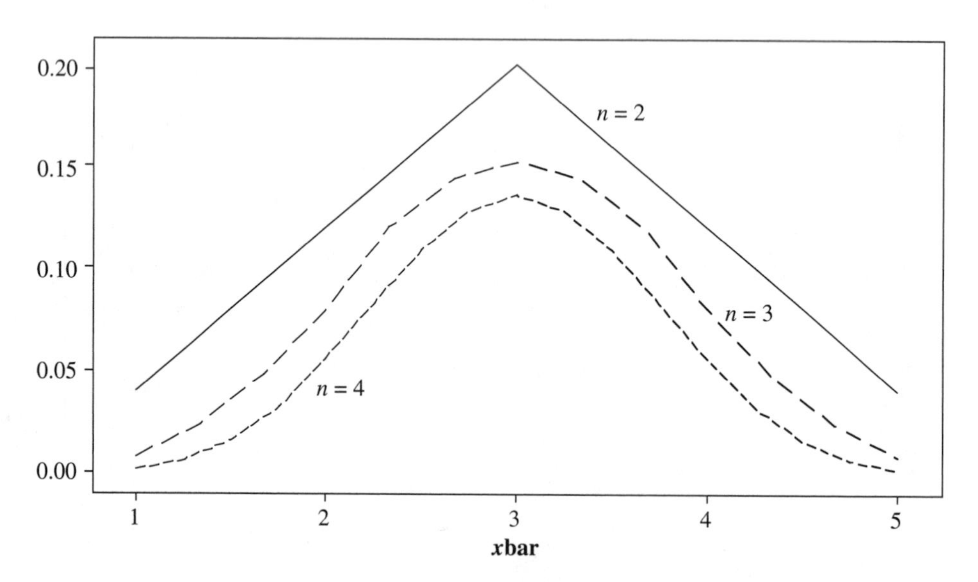

Figure Solutions-19 Distribution of *x*bar for $n = 2$, 3, and 4, overlaid on the same graph.

3.

Calculate the probability of finding a sample mean of 3.1 or larger, under the assumption that the population mean is 2.5. The standard error of the mean is $1/\sqrt{35} = 0.169$. The central limit theorem tells us that $\overline{X}$ is normally distributed with mean 2.5 and standard error 0.169. The probability that $\overline{X} \geq 3.1$ is given by the EXCEL command =1-NORMDIST (3.1,2.5,0.169,1), which equals 0.00019. Since this probability is less than 0.05, it is concluded that the population mean fluoride is greater than 2.5. What this result says is that if the mean of the population is really 2.5 and a sample mean is computed, the probability is only 0.00019 that the sample mean would be as large as 3.1 or larger. Because the sample size is large ($n = 35$), we know from the central limit theorem that $\overline{X}$ has a normal distribution with mean equal to the population mean and standard deviation equal to the standard error. *The student really needs to make sure he/she understands the logic here, and not just learn the mechanics.*

4.

The Anderson-Darling normality test gives a p-value of 0.778, the Ryan-Joiner normality test gives a p-value > 0.1, and the Kolmogorov-Smirnov normality test gives a p-value > 0.15. No matter which of these tests are used, normality is not rejected.

Proceeding, the calculated value for the sample mean, standard error, and standard deviation are:

Descriptive Statistics: Fluoride

```
Variable   Mean   SE Mean   SD
Fluoride  2.740     0.338  1.069
```

$$t_9 = \frac{\bar{x} - \mu}{\frac{s}{\sqrt{10}}} = \frac{2.74 - 2.5}{0.338} = 0.71$$

The probability of getting a t_9 equal to 0.71 or larger is found using MINITAB.

Cumulative Distribution Function

```
Student's t distribution with 9 DF

  x     P(X <= x)
0.71     0.752156
```

$$P(t_9 > 0.71) = 1 - 0.75 = 0.25$$

There is a good chance (25%) of getting such a sample if the population mean is 2.5. The population mean, 2.5, is not rejected.

5.

The two rules usually give the same answer, but not always. I would use the simpler of the two rules and stick with it. That is the second of the two rules.

(a) $n = 150,\ \hat{p} = 0.10$

$$n\hat{p} = 15 \text{ and } n\hat{q} = 135 \quad \text{Yes}$$

$$\hat{p} \pm 3\sqrt{\frac{\hat{p}\hat{q}}{n}} \text{ or } 0.10 \pm 3\sqrt{\frac{0.10(0.90)}{150}} \text{ or } 0.10 \pm 0.07 \ (0.03, 0.17) \quad \text{Yes}$$

(b) $n = 500,\ \hat{p} = 0.05$

$$n\hat{p} = 25 \text{ and } n\hat{q} = 475 \quad \text{Yes}$$

$$\hat{p} \pm 3\sqrt{\frac{\hat{p}\hat{q}}{n}} \text{ or } 0.05 \pm 3\sqrt{\frac{0.05(0.95)}{500}} \text{ or } 0.05 \pm 0.03 \ (0.02, 0.08) \quad \text{Yes}$$

(c) $n = 50,\ \hat{p} = 0.15$

$$n\hat{p} = 7.5 \text{ and } n\hat{q} = 42.5 \quad \text{Yes}$$

$$\hat{p} \pm 3\sqrt{\frac{\hat{p}\hat{q}}{n}} \text{ or } 0.15 \pm 3\sqrt{\frac{0.15(0.85)}{50}} \text{ or } 0.15 \pm 0.151 \ (-0.001, 0.301) \quad \text{No}$$

(d) $n = 1000,\ \hat{p} = 0.001$

$$n\hat{p} = 1 \text{ and } n\hat{q} = 999 \quad \text{No}$$

$$\hat{p} \pm 3\sqrt{\frac{\hat{p}\hat{q}}{n}} \text{ or } 0.001 \pm 3\sqrt{\frac{0.001(0.999)}{1000}} \text{ or } 0.001 \pm 0.003 \ (-0.002, 0.004) \quad \text{No}$$

6.

Set the approximate 95% error of estimate equal to 0.05, and solve for n.

$$2\sigma_{\hat{p}} = 2\sqrt{\frac{\hat{p}\hat{q}}{n}} = 2\sqrt{\frac{0.15(0.85)}{n}} = 0.05$$

The solution is $n = 204$. The engineer would need to sample an additional 104 units, combine the two results, and perform the calculations again.

7.

				PHAT				
A	B	C	D	1		PHAT	Prob	G*H
A	B	C	E	0.75		0.5	0.4	0.2
A	B	C	F	0.75		0.75	0.53	0.3975
A	B	D	E	0.75		1	0.07	0.07
A	B	D	F	0.75			Mean of PHAT	0.6675
A	B	E	F	0.5				
A	C	D	E	0.75				
A	C	D	F	0.75				
A	C	E	F	0.5				
A	D	E	F	0.5				
B	C	D	E	0.75				
B	C	D	F	0.75				
B	C	E	F	0.5				
B	D	E	F	0.5				
C	D	E	F	0.5				

Note that $\mu_{\text{PHAT}} = p$. This population is small, and the distribution of PHAT is not approximately normal.

8.

(a) $\sigma^2 = 1^2(0.2) + 2^2(0.2) + 3^2(0.2) + 4^2(0.2) + 5^2(0.2) - 3^2 = 2$ and $\sigma = 1.414$

(b)

S^2	$p(S^2)$
0	0.2
0.5	0.32
2	0.24
4.5	0.16
8	0.08

$E(S^2) = 0(0.2) + 0.5(0.32) + 2(0.24) + 4.5(0.16) + 8(0.08) = 2$

S^2 is an unbiased estimator of σ^2.

S	$p(S)$
0	0.2
0.707	0.32
1.414	0.24
2.121	0.16
2.828	0.08

$$E(S) = 0(0.2) + 0.707(0.32) + 1.414(0.24) + 2.121(0.16) + 2.828(0.08) = 1.131$$

S is not an unbiased estimator of σ.

(c)

S^2	$p(S^2)$
0.000	0.040
0.333	0.192
1.000	0.144
1.333	0.144
2.333	0.192
3.000	0.096
4.000	0.048
4.333	0.096
5.333	0.048

$$E(S^2) = 0(0.040) + 0.333\,(0.192) + 1(0.144) + 1.333(0.144)$$
$$+ 2.333(0.192) + 3(0.096) + 4(0.048) + 4.333(0.096) + 5.333(0.048) = 2$$

S^2 is an unbiased estimator of σ^2.

S	$p(S)$
0.000	0.040
0.577	0.192
1	0.144
1.155	0.144
1.527	0.192
1.732	0.096
2	0.048
2.082	0.096
2.309	0.048

$$E(S) = 0(0.040) + 0.577\,(0.192) + 1(0.144) + 1.155(0.144) + 1.572(0.192) + 1.732(0.096) + 2(0.048) + 2.082(0.096) + 2.309(0.048) = 1.287$$

S is not an unbiased estimator of σ.

(d)

S^2	$p(S^2)$
0.00000	0.0080
0.25000	0.0512
0.33333	0.0384
0.66667	0.0576
0.91667	0.1152
1.00000	0.0384
1.33333	0.0288
1.58333	0.0768
1.66667	0.0768
2.00000	0.0768
2.25000	0.1024
2.66667	0.0192
2.91667	0.0768
3.00000	0.0576
3.33333	0.0384
3.58333	0.0384
3.66667	0.0384
4.00000	0.0128
4.25000	0.0384
5.33333	0.0096

$E(S^2) = 2$, S^2 is an unbiased estimator of σ^2.

S	$p(S)$
0	0.0080
0.5	0.0512
0.577347	0.0384
0.816499	0.0576
0.957429	0.1152
1	0.0384

1.154699	0.0288
1.258304	0.0768
1.290996	0.0768
1.414214	0.0768
1.5	0.1024
1.632994	0.0192
1.707826	0.0768
1.732051	0.0576
1.825741	0.0384
1.892969	0.0384
1.914855	0.0384
2	0.0128
2.061553	0.0384
2.3094	0.0096

$E(S) = 1.340$, S is not an unbiased estimator of σ.

9.

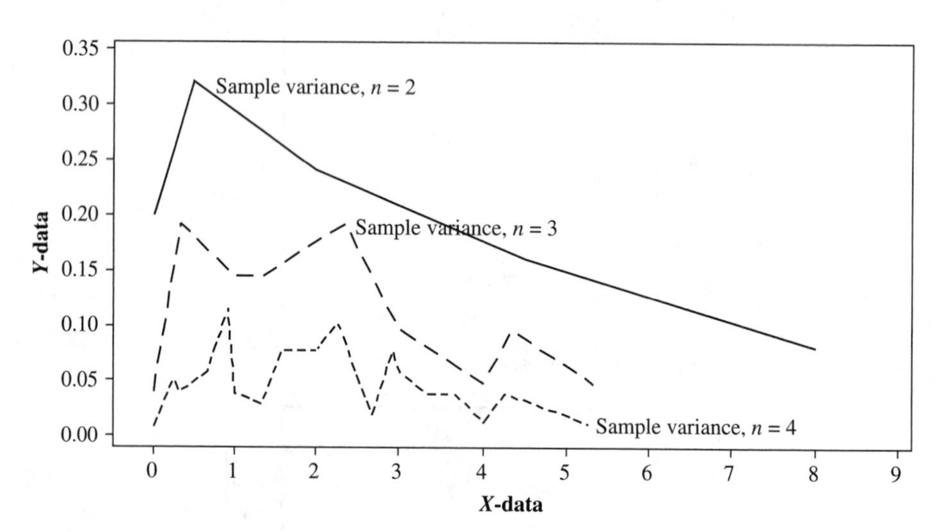

Figure Solutions-20 Sampling distribution of sample variance for $n = 2$, 3, and 4.

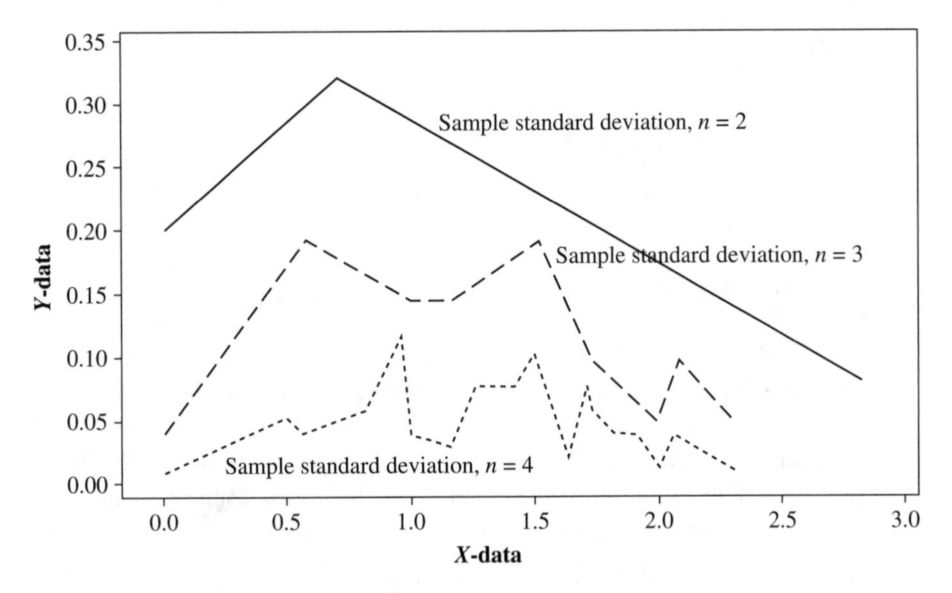

Figure Solutions-21 Sampling distribution of sample standard deviation for $n = 2$, 3, and 4.

10.

Suppose we use the package STATISTIX to aid in our analysis.

First, test to determine it is reasonable to assume normality of the data. The results of the Shapiro-Wilk test are:

Shapiro-Wilk Normality Test

Variable	N	W	p
Length	10	0.9318	0.4663

The large p-value allows us to not reject normality.

Descriptive Statistics

Variable	N	Mean	SD
Length	10	99.270	3.3556

The standard deviation of the 10 values is 3.3556. The value of $(n - 1)S^2/\sigma^2$ is $9(11.26)/12.25 = 8.27$. This is very close to the mean value of a chi-square distribution with 9 degrees of freedom, which is 9. Therefore, assuming that the population has a standard deviation equal to 3.5 does not yield any unusual value for a chi-square random variable. It is reasonable to assume the process is in control.

Chapter 6

1.

(a) The maximum error of estimate is given by

$$E = z_{\alpha/2} \frac{\sigma}{\sqrt{n}}$$

The sample standard deviation is used in place of σ. The sample size is $n = 75$. The level of confidence is $1 - \alpha = 0.98$, or $\alpha = 0.02$, or $\alpha/2 = 0.01$. We use MINITAB to find $z_{0.01}$. The pull-down menu **Calc ⇒ Probability Distributions ⇒ Normal** gives the dialog box shown in Figure 10-22 which is filled out as shown. Remember the inverse cumulative probability is used because you are given an area and are asked to find a point along the z-axis. This is the inverse of what you usually do. Usually, you are given a point along the z-axis, and you find the area to the left of that point.

The dialog box produces the following output:

Inverse Cumulative Distribution Function

```
Normal with mean = 0 and standard deviation = 1

P(X <= x)        x
     0.99  2.32635
```

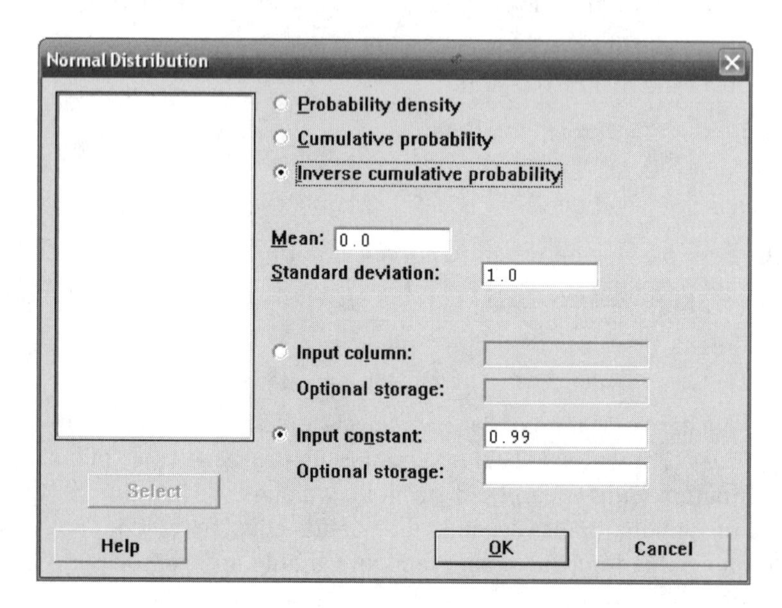

Figure Solutions-22 Dialog box for finding $z_{0.01}$.

Therefore, the area to the right of 2.32635 is 0.01.

$$E = (2.32635)(1.8)/8.66 = 0.48$$

(b) $n = \left[\frac{z_{\alpha/2}\sigma}{E}\right]^2$ is the formula for the sample size needed. $E = 0.25$ and σ is estimated by 1.8. The level of confidence is $1 - \alpha = 0.99$, or $\alpha = 0.01$, or $\alpha/2 = 0.005$. Using EXCEL, rather than MINITAB, $z_{0.005}$ =NORMSINV(0.995) = 2.575829.

$$n = \left[\frac{2.575829(1.8)}{0.25}\right]^2 = 343.95$$

Rounding up, we get 344. You will need to sample an additional 344 – 75 = 269 locations.

2.

EXCEL

	A	B	C	D
1	Temp			
2	65.7		67.895	AVERAGE(A2:A21)
3	64.9		1.951915	STDEV(A2:A21)
4	69.8		0.436462	C3/sqrt(20)
5	70.3		2.093024	TINV(0.05,19)
6	67.3			
7	64.9		66.98148	C2-C4*C5
8	67.7		68.80852	C2+C4*C5
9	71.5			
10	68.7			
11	67.5			
12	68.4			
13	66.4			
14	69.2			
15	66.6			
16	69.9			
17	70			
18	68.5			
19	65.7			
20	69.2			
21	65.7			

MINITAB

One-Sample *t*: Temp

```
Variable   N     Mean     SD      SE Mean       95% CI
Temp      20  67.8998  1.9525   0.4366   (66.9860, 68.8136)
```

SAS

```
    Sample Statistics for Temp
           N           Mean             SD       Std. Error
       -----------------------------------------------------
         20          67.90            1.95            0.44
    95 % Confidence Interval for the Mean
           Lower limit:            66.98
           Upper limit:            68.81
```

SPSS

One-Sample Statistics

	N	Mean	Std. Deviation	Std. Error Mean
temp	20	67.895	1.9519	.4365

One-Sample Test

	Test Value = 0					
					95% Confidence Interval of the Difference	
	t	df	Sig. (2-tailed)	Mean Difference	Lower	Upper
temp	155.558	19	.000	67.8950	66.981	68.809

STATISTIX

```
Statistix 8.0
Descriptive Statistics

Variable          N   Lo 95% CI        Mean   Up 95% CI         SD
Temp             20      66.981      67.895      68.809     1.9519
```

3.

(a) α=BINOMDIST(2,36,0.167,1)+(1-BINOMDIST(9,36,0.167,1)) = 0.112586

(b) $\beta = P$(not reject null for different values of p)

p	β	$1-\beta$	p	β	$1-\beta$
0.01	0.005581	0.994419	0.16	0.890654	0.109346
0.02	0.034983	0.965017	0.17	0.884663	0.115337
0.03	0.092775	0.907225	0.18	0.869847	0.130153
0.04	0.173367	0.826633	0.19	0.846847	0.153153
0.05	0.267914	0.732086	0.2	0.816397	0.183603

0.06	0.367779	0.632221	0.21	0.779343	0.220657
0.07	0.46598	0.53402	0.22	0.736647	0.263353
0.08	0.557565	0.442435	0.23	0.689373	0.310627
0.09	0.639433	0.360567	0.24	0.638659	0.361341
0.1	0.709939	0.290061	0.25	0.58568	0.41432
0.11	0.76846	0.23154	0.26	0.531605	0.468395
0.12	0.815018	0.184982	0.27	0.477548	0.522452
0.13	0.849981	0.150019	0.28	0.424536	0.575464
0.14	0.873869	0.126131	0.29	0.373468	0.626532
0.15	0.887241	0.112759	0.3	0.325093	0.674907

The values of p are entered in A1:A30. The function =BINOMDIST (9,36,A1,1)-BINOMDIST(2,36,A1,1) is entered into B1 and a click-and-drag is performed from B1 to B30. The expression 1 – B1 is entered in C1 and a click-and-drag is performed from C1 to C30.

Note that in Figures Solutions-23 and Solutions-24 the values of p are between 0 and 1. The plots were stopped at $p = 0.3$.

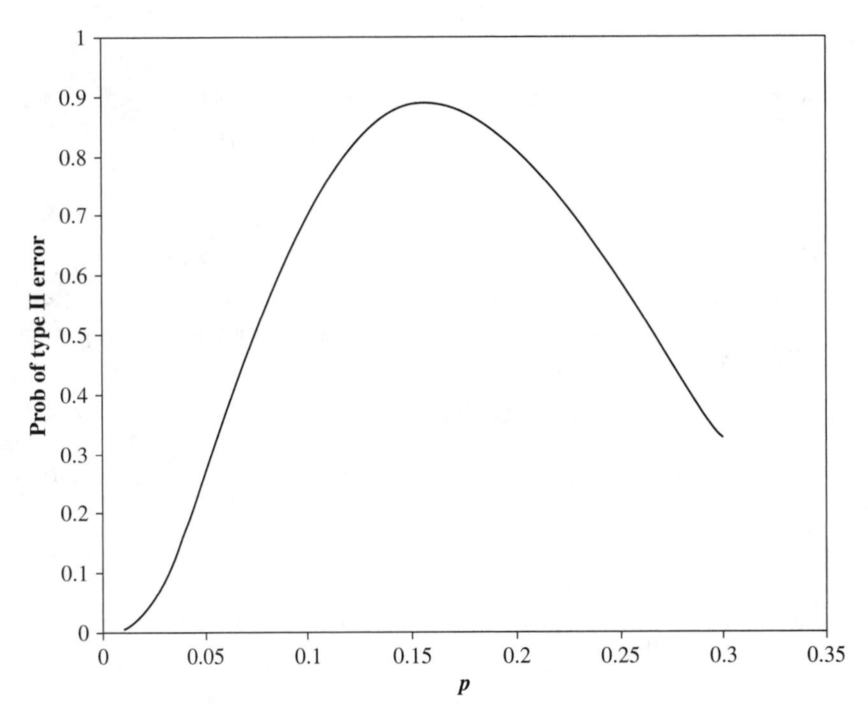

Figure Solutions-23 Graph of type II errors.

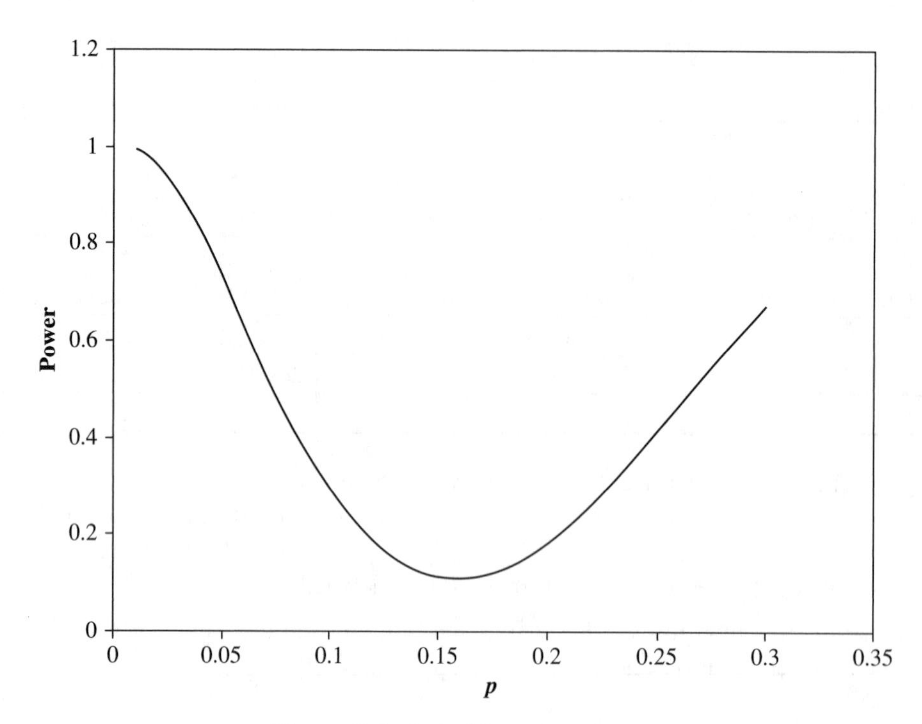

Figure Solutions-24 Power curve.

(c) p-value = $P(X \geq 9$ or $X \leq 3$, when $p = 0.167)$ =BINOMDIST(3,36, 0.167,1)+(1-BINOMDIST(8,36,0.167,1)) = 0.261.

4.

(a) The test statistic is computed as =(3.4-3)/(1.5/SQRT(50)), and the value is 1.886. The critical values are calculated as =NORMSINV(0.025) = -1.96 and =NORMSINV(0.975) = 1.96. The rejection region is $Z < -1.96$ and $Z > 1.96$. The computed test statistic does not fall in the rejection region, and the null hypothesis is not rejected.

(b) The 95% confidence interval is

$$\bar{x} - z_{\alpha/2}\frac{s}{\sqrt{n}} < \mu < \bar{x} + z_{\alpha/2}\frac{s}{\sqrt{n}}, \quad \text{or} \quad 3.4 - 1.96\frac{1.5}{\sqrt{50}} < \mu < 3.4 + 1.96\frac{1.5}{\sqrt{50}}$$

Using EXCEL, the lower confidence limit is =3.4-1.96*1.5/SQRT(50) = 2.98 and the upper confidence limit is =3.4+1.96*1.5/SQRT(50) = 3.82. The null hypothesis value for the mean is 3, and 3 falls inside the interval (2.98, 3.82) so that we are unable to reject the null.

(c) The p-value is twice the area to the right of 1.886, or =2*(1-NORMSDIST (1.886)), which equals 0.059 which is larger than the preset α (0.05) so that we are unable to reject the null.

Therefore, we are unable to reject the null, no matter which of the three methods we use, as will always be the case. The three methods are equivalent.

5.

(a) The test statistic is computed as =(3.9 – 3)/(1.5/SQRT(13)), and the value is 2.163. The critical values are calculated as =TINV(0.05,12), which equals 2.179. The rejection region is $t < -2.179$ and $t > 2.179$. The computed test statistic does not fall in the rejection region and the null hypothesis is not rejected.

(b) The 95% confidence interval is

$$\bar{x} - t_{\alpha/2}\frac{s}{\sqrt{n}} < \mu < \bar{x} + t_{\alpha/2}\frac{s}{\sqrt{n}}, \quad \text{or} \quad 3.9 - 2.179\frac{1.5}{\sqrt{13}} < \mu < 3.9 + 2.179\frac{1.5}{\sqrt{13}}$$

Using EXCEL, the lower confidence limit is =3.9-2.179*1.5/SQRT(13) = 2.993 and the upper confidence limit is =3.9 + 2.179*1.5/SQRT(13) = 4.807. The null hypothesis value for the mean is 3, and 3 falls inside the interval (2.993, 4.807) so that we are unable to reject the null.

(c) The p-value is twice the area to the right of 2.163, or =TDIST(2.163,12,2), which equals 0.051 which is larger than the preset α (0.05) so that we are unable to reject the null.

6.

She uses the pull-down **Stat ⇒ Basic Statistics ⇒ 2-Sample *t*** in MINITAB, and in the dialog box, she chooses summarized data. She may or may not choose equal variances since that assumption is not needed for large samples. She also does not need to worry about the normality assumption because of the large samples.

Two-Sample *t*-Test and CI

```
Sample   N    Mean     SD     SE Mean
1        65   16.100   0.250    0.031
2        65   16.350   0.550    0.068

Difference = mu (1) - mu (2)
Estimate for difference:  -0.250000
95% CI for difference:  (-0.398896, -0.101104)
t-test of difference = 0 (vs not =): t-value = -3.34  p-value = 0.001  DF = 89
```

(a) Because of the large samples, the t and z distributions may be used interchangeably. Therefore, the rejection region is $z < -1.96$ and $z > 1.96$. The computed z value is equal to -3.34. Since the computed test statistic is in the rejection region, the null hypothesis is rejected.

(b) The 95% confidence interval may be read directly from the output as `(-0.399,-0.101)`, and since the confidence interval does not contain $D_o = 0$, the null may be rejected if this method of testing is used.

(c) If the p-value method of testing is used, the null hypothesis is rejected since the p-value $< \alpha$.

7.

She uses the pull-down **Stat ⇒ Basic Statistics ⇒ 2-Sample *t*** in MINITAB, and in the dialog box, she chooses summarized data. She also chooses equal variances in the dialog box.

Two-Sample *t*-Test and CI

```
Sample   N    Mean      SD     SE Mean
1        10   16.100   0.250   0.079
2        15   16.350   0.550   0.14

Difference = mu (1) - mu (2)
Estimate for difference:  -0.250000
95% CI for difference:  (-0.635706, 0.135706)
t-test of difference = 0 (vs not =): t-value = -1.34  p-value = 0.193  DF = 23
Both use Pooled SD = 0.4567.
```

(a) The computed test statistic is $t = -1.34$. The degrees of freedom is $\nu = 10 + 15 - 2 = 23$.

The rejection region is determined as follows:

Inverse Cumulative Distribution Function

```
Student's t distribution with 23 DF
P(X <= x)       x
   0.025     -2.06866
```

The rejection region is $t < -2.069$ and $t > 2.069$. The computed $t = -1.34$ does not fall in the rejection region.

(b) The 95% confidence interval is `(-0.635706, 0.135706)`. Since this interval does contain $D_o = 0$, the null is not rejected.

(c) The p-value is 0.193 which is not less than $\alpha = 0.05$. The null hypothesis is not rejected.

8.

The differences pass the normality test. The p-value > 0.15 for the Kolmogorov-Smirnov test. Thus we may assume that the differences are normally distributed.

EXCEL

t-Test: Paired Two Sample for Means		
	Variable 1	Variable 2
Mean	26.33333	31.22222
Variance	12.25	9.444444
Observations	9	9
Pearson correlation	0.79412	
Hypothesized mean difference	5	
DF	8	
t stat	−13.8155	
P(*T* <= *t*) one-tail	3.64E-07	
t critical one-tail	1.859548	
P(*T* <= *t*) two-tail	7.28E-07	
t critical two-tail	2.306004	

The above analysis is incorrect. The information entered into the dialog box was that the range of variable 1 was A1:A9 and the range for variable 2 was B1:B9, and that the mean difference was 5 which in fact would have been −5. However, EXCEL will not allow a negative value to be put in for D_o. What you have to do is tell the software that the range of variable 1 is B1:B9 and that the range of variable 2 is A1:A9. The mean difference is 5. The output from EXCEL is then shown below. This is the correct output. The last five rows are now correct. You must peruse your output carefully and ask the question: "Is this output reasonable?" If the answer is **No**, then you have to ask: "What have I done that may not be in accord with the programmer who wrote the code?"

t-Test: Paired Two Sample for Means		
	Variable 1	Variable 2
Mean	31.22222	26.33333
Variance	9.444444	12.25
Observations	9	9
Pearson correlation	0.79412	
Hypothesized mean difference	5	
df	8	
t stat	−0.15523	
P(*T* <= *t*) one-tail	0.440243	
t critical one-tail	1.859548	
P(*T* <= *t*) two-tail	0.880485	
t critical two-tail	2.306004	

The MINITAB output is

Paired *t*-Test and CI: Highway, Stop-and-go

```
Paired t for Highway - Stop-and-go

                 N     Mean       SD     SE Mean
Highway          9  31.2222   3.0732    1.0244
Stop-and-go      9  26.3333   3.5000    1.1667
Difference       9  4.88889  2.14735   0.71578

95% upper bound for mean difference: 6.21992
t-test of mean difference = 5 (vs < 5): t-value = -0.16  p-value = 0.440
```

Note that the t-value in the MINITAB output is –0.16. In the EXCEL output it is called t Stat and is equal to –0.15523. The p-value in MINITAB is called p-value and is equal to 0.440. In EXCEL it is written as $P(T <= t)$ one-tail and is 0.440243.

```
Statistix 8
Paired t-Test for Highway - Stop

Null hypothesis: difference = 5
Alternative hypothesis: difference < 5

Mean              4.8889
Std. Error        0.7158
Mean - H0        -0.1111
Lower 95% CI     -1.7617
Upper 95% CI      1.5395
t                  -0.16
DF                     8
P                 0.4402
```

9.

$$v = \frac{(s_1^2/n_1 + s_2^2/n_2)^2}{\dfrac{(s_1^2/n_1)^2}{n_1 - 1} + \dfrac{(s_2^2/n_2)^2}{n_2 - 1}}$$

is used to compute the degrees of freedom.

EXCEL Computation of the Degrees of Freedom

Process 1	Process 2
84.3	77
75.3	72.3
76	90.4
75.2	101.9
78.1	60.8
74.3	89.1
78.5	70
70.3	20.8
75	26
77.4	91.4

1.300933	VAR(A2:A11)/10
75.02334	VAR(B2:B11)/10
5825.395	(D2+D3)^2
625.5772	(D2^2/9+D3^2/9)
9.312033	D4/D5

$$v = \frac{(s_1^2 / n_1 + s_2^2 / n_2)^2}{\frac{(s_1^2 / n_1)^2}{n_1 - 1} + \frac{(s_2^2 / n_2)^2}{n_2 - 1}}$$

The MINITAB computation gives

Two-Sample *t*-Test and CI: Process 1, Process 2

```
Two-sample t for Process 1 vs Process 2

           N    Mean     SD    SE Mean
Process 1  10   76.44   3.61    1.1
Process 2  10   70.0    27.4    8.7

Difference = mu (process 1) - mu (process 2)
Estimate for difference:  6.47377
95% CI for difference:  (-13.29480, 26.24233)
t-test of difference = 0 (vs not =): t-value = 0.74  p-value = 0.478  DF = 9
```

If the 9.312033 calculated using EXCEL is rounded down to the next integer as is recommended, the MINITAB, DF = 9 is obtained.

10.

EXCEL Computation of Pooled Standard Deviation

Type 1	Type 2
2164	1641
2040	1827
1926	1847
1911	1800
1973	1932
2064	1937
2018	2023
2097	1836
2105	1891
1972	1715

6774.444	VAR(A2:A11)
12309.21	VAR(B2:B11)
9541.828	(D2+D3)/2
97.68228	SQRT(D4)

$$S_p^2 = \frac{(n_1 - 1)S_1^2 + (n_2 - 1)S_2^2}{n_1 + n_2 - 2}$$

When the sample sizes are equal, the pooled variance is the average of the two sample variances.

Two-Sample *t*-Test and CI: Type 1, Type 2

```
Two-Sample t for Type 1 vs Type 2

                                 SE
           N    Mean      SD    Mean
Type 1    10    2027.0    82.3    26
Type 2    10    1845      111     35

Difference = mu (type 1) - mu (type 2)
Estimate for difference:  182.100
95% CI for difference:  (90.322, 273.878)
t-test of difference = 0 (vs not =): t-value = 4.17  p-value = 0.001  DF = 18
Both use Pooled SD = 97.6823.
```

Chapter 7

1.

EXCEL Output

A	B	C	D
Diameter	0.156347	Standard deviation	=STDEV(A2:A11)
1.6	0.024444	Variance	=VAR(A2:A11)
1.5	22	Test statistic	=9*B2/0.01
1.8	0.008879	Prob	=CHIDIST(22,9)
1.5	0.013003	90% lower limit	=9*B2/CHIINV(0.05,9)
1.4	0.066163	90% upper limit	=9*B2/CHIINV(0.95,9)
1.5			
1.6			
1.7			
1.9			
1.5			

Method 1

The rejection region is $\chi^2 >$ =CHIINV(0.1,9) = 14.684. Since the test statistic exceeds this, the null is rejected and the process is shut down.

Method 2

The *p*-value is 0.008879 which is less than 0.1, and the process is shut down.

Method 3

The interval (0.013, 0.066) does not contain 0.01 and the process is shut down.

2.

Value	Function
2.558212	=CHIINV(0.99,10)
3.940299	=CHIINV(0.95,10)
18.30704	=CHIINV(0.05,10)
23.20925	=CHIINV(0.01,10)

Inverse Cumulative Distribution Function

```
Chi-Square with 10 DF

   P(X <= x)     x
        0.01   2.5582
        0.05   3.9403
        0.95  18.3070
        0.99  23.2093
```

Note that the EXCEL function gives areas to the right, and MINITAB gives area to the left of the point.

3.

3.178893	=FINV(0.05,9,9)
0.314575	=FINV(0.95,9,9)
2.147727	=3.78/1.76

The rejection region is $F < 0.315$ and $F > 3.179$. The computed test statistic is $F = 2.15$. We are unable to reject that $\frac{\sigma_1^2}{\sigma_2^2} = 1$, or that the two population variances are equal.

4.

The data is in summarized form. All you need to give in the MINITAB dialog box are the sample sizes and the sample variances. The results are shown in Figure Solutions-25. Note that the F-value (2.15) is given in Figure Solutions-25 along with the p-value (0.270).

5.

The MINITAB output for the F distribution is shown in Figure Solutions-26

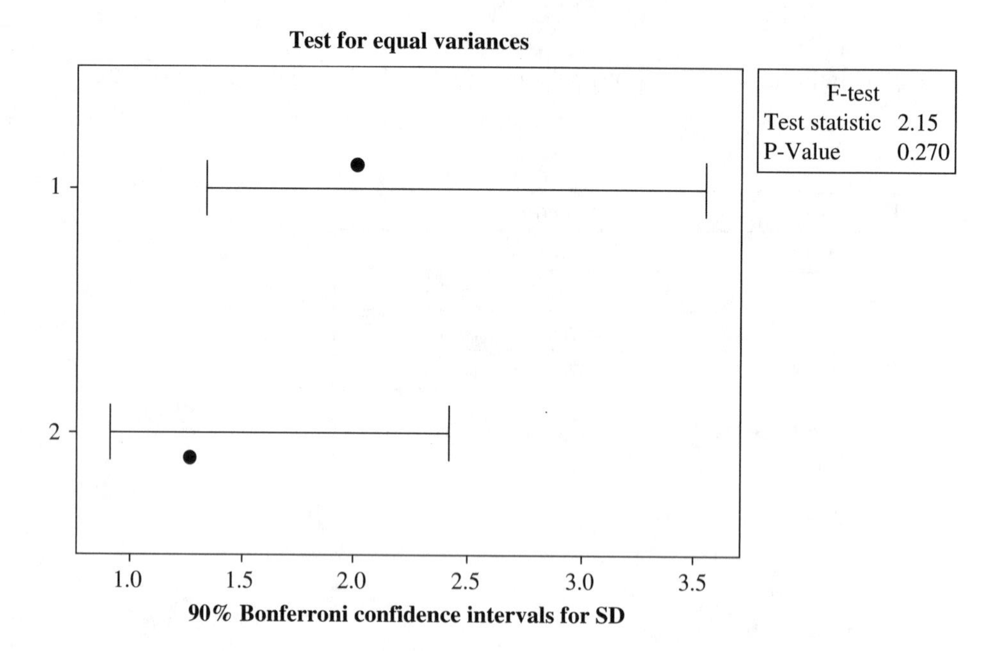

Figure Solutions-25 MINITAB comparison of variances of lifetimes.

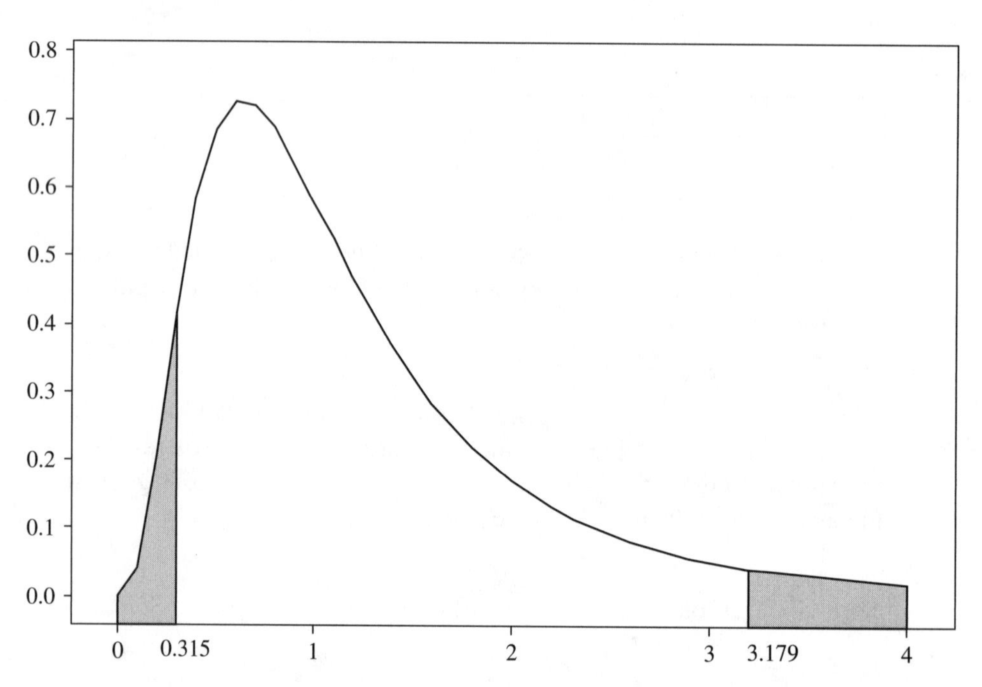

Figure Solutions-26 F distribution with right and left rejection regions shown.

6.

Type 1	Type 2
43	64
52	36
46	50
53	42
62	52
62	52
54	57
41	64
51	52
42	70
	68
	67

58.26667	=VAR(A2:A11)
115.4242	=VAR(B2:B13)
0.504804	=C1/C2
0.314187	=2*(1-FDIST(C3,9,11))

The first sample variance is 58.267 and the second sample variance is 115.4242. The computed test statistic is $F = 0.505$. The p-value is 0.314. It is assumed that the samples are taken from normally distributed populations.

7.

7.846645	FINV(0.01,4,7)
4.120312	FINV(0.05,4,7)
0.16409	FINV(0.95,4,7)
0.066775	FINV(0.99,4,7)

Inverse Cumulative Distribution Function

```
F Distribution with 4 DF in Numerator and 7 DF in Denominator

   P(X <= x)     x
        0.01   0.06677
        0.05   0.16409
        0.95   4.12031
        0.99   7.84665
```

8.

A	B	C	D
Type 1	Type 2	58.26667	=VAR(A2:A11)
43	64	115.4242	=VAR(B2:B13)
52	36	3.587899	=FINV(0.025,9,11)
46	50	3.912074	=FINV(0.025,11,99)
53	42	0.140696	=(C1/C2)/C3
62	52	1.974832	=(C1/C2)*C4
62	52		
54	57	0.375095	=SQRT(C5)
41	64	1.405287	=SQRT(C6)
51	52		
42	70		
	68		
	67		

$$0.1410 < \frac{\sigma_1^2}{\sigma_2^2} < 1.9748 \qquad 0.3751 < \frac{\sigma_1}{\sigma_2} < 1.4053$$

9.

$S_1^2 = 58.26667, S_2^2 = 115.4242, S_1 = 7.6333, S_2 = 10.744$, null hypothesis Variance 1/Variance 2 = 1, alternative hypothesis Variance 1/ Variance 2 ^= 1, computed test statistic $F = 0.50$, two-tailed p-value Pr > F is 0.3142, 95% confidence interval for $\frac{\sigma_1^2}{\sigma_2^2}$ and compare the 95% confidence interval for $\frac{\sigma_1^2}{\sigma_2^2}$ with the one obtained using EXCEL in problem 8 of this chapter.

SAS

```
Lower Limit     Upper Limit
-----------     -----------
0.1407          1.9748
```

EXCEL

$$0.1410 < \frac{\sigma_1^2}{\sigma_2^2} < 1.9748$$

10.

$$S_1^2 = 58.2673, S_2^2 = 115.4249, S_1 = 7.6333, S_2 = 10.7436, F = \frac{S_1^2}{S_2^2} = 0.505, p - \text{value} = 0.314$$

This is a two-tailed p-value.

Chapter 8

1.

(a) $\hat{p} = \frac{X}{n} = \frac{8}{35} = 0.22857$

(b) Difference = $\hat{p} - p_o = 0.22857 - 0.1 = 0.12857$

(c) $\sqrt{\frac{\hat{p}\hat{q}}{n}} = \sqrt{\frac{0.22857(0.77143)}{35}} = 0.07098$

(d) $Z = \frac{\hat{p} - p_o}{\sqrt{\frac{p_o q_o}{n}}} = \frac{0.22857 - 0.1}{\sqrt{\frac{.1(.9)}{35}}} = 2.54$

(e)

(f)

$$\left(\hat{p} - 1.96\sqrt{\frac{\hat{p}\hat{q}}{n}}, \hat{p} + 1.96\sqrt{\frac{\hat{p}\hat{q}}{n}}\right)$$

or $\left(.22857 - 1.96\sqrt{\frac{0.22857(0.77143)}{35}}, .22857 + 1.96\sqrt{\frac{0.22857(0.77143)}{35}}\right)$

or (0.08946, 0.36769)

2.

Method 1

The rejection region is $Z > 1.645$. The test statistic is 2.54. Since the test statistic is in the rejection region, reject the null hypothesis.

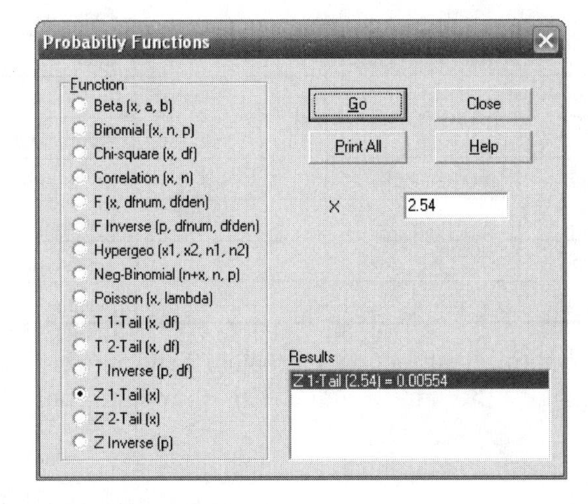

Figure Solutions-27 *p*-value is the area to the right of 2.54.

Method 2

The p-value is 0.00554 and since this is less than 0.05, the null is rejected.

Method 3

The one-sided confidence interval is (0.111832, 1). Since this interval does not contain the value stated for p in the null hypothesis, the null is rejected.

3.

```
Sample   X     N  Sample p
1       10   100  0.100000
2        8   125  0.064000

Difference = p (1) - p (2)
Estimate for difference:  0.036
95% CI for difference:   (-0.0367892, 0.108789)
Test for difference = 0 (vs not = 0):   Z = 0.99  p-value = 0.323
```

The MINITAB test is based on the standard normal distribution. The Z (standard normal value) is 0.99. The p-value is 0.323. There is no difference in the proportions.

4.

A	B	C	D
10	8	18	Observed
90	117	207	data
100	125	225	
8	10		Expected
92	115		data
0.5	0.4		4 contributions
0.04347826	0.034783		to the chi-square
			test statistic
0.97826087			Test statistic
0.32262855			p-value

Note that the normal distribution in problem 3 and the chi-square distribution in problem 4 give the same p-value. This shows the equivalence of the two tests.

5.

A	B	C	D	E	F
68	75	80	60	283	Observed
57	50	45	65	217	data
125	125	125	125	500	
70.75	70.75	70.75	70.75		Expected
54.25	54.25	54.25	54.25		data
0.10689	0.2553	1.209364	1.633392		
0.139401	0.332949	1.577189	2.130184		
7.38467					Test statistic
0.060597					*p*-value

The computed test statistic is 7.38467 and the *p*-value is 0.060597. You can not reject the null that the proportions are same at $\alpha = 0.05$.

6.

(a) $k = 4$

(b) $o_{11} = 68$, $e_{11} = 70.75$, and $(o_{11} - e_{11})^2 / e_{11} = 0.107$

(c)

$$\chi^2 = \sum_{i=1}^{2} \sum_{j=1}^{k} \frac{(o_{ij} - e_{ij})^2}{e_{ij}} = 7.385$$

(d) 0.061

7.

Row	Col	Counts
1	1	68
2	1	57
1	2	75
2	2	50
1	3	80
2	3	45
1	4	60
2	4	65

8.

Chi-square Test: C1, C2, C3

```
Expected counts are printed below observed counts.
Chi-square contributions are printed below expected counts.

           C1      C2      C3   Total
    1      15       8       5      28
         7.39   12.54    8.06
        7.830   1.646   1.164

    2      10      35       6      51
        13.46   22.85   14.69
        0.891   6.463   5.139

    3       8      13      25      46
        12.14   20.61   13.25
        1.414   2.809  10.425

Total      33      56      36     125

Chi-sq = 37.782, DF = 4, p-value = 0.000
```

The computed test statistic is 37.782. The distribution is chi-square with degrees of freedom = (3 – 1)(3 – 1) = 4. The *p*-value is 0 to three decimal places. Reject the null hypothesis of independence.

9.

The frequency for cell (1, 1) is 15. There are 15 welds with appearance = 1, x-ray = 1.

The expected for cell (1, 1) is 7.392. This expected value is arrived at as (33)(28)/125.

The deviation for cell (1, 1) is (observed – expected) = (15 – 7.392) = 7.608.

The percent for cell (1, 1) is (15/125) × 100 = 12.

The row percent for cell (1, 1) is (15/28) × 100 = 53.57.

The col percent for cell (1, 1) is (15/33) × 100 = 45.45.

These same 6 quantities are calculated and printed for all 9 cells.

Chi-square DF = (3 – 1)(3 – 1) = 4.

$$\chi^2 = \sum_{i=1}^{2}\sum_{j=1}^{k}\frac{(o_{ij} - e_{ij})^2}{e_{ij}} = 37.7817$$

The area to the right of 37.7817 is less than 0.0001.

10.

Obs	X-ray	Appear	Count
1	1	1	15
2	2	1	10
3	3	1	8
4	1	2	8
5	2	2	35
6	3	2	13
7	1	3	5
8	2	3	6
9	3	3	25

FINAL EXAMINATION ONE

(Answers with comments)

1. b

   The mode is the number that occurs most frequently. Mode = 10.

2. a

   The range = maximum – minimum = 20 – 2 = 18.

3. c

   (mean – 2 standard deviations, mean + 2 standard deviations) = (3.773, 16.453)

   There are 76 out of 80 that fall in the interval.

4. d

   It has a bell-shape, or what is called a normal distribution.

5. c

   $10.113 \pm 1.96(0.354)$

6. a

$$Z = \frac{\bar{X} - \mu_o}{\sigma_{\bar{x}}} = \frac{10.113 - 11.5}{3.170/\sqrt{80}} = \frac{-1.387}{0.354} = -3.92$$

7. d 10/80
8. d 6/28

9. b

The two events **less than 8** or **greater than 16** are mutually exclusive, so add the probabilities of the two events. 17/80 + 1/80 = 18/80.

10. c

$$\int_0^1 0.75(1-x^2)\,dx = 0.5$$

11. c

$$\mu = \int_{-1}^{1} xf(x)\,dx = 0.75\int_{-1}^{1} x(1-x^2)\,dx = 0.75\int_{-1}^{1} (x-x^3)\,dx$$

$$= 0.75\left\{\frac{x^2}{2} - \frac{x^4}{4}\right\}\Bigg|_{-1}^{1} = 0.75(0) = 0$$

12. c

$$\sigma = \sqrt{\int_{-1}^{1} x^2 f(x) - \mu^2} = \sqrt{0.75\int_{-1}^{1} x^2(1-x^2)\,dx - 0^2} = \sqrt{0.75\int_{-1}^{1} (x^2 - x^4)\,dx}$$

$$= \sqrt{0.75\left(\frac{1}{3} + \frac{1}{3} - \frac{1}{5} - \frac{1}{5}\right)} = 0.447$$

or using MAPLE, the answer is

```
> evalf(int(0.75*(x^2-x^4),x=-1..1));   0.2000000000
```

Taking the square root of 0.2 gives 0.447 as the standard deviation.

13. b

$$\int_{-1}^{M} 0.75(1-x^2)\,dx = 0.5$$

This equation is solved for M. The resulting equation is $0.75(M - M^3/3) = 0$. The solution is $M = 0$. The median is 0. Notice that the median is such that 0.5 of the area under the pdf is less than the median.

Using MAPLE, the solution is

```
> evalf(int(0.75*(1-x^2),x=-1..M)=0.5);
0.7500000000M + 0.5000000000 –0.2500000000M³ =.0.5

> solve(.7500000000*M+.5000000000-.2500000000*M^3=.5,M);
0.,1.732050808,-1.732050808
```

The 0 in the 3 solutions is the only one between –1 and 1. The median is 0.

14. d

    To find the probability that $X < 2$, integrate from –1 to 1. This integral equals 1.

15. d

    The chi-square value is printed out. It is shown to be 3.684.

16. b

    The degrees of freedom is $(4 - 1)(2 - 1) = 3$.

17. a

    The test is upper-tailed only. If we put 0.05 in the upper tail of the chi-square distribution having 3 degrees of freedom, the critical value is =CHIINV(0.05, 3).

18. d

    The p-value is the area to the right of the computed test statistic. This is given by =CHIDIST(3.684,3).

19. a

    The 95% confidence interval for p_4 is

$$\left(\hat{p}_4 - 1.96\sqrt{\frac{\hat{p}_4\hat{q}_4}{n}}, \hat{p}_4 + 1.96\sqrt{\frac{\hat{p}_4\hat{q}_4}{n}}\right) \text{ or } 0.1 \pm 1.96(0.0212)$$

20. c

$$\int_0^{0.2} f(x)\,dx = \int_0^{0.2} 4x^3\,dx = x^4 \Big|_0^{0.2} = 0.2^4 - 0 = 0.0016$$

21. c

$$\mu = \int_0^1 xf(x)\,dx = \int_0^1 4x^4\,dx = \frac{4}{5}x^5 \Big|_0^1 = 0.8$$

22. d

$$\sigma^2 = \int_0^1 x^2 f(x) - \mu^2 = \int_0^1 x^2 4x^3\,dx - 0.8^2 = \int_0^1 4x^5\,dx - 0.64 = \frac{4}{6}x^6 \Big|_0^1 - 0.64$$

$$= 0.667 - 0.64 = 0.027$$

23. a

$$\int_1^2 \frac{c}{x^2}\,dx = -\frac{c}{x}\Big|_1^2 = -\frac{c}{2} + \frac{c}{1} = 1$$

Solving for c we obtain, $c = 2$.

24. a

The value is listed as the chi-square value or 10.067.

25. d

The degrees of freedom is $(4 - 1)(3 - 1) = 6$.

26. a

The critical value is =CHIINV(0.05,6) = 12.59159

27. a

The p-value is given by =CHIDIST(10.067,6) = 0.121859.

28. c

=HYPGEOMDIST(2,2,2,15) = 0.0095

29. d

=BINOMDIST(5,100,0.1,1) = 0.0576

30. a

=BINOMDIST(10,100,0.1,1)-BINOMDIST(4,100,0.1,1) = 0.5594

31. d

The geometric distribution is $g(x) = q^{x-1}p$ for $x = 1, 2, 3, \ldots$ The mean is

$$\mu = \frac{1}{p} = \frac{1}{0.0005} = 2000$$

32. a

$$P(X > 500) = .9995^{500}(0.0005) + .9995^{501}(0.0005) + \cdots$$

$$= \frac{a}{1-r} = \frac{.9995^{500}(0.0005)}{1-0.9995} = 0.7788$$

33. c

The variance of a geometric random variable is

$$\sigma^2 = \frac{1-p}{p^2} = \frac{.9995}{0.00000025} = 3998000$$

The standard deviation is 1999.5.

34. b

Using Chebyshev's theorem, the probability is at least 0.75 that *X* will be within two standard deviations of the mean. Such an interval is (0, 5999).

35. b

In Chapter 4, it is shown that the mean of a beta distribution is $\mu = \frac{\alpha}{\alpha + \beta} = \frac{1}{10}$.

36. c

Making the substitution $y = 1 - x$ changes the integral as follows:

$$\int_0^{0.15} 9(1-x)^8\,dx = \int_1^{0.85} 9y^8(-dy) = 1 - 0.232 = 0.768$$

37. a

The standard deviation of a uniform variable over (a, b) is $\sigma = \frac{(b-a)}{\sqrt{12}}$ or 0.577.

38. d

$$\int_{2.5}^{3} 0.5\,dx = (3 - 2.5)0.5 = .25 \text{ or } 25\%$$

39. b

=1-NORMDIST(20,15,2.5,1) or 2.275%.

40. d

The sample mean has mean $\mu_{\bar{x}} = \mu = 15$ and standard error $\sigma_{\bar{x}} = \frac{\sigma}{\sqrt{n}} = \frac{2.5}{5} = 0.5$. Because the sample came from a normal distribution, the sample mean is normally distributed no matter what the sample size is. $P(\overline{X} < 14.5)$ =NORMDIST(14.5,15,0.5,1) = 0.1587.

41. d

$$S_1^2 = 30.25, S_2^2 = 69.7225, F = \frac{S_1^2}{S_2^2} = 0.43$$

42. a

=FINV(0.025,34,34) = 1.9811 and =FINV(0.975,34,34) = 0.5048, reject the null since 0.43 is less than 0.5048.

43. b

=2*(1-FDIST(0.43,34,34)) = 0.016

44. d

The test statistic can be either the T or the Z since both samples are greater than 30.

$$Z = \frac{\bar{X}_1 - \bar{X}_2 - D_o}{\sqrt{\frac{\sigma_1^2}{n_1} + \frac{\sigma_2^2}{n_2}}} = \frac{4.5 - 3.9 - 0}{\sqrt{\frac{2.25}{35} + \frac{7.84}{35}}} = 1.12$$

45. b

=2*(1-NORMSDIST(1.12)) = 0.263

46. a

$$(\bar{X}_1 - \bar{X}_2) \pm z_{\alpha/2}\sqrt{\frac{\sigma_1^2}{n_1} + \frac{\sigma_2^2}{n_2}}$$

gives $(4.5 - 3.9) \pm 2.58\sqrt{\frac{2.25}{35} + \frac{7.84}{35}}$ or (–0.8356, 2.0356).

47. c

It is assumed that the differences are normally distributed.

48. d

All three.

49. d

$$t = \frac{\bar{d} - 0}{s_d / \sqrt{n}} = \frac{-5.524}{21.472 / \sqrt{20}} = -1.15$$

50. a

=TDIST(1.15,19,1) = 0.132 is the p-value.

FINAL EXAMINATION TWO

1. 0.2
2. 78.0
3. Right
4. 50
5. 52
6. 72
7. 84
8. Right
9. 61.1
10. 75.2
11. 78.0
12. Three
13. Skewed

14.

$$n = 100, p = 25, np/100 = 25$$

Average the 25th and 26th values in the sorted group.

$$(5.2 + 5.4)/2 = 5.3$$

The EXCEL solution is 5.35.

15.

$$n = 100, p = 75, np/100 = 75$$

Average the 75th and 76th values in the sorted group.

$$(27.1 + 27.1)/2 = 27.1$$

The EXCEL solution is 27.1.

16.

$$n = 100, p = 90, np/100 = 90$$

Average the 90th and 91st values in the sorted group.

$$(43.1 + 43.2)/2 = 43.15$$

The EXCEL solution is 43.11.

17.

$$n = 100, p = 95, np/100 = 95$$

Average the 95th and 96th values in the sorted group.

$$(50.3 + 54.6)/2 = 52.45$$

The SAS solution is 52.45.

18.

$$n = 100, p = 50, np/100 = 50$$

Average the 50th and 51st values in the sorted group.

$$(11.5 + 11.8)/2 = 11.65$$

The SAS solution is 11.65.

19.

$$n = 100, p = 10, np/100 = 10$$

Average the 10th and 11th values in the sorted group.

$$(1.9 + 2.2)/2 = 2.05$$

The SAS solution is 2.05. SAS is using the algorithm given in the book. EXCEL is using another algorithm.

Questions 20 through 24:

Category	Angle of the slice
Mechanical	**20** 0.26(360) = 93.6 degrees
Chemical	**21** 0.1(360) = 36 degrees
Computer	**22** 0.1(360) = 36 degrees
Industrial	**23** 0.2(360) = 72 degrees
Civil	**24** 0.34(360) = 122.4 degrees

25.

$$P(\text{Mechanical or Civil}) = 0.26 + 0.34 = 0.6$$

26.

$$1 - P(\text{Civil}) = 1 - 0.34 = 0.66$$

27.

$$P(\text{Civil} \cap \text{Civil} \cap \text{Civil}) = \frac{17}{50}\frac{16}{49}\frac{15}{48} = \frac{4080}{117600} = 0.03469$$

28.

$$P(\text{Chemical} \cap \text{Chemical} \cap \text{Chemical}) = \frac{5}{50}\frac{4}{49}\frac{3}{48} = \frac{60}{117600} = 0.00051$$

29.

P(computer ∩ industrial ∩ mechanical or computer ∩ mechanical ∩ industrial or industrial ∩ computer ∩ mechanical or industrial ∩ mechanical ∩ computer or mechanical ∩ industrial ∩ computer or mechanical ∩ computer ∩ industrial)

The six probabilities are all the same numerically.

$$6\frac{5}{50}\frac{10}{49}\frac{13}{48} = \frac{3900}{117600} = 0.03316$$

30.

P(second and third were computer engineers | first selected was a civil engineer)

$$= P(A \mid B) = P(A \cap B)/P(B) = \frac{\frac{17}{50}\frac{5}{49}\frac{4}{48}}{\frac{17}{50}} = \frac{20}{2352} = 0.008503$$

31.

If the first two selected are known to be mechanical engineers, the space to pull the third one from consists of 48, of which 11 are mechanical, 5 are chemical, 5 are computer, 10 are industrial, and 17 are civil. The probability of selecting an industrial engineer from this space is 10/48, or 0.20833.

32.

$$P(X = 0) = 4/10 = 0.4$$

33.

$$P(\text{at least two defectives}) = P(X = 2 \text{ or } X = 3) = 2/10 + 1/10 = 3/10 = 0.3$$

34.

$$P(\text{at most one defective}) = P(X = 1 \text{ or } X = 0) = 3/10 + 4/10 = 7/10 = 0.7$$

35.

$$\mu = \sum x_i\, p(x_i) = 0(0.4) + 1(0.3) + 2(0.2) + 3(0.1) = 1$$

36.

$$\sigma = \sqrt{\sum (x_i - \mu)^2\, p(x_i)}$$
$$= \sqrt{(0-1)^2(0.4) + (1-1)^2(0.3) + (2-1)^2(0.2) + (3-1)^2(0.1)} = \sqrt{1} = 1$$

37.

$$P(X > 0.3) = \int_{0.3}^{0.5} 80x^4 dx = \frac{80}{5}x^5 \Big|_{0.3}^{0.5} = 16(0.5)^5 - 16(0.3)^5 = 0.46112$$

38.

$$P(-0.2 < X < 0.2) = \int_{-0.2}^{0.2} 80x^4 dx = \frac{80}{5}x^5 \Big|_{-0.2}^{0.2} = 16(0.2)^5 - 16(-0.2)^5 = 0.01024$$

39.

$$P(X < -0.3 \text{ or } X > 0.4) = \int_{-0.5}^{-0.3} 80x^4 dx + \int_{0.4}^{0.5} 80x^4 dx = 16x^5 \Big|_{-0.5}^{-0.3} + 16x^5 \Big|_{0.4}^{0.5} = 0.79728$$

40.

$$\mu = \int_{-0.5}^{0.5} 80x^4 x\, dx = \frac{80}{6}x^6 \Big|_{-0.5}^{0.5} = 0$$

41.

$$\sigma^2 = \int_{-0.5}^{0.5} x^2 f(x)\,dx - \mu^2 = \int_{-0.5}^{0.5} 80x^6\,dx - 0^2 = \frac{80}{7}x^7 \Big|_{-0.5}^{0.5} = \frac{80}{7}(0.5^7 + 0.5^7) = 0.17857$$

$$\sigma = 0.42258$$

42.

$$NB(x) = \binom{x-1}{r-1}(p)^r (q)^{(x-r)} = \binom{50-1}{3-1}(0.05)^3(0.95)^{50-3}$$

$$= 1176(0.000125)(0.089745) = 0.01319$$

43.

$$\mu = \frac{r}{p} = \frac{3}{0.05} = 60$$

On the average, the third flaw occurs on the 60th trial.

44.

$$0.95^{30}(0.05) + 0.95^{31}(0.05) + \cdots = \frac{a}{1-r} = \frac{0.95^{30}(0.05)}{1-0.95} = 0.95^{30} = 0.214639$$

45.

=NORMDIST(5,8,2,1) = 0.066807

46.

=NORMINV(0.25,8,2) = 6.65102

47.

$$\mu = e^{\alpha+\beta^2/2} = e^{2+0.5} = e^{2.5} = 12.1825$$

48.

$$\sigma^2 = e^{2\alpha+\beta^2}\left(e^{\beta^2} - 1\right) = e^{2.5}(e^1 - 1) = 12.1825(2.7183 - 1) = 20.9332 \quad \sigma = 4.5753$$

49.

$P(5 < X < 15) = P(\ln(5) < \ln(X) < \ln(15)) = P(1.609 < \ln(X) < 2.708)$

=NORMDIST(2.708,2,1,1) - NORMDIST(1.609,2,1,1) = 0.4126

50.

$$Z = \frac{\bar{X}_1 - \bar{X}_2 - D_o}{\sqrt{\frac{\sigma_1^2}{n_1} + \frac{\sigma_2^2}{n_2}}} = \frac{257 - 185 - 0}{\sqrt{\frac{11025}{35} + \frac{9025}{35}}} = \frac{72}{23.9344} = 3.01$$

51.

Critical values =NORMSINV(0.005) = –2.5758 and =NORMSINV(0.995) = 2.5758

52.

p-value =2*(1-NORMSDIST(3.01)) = 0.002612

53.

$$(\bar{X}_1 - \bar{X}_2) \pm z_{\alpha/2}\sqrt{\frac{\sigma_1^2}{n_1} + \frac{\sigma_2^2}{n_2}} \quad \text{or} \quad 72 \pm 2.5758\sqrt{\frac{11025}{35} + \frac{9025}{35}}$$

or $72 \pm 2.5758(23.9344)$ or 72 ± 61.6502 or (10.3498, 133.6502)

54.

The null hypothesis is rejected.

55.

Chi-square

56.

3 degrees of freedom

57.

p-value =CHIDIST(30.245,3) = 0.000001226

58.

The 9 expected values are shown in bold. These are computed assuming independence.

```
Expected counts are printed below observed counts.
Chi-square contributions are printed below expected counts.
        Above  Average  Below  Total
    1      15       35     17     67
        15.14    35.88  15.98
        0.001    0.022  0.065

    2      13       38     16     67
        15.14    35.88  15.98
        0.302    0.125  0.000

    3      26       55     24    105
        23.72    56.23  25.04
        0.218    0.027  0.043

Total      54      128     57    239

Chi-sq = 0.804, DF = 4, p-value = 0.938
```

59.

The value of

$$\chi^2 = \sum_{j=1}^{c}\sum_{i=1}^{r}\frac{[o_{ij} - e_{ij}]^2}{e_{ij}}$$

when the sum is taken over the 3 × 3 = 9 row-column combinations is 0.804.

60.

This is an upper-tailed test, that is, we reject the null only for large values of the test statistic. The value for which there is only a 5% chance of exceeding that value is given by =CHIINV(0.05,4), or 9.487729.

61.

The p-value is given by =CHIDIST(0.804,4), which equals 0.937911.

62.

Since the p-value is very large, there is no evidence supporting a dependency.

63.

Half of the outcome space for rolling three dice and recording the sum is given in Figure Solutions-28a and the other half is given in Figure Solutions-28b. These figures are very easy to form using EXCEL and by using click-and-drag techniques. By summarizing the 216 sums, the distribution of X, the sum of the three dice can be found. Figure Solutions-29 gives the distribution of X summarized from Figures Solutions-28a and Solutions-28b.

64.

$$\mu = \sum x_i p(x_i), \text{ where } i \text{ goes from 1 to 16 } \mu = 10.5$$

65.

$$\sigma^2 = \sum x_i^2 p(x_i) - \mu^2 = \frac{25704}{216} - 110.25 = 8.75$$

66.

$P(8 \le X \le 11) = 21/216 + 25/216 + 27/216 + 27/216 = 100/216 = 0.463$

67.

A normal distribution, because the sample size exceeds 30.

68.

$$\mu_{\bar{x}} = \mu = 10.5$$

Die 1	Die 2	Die 3	Sum	Die 1	Die 2	Die 3	Sum	Die 1	Die 2	Die 3	Sum
1	1	1	3	2	1	1	4	3	1	1	5
1	1	2	4	2	1	2	5	3	1	2	6
1	1	3	5	2	1	3	6	3	1	3	7
1	1	4	6	2	1	4	7	3	1	4	8
1	1	5	7	2	1	5	8	3	1	5	9
1	1	6	8	2	1	6	9	3	1	6	10
1	2	1	4	2	2	1	5	3	2	1	6
1	2	2	5	2	2	2	6	3	2	2	7
1	2	3	6	2	2	3	7	3	2	3	8
1	2	4	7	2	2	4	8	3	2	4	9
1	2	5	8	2	2	5	9	3	2	5	10
1	2	6	9	2	2	6	10	3	2	6	11
1	3	1	5	2	3	1	6	3	3	1	7
1	3	2	6	2	3	2	7	3	3	2	8
1	3	3	7	2	3	3	8	3	3	3	9
1	3	4	8	2	3	4	9	3	3	4	10
1	3	5	9	2	3	5	10	3	3	5	11
1	3	6	10	2	3	6	11	3	3	6	12
1	4	1	6	2	4	1	7	3	4	1	8
1	4	2	7	2	4	2	8	3	4	2	9
1	4	3	8	2	4	3	9	3	4	3	10
1	4	4	9	2	4	4	10	3	4	4	11
1	4	5	10	2	4	5	11	3	4	5	12
1	4	6	11	2	4	6	12	3	4	6	13
1	5	1	7	2	5	1	8	3	5	1	9
1	5	2	8	2	5	2	9	3	5	2	10
1	5	3	9	2	5	3	10	3	5	3	11
1	5	4	10	2	5	4	11	3	5	4	12
1	5	5	11	2	5	5	12	3	5	5	13
1	5	6	12	2	5	6	13	3	5	6	14
1	6	1	8	2	6	1	9	3	6	1	10
1	6	2	9	2	6	2	10	3	6	2	11
1	6	3	10	2	6	3	11	3	6	3	12
1	6	4	11	2	6	4	12	3	6	4	13
1	6	5	12	2	6	5	13	3	6	5	14
1	6	6	13	2	6	6	14	3	6	6	15

Figure Solutions-28a Half of the outcome space for tossing three dice.

Die 1	Die 2	Die 3	Sum	Die 1	Die 2	Die 3	Sum	Die 1	Die 2	Die 3	Sum
4	1	1	6	5	1	1	7	6	1	1	8
4	1	2	7	5	1	2	8	6	1	2	9
4	1	3	8	5	1	3	9	6	1	3	10
4	1	4	9	5	1	4	10	6	1	4	11
4	1	5	10	5	1	5	11	6	1	5	12
4	1	6	11	5	1	6	12	6	1	6	13
4	2	1	7	5	2	1	8	6	2	1	9
4	2	2	8	5	2	2	9	6	2	2	10
4	2	3	9	5	2	3	10	6	2	3	11
4	2	4	10	5	2	4	11	6	2	4	12
4	2	5	11	5	2	5	12	6	2	5	13
4	2	6	12	5	2	6	13	6	2	6	14
4	3	1	8	5	3	1	9	6	3	1	10
4	3	2	9	5	3	2	10	6	3	2	11
4	3	3	10	5	3	3	11	6	3	3	12
4	3	4	11	5	3	4	12	6	3	4	13
4	3	5	12	5	3	5	13	6	3	5	14
4	3	6	13	5	3	6	14	6	3	6	15
4	4	1	9	5	4	1	10	6	4	1	11
4	4	2	10	5	4	2	11	6	4	2	12
4	4	3	11	5	4	3	12	6	4	3	13
4	4	4	12	5	4	4	13	6	4	4	14
4	4	5	13	5	4	5	14	6	4	5	15
4	4	6	14	5	4	6	15	6	4	6	16
4	5	1	10	5	5	1	11	6	5	1	12
4	5	2	11	5	5	2	12	6	5	2	13
4	5	3	12	5	5	3	13	6	5	3	14
4	5	4	13	5	5	4	14	6	5	4	15
4	5	5	14	5	5	5	15	6	5	5	16
4	5	6	15	5	5	6	16	6	5	6	17
4	6	1	11	5	6	1	12	6	6	1	13
4	6	2	12	5	6	2	13	6	6	2	14
4	6	3	13	5	6	3	14	6	6	3	15
4	6	4	14	5	6	4	15	6	6	4	16
4	6	5	15	5	6	5	16	6	6	5	17
4	6	6	16	5	6	6	17	6	6	6	18

Figure Solutions-28b Other half of the outcome space for tossing three dice.

x	3	4	5	6	7	8	9	10
$f(x)$	1/216	3/216	6/216	10/216	15/216	21/216	25/216	27/216

x	11	12	13	14	15	16	17	18
$f(x)$	27/216	25/216	21/216	15/216	10/216	6/216	3/216	1/216

Figure Solutions-29 Probability distribution of X = sum of the three dice.

69.

$$\sigma_{\bar{x}} = \frac{\sigma}{\sqrt{n}} = \frac{\sqrt{8.75}}{\sqrt{36}} = 0.493$$

70.

$$Z = \frac{\bar{X} - \mu_o}{\sigma_{\bar{x}}} = \frac{551.6 - 555}{\frac{5.7}{\sqrt{35}}} = \frac{-3.4}{0.963} = -3.53$$

71.

=NORMSINV(0.05) = −1.64485

72.

=NORMSINV(0.95) = 1.64485

73.

=2*NORMSDIST(−3.53) = 0.000416

74.

$$\chi^2 = \frac{(n-1)S^2}{\sigma_o^2} = \frac{34(5.7)^2}{9} = 122.74$$

75.

=CHIINV(0.025,34) = 51.966

76.

=CHIINV(0.975,34) = 19.80625

77.

p-value =2*CHIDIST(122.74,34) = 1.15628E-11

78.

$$\frac{(n-1)S^2}{\chi^2_{\alpha/2}} < \sigma^2 < \frac{(n-1)S^2}{\chi^2_{1-\alpha/2}}$$

or $\frac{34(5.7)^2}{51.966} < \sigma^2 < \frac{34(5.7)^2}{19.80625}$ or $21.257 < \sigma^2 < 55.773$ or $4.61 < \sigma < 7.468$

79.

Using MAPLE,

```
> evalf(int(int(int(c/(x*y*z),x=1..2),y=2..3),z=3..4));
```

0.0808521819*c*

Setting this equal to 1 gives the following:

$$0.08085c = 1 \quad \text{or} \quad c = 12.3686$$

Or doing the integration by finding antiderivatives,

$$c\int_1^2\int_2^3\int_3^4 \frac{1}{xyz}\,dx\,dy\,dz = c\ln(x)\Big|_1^2 \ln(y)\Big|_2^3 \ln(z)\Big|_3^4 = 1$$

$$\text{or } c\ln(2)[\ln(3)-\ln(2)][\ln(4)-\ln(3)] = 1$$

Solving for c gives

$$c = \frac{1}{0.69315[1.09861-0.69315][1.38629-1.09861]}$$

$$= \frac{1}{0.69315[0.40546][0.28768]} = 12.3686$$

80.

Integrate Y and Z out to get the marginal of X. Using MAPLE, we find the marginal of X to be

```
> evalf(int(int(12.3686/(x*y*z),y=2..3),z=3..4));
```

$$\frac{1.442735874}{x}$$

Verify the above by doing the integration by hand.

81.

Integrate X and Z out to get the marginal of Y. Using MAPLE, we find the marginal of Y to be

```
> evalf(int(int(12.3686/(x*y*z),x=1..2),z=3..4));
```

$$\frac{2.466373267}{y}$$

Verify the above by doing the integration by hand.

82.

Integrate X and Y out to get the marginal of Z. Using MAPLE, we find the marginal of Z to be

```
> evalf(int(int(12.3686/(x*y*z),x=1..2),y=2..3));
```

$$\frac{3.476157881}{z}$$

Verify the above by doing the integration by hand.

Note that the product of the three marginal densities gives the joint density of the three variables. The three variables are said to be independent.

83.

$$12.3686\int_1^{1.5}\int_2^{2.5}\int_3^{3.5} \frac{1}{xyz}\,dx\,dy\,dz = 12.3686\ln(x)\Big|_1^{1.5} \ln(y)\Big|_2^{2.5} \ln(z)\Big|_3^{3.5}$$

$$12.3686\ln(1.5)[\ln(2.5)-\ln(2)][\ln(3.5)-\ln(3)]$$

$$= 12.3686[0.40547][0.22314][0.15415] = 0.1725$$

Or, using MAPLE

```
evalf(int(int(int(12.3686/(x*y*z),x=1..(1.5)),y=2..(2.5)),z=3..(3.5)));
0.1725058458
```

84.

The null hypothesis is that the data comes from a normally distributed population. This hypothesis is rejected because the p-value for the Anderson-Darling test of normality is less than 0.005.

85.

The null hypothesis is that the data comes from a normally distributed population. This hypothesis is rejected because the p-value for the Ryan-Joiner test of normality is less than 0.01.

86.

The null hypothesis is that the data comes from a normally distributed population. This hypothesis is rejected because the p-value for the Kolmogorov-Smirnov test of normality is less than 0.01.

87.

Substituting $\alpha = 3$ and $\beta = 2$ in the beta distribution, we find

$$f(x) = \begin{cases} \dfrac{\Gamma(\alpha+\beta)}{\Gamma(\alpha)\Gamma(\beta)} x^{\alpha-1}(1-x)^{\beta-1}, & 0 < x < 1, \alpha > 0, \beta > 0 \\ 0, & \text{otherwise} \end{cases}$$

$$f(x) = \begin{cases} \dfrac{\Gamma(5)}{\Gamma(3)\Gamma(2)} x^{2}(1-x)^{1}, & 0 < x < 1 \\ 0, & \text{otherwise} \end{cases} = \begin{matrix} \dfrac{4!}{2!1!}(x^2 - x^3), & 0 < x < 1 \\ 0, & \text{otherwise} \end{matrix}$$

Simplifying further,

$$f(x) = 12(x^2 - x^3),\ 0 < x < 1$$

If this function is integrated from 0 to 1, we obtain 1, indicating that it is a density.

88.

The mean for the beta distribution is

$$\mu = \frac{\alpha}{\alpha+\beta} = \frac{3}{5} = 0.6$$

The average is 60% daily.

89.

The probability that the daily proportion is greater than 0.5 is $P(X > 0.5) =$
$\int_{0.5}^{1} 12(x^2 - x^3)\,dx = 12\left(\frac{x^3}{3} - \frac{x^4}{4}\right)\Bigg|_{0.5}^{1} = 12(1/3 - 1/4) - 12(0.125/3 - 0.0625/4) = 0.6875$

90.

Normality assumption, that is, it is assumed that you are sampling from a normally distributed population.

91.

The t value for the paired t-test is

$$t = \frac{\bar{d} - 0}{s_d/\sqrt{n}} = \frac{1.9 - 0}{2.33095/\sqrt{10}} = \frac{1.9}{0.73711} = 2.58$$

92.

p-value =TDIST(2.58,9,1) = 0.014848.

93.

Reject the null hypothesis and conclude that the accidents have gone down since the p-value < 0.05.

94.

95% lower bound for mean difference: 0.54879. The interval is (0.54879, ∞), and since the interval does not include the number 0, it is concluded that the number of accidents has gone down with the installations of the traffic device.

95.

The expression

$$f(x) = \begin{cases} 0, & x < a \\ \dfrac{1}{b-a}, & a \le x \le b \\ 0, & x > b \end{cases}$$

is the density for a uniform random variable over the interval $[a, b]$. In this case, $a = 10$ and $b = 12$. The mean is

$$\mu = \frac{a+b}{2} = \frac{10+12}{2} = 11$$

96.

In problem 95, the variance of a uniform random variable is $\sigma^2 = \frac{(b-a)^2}{12}$ and the standard deviation is

$$\sigma = \sqrt{\frac{(b-a)^2}{12}} = \frac{b-a}{\sqrt{12}} = 0.289(b-a) = 0.289(2) = 0.578$$

97.

Assuming the null hypothesis is true, the variable $Z = \frac{\hat{p} - p_o}{\sqrt{\frac{p_o q_o}{n}}}$ will have a standard normal distribution. The expression $Z = \frac{\hat{p} - p_o}{\sqrt{\frac{p_o q_o}{n}}}$ is the test statistic and it equals

$$Z = \frac{\hat{p} - p_o}{\sqrt{\frac{p_o q_o}{n}}} = \frac{37.5 - 30}{\sqrt{\frac{30(70)}{1000}}} = \frac{7.5}{\sqrt{2.1}} = 5.18$$

98.

Using EXCEL, the p-value is given by =1-NORMSDIST(5.18) = 1.10943E-07 = 0.000000110943, or practically zero.

99.

Because 0.000000110943 is much smaller than 0.1, the null hypothesis is rejected.

100.

A $(1 - \alpha)$ confidence interval for p is

$$\left(\hat{p} - z_{\alpha/2}\sqrt{\frac{\hat{p}\hat{q}}{n}}, \hat{p} + z_{\alpha/2}\sqrt{\frac{\hat{p}\hat{q}}{n}}\right)$$

From the data we know that

$$\hat{p} = \frac{x}{n} = \frac{375}{1000} = 37.5\%, \hat{q} = 62.5, z_{0.05} = 1.645$$

The value of $1.645\sqrt{\frac{37.5(62.5)}{1000}}$ is 2.518% and the interval is (37.5% ± 2.5%), or (35%, 40%).

Bibliography

1. Richard Johnson, Irwin Miller, John Freund, ***Miller & Freund's Probability and Statistics for Engineers***, 7th edition, Prentice Hall, Upper Saddle River, 2004.
2. David M. Levine, Patricia P. Ramsey, Robert K. Smidt, ***Applied Statistics For Engineers and Scientists Using Microsoft Excel and MINITAB (With CD-ROM)***, Prentice Hall, Upper Saddle River, New Jersey, 2001.
3. William Mendenhall, Terry L. Sincich, ***Statistics for Engineering and the Sciences***, 5th edition, Prentice Hall, Upper Saddle River, 2006.
4. William Navidi, ***Statistics for Engineers and Scientists***, McGraw-Hill, New York, 2005.
5. Joseph Petruccelli, Balgobin Nandram, Minghui Chen, ***Applied Statistics for Engineers and Scientists***, 4th edition, Prentice Hall, Upper Saddle River, 1999.
6. Murray R. Spiegel and Larry J. Stephens, ***Schaum's Outline of Statistics***, 3rd edition, McGraw-Hill, New York, 1999.
7. Larry J. Stephens, ***Advanced Statistics Demystified***, McGraw-Hill, New York, 2004.
8. Ronald E. Walpole, Raymond H. Myers, Sharon L. Myers, Keying Ye, ***Probability & Statistics for Engineers & Scientists***, 8th edition, Prentice Hall, Upper Saddle River, 2007.

INDEX

E

F

G

N

Q

R

S

T

U

V

W

X

Z